Peter Hackl · Anders H. Westlund (Eds.)

# Economic Structural Change

## Analysis and Forecasting

With 101 Figures and 56 Tables

Springer-Verlag Berlin Heidelberg GmbH

Professor Dr. Peter Hackl
Wirtschaftsuniversität Wien, Institut für Statistik,
Augasse 2, A-1090 Wien

Professor Dr. Anders Holger Westlund
Stockholm School of Economics,
Box 6501, S-113 83 Stockholm

ISBN 978-3-662-06826-7 ISBN 978-3-662-06824-3 (eBook)
DOI 10.1007/978-3-662-06824-3



Originally published by International Institute for Applied Systems Analysis, Laxenburg, Austria in 1991.

Softcover reprint of the hardcover 1st edition 1991



Typesetting: Camera ready by author;
Offsetprinting: Color-Druck Dorfi GmbH, Berlin;
42/3020-543210 – Printed on acid-free paper

# Preface

Structural change is a fundamental concept in economic model building. Statistics and econometrics provide the tools for identification of change, for estimating the onset of a change, for assessing its extent and relevance. Statistics and econometrics also have developed models that are suitable for picturing the data-generating process in the presence of structural change by assimilating the changes or due to the robustness to its presence. Important subjects in this context are forecasting methods.

The need for such methods became obvious when, as a consequence of the oil price shock, the results of empirical analyses suddenly seemed to be much less reliable than before. Nowadays, economists agree that models with fixed structure that picture reality over longer periods are illusions. An example for less dramatic causes than the oil price shock with similarly profound effects is economic growth and its impacts on the economic system. Indeed, economic growth was a motivating concept for this volume.

In 1983, the International Institute for Applied Systems Analysis (IIASA) in Laxenburg/Austria initiated an ambitious project on "Economic Growth and Structural Change". These two economic issues and their interrelation are crucial determinants and pose a great challenge for economic theory. An outstanding effort, guided and supervised by Wilhelm Krelle, was started in the form of a joint IIASA-University of Bonn Project. The empirical basis of the project covered all important countries and regions. Excellent economists and econometricians participated in the project, partly as members of a central group in Bonn and partly as members of country or regional groups. The heart of the work was a highly aggregated world model established by the central group, which related results of country and regional groups to guarantee consistency. Several conferences on these topics took place between 1984 and 1986.

At the same time when the IIASA-University of Bonn Project was started, an IIASA Working Group on "Statistical Analysis and Forecasting of Economic Structural Change" was established. In the planning stage of the IIASA-University of Bonn Project, it became clear that many questions would require special statistical techniques. Such questions are, e.g., how to identify structural changes, what type of models are adequate in the presence of structural change, how to take into account structural changes when forecasting future developments? Some 50 economists and econometricians from about 15 countries from the East and West contributed to the aims of this IIASA Working Group. A bibliography of related books and papers was compiled [Hackl and Westlund (1989)]. In 1985 and 1986, three workshops took place where available and new statistical methods were presented and discussed.

The activities of the IIASA-University of Bonn Project and the IIASA Working Group resulted in the publication of the main outcomes in two books. *The Future of the World Economy: Economic Growth and Structural Change*, edited by Wilhelm Krelle, contains a comprehensive description of the model and results from the IIASA-University of Bonn Project. The main part of the volume concerns the work done by the central group in Bonn. Other chapters cover results of the collaborating country and regional groups. The model, some data, and some results are given by Krelle in Chapter 17 of this volume. *Statistical Analysis and Forecasting of Economic Structural Change*, edited by Peter Hackl, is a multi-authored volume, consisting of papers that were presented at the Working Group's meetings or were specially invited as surveying papers. Both volumes were published in 1989.

Interest in statistical methods related to economic structural change was not restricted to IIASA activities during the last ten years. Hackl and Westlund's (1989) bibliography of the literature in this area contains about 400 entries, among them some 100 that appeared between 1985 and 1988. In addition, some systematic surveys have been published recently: Schulze (1987) deals with multiphase regression models; Broemeling and Tsurumi (1987) present a comprehensive survey on Bayesian econometric methods. Somewhat older is Poirier's (1976) book that discusses models based on spline functions. A special issue of *Empirical Economics* on econometric methods has been edited by Krämer (1989). Moreover, since 1989, a new journal entitled *Structural Change and Economic Dynamics* attempts to provide a forum for methodological discussions.

The growing interest and activities indicated led to organizing another workshop on this topic. Many of the participants in the IIASA Working Group were still engaged in related research. Those as well as other statisticians and econometricians were invited to the international conference on "Economic Structural Change: Analysis and Forecasting" that took place in May 1989 in Stockholm. The program presented 16 lectures; invited discussants stimulated the discussion of each lecture. Most of the papers that were presented at the Stockholm conference are part of this volume.

From the beginning it was clear that it would not be possible to cover the whole area of such a complex topic at the Stockholm conference. A wide range of statistical methods interfers in this topic, such as time series analysis, regression analysis, econometrics, filtering techniques, etc. Demands from applications in areas other than economics, like engineering, biology, hydrology, lead to interesting impacts and contributions. This is especially true for themes that are not of central interest in the discussion of statistical analysis of economic structural change, like special modeling approaches, the relation to disequilibrium modeling, etc. When designing the plan for this volume, we invited scientists who are competent in a series of related areas.

We would like to acknowledge the help and support that made this book possible. We wish to thank the Prince Bertil Foundation at the Stockholm School of Economics, Professor Hans Landberg (The Swedish Council for Planning and Coordination of Research), and Professor Robert H. Pry (International Institute for Applied Systems Analysis) for their financial support and their interest; the contributors to the volume and the referees, in particular Åke Anderson, Wolfgang Polasek, and Howell Tong, for their very fruitful and uncomplicated collaboration; and the Publications Department at IIASA for providing its very professional know-how in producing this volume.

We hope that this volume will contribute to stimulate the interest of statisticians and econometricians in this topic and to improve models for analyzing real-world phenomena and reliability of results.

*Peter Hackl*
Department of Statistics
University of Economics
Vienna, Austria

*Anders H. Westlund*
Department of Economic Statistics
Stockholm School of Economics
Stockholm, Sweden

## References

Broemeling, L.D. and Tsurumi, H. (1987), *Econometrics and Structural Change.* New York: Marcel Dekker.
Hackl, P. (ed.) (1989), *Statistical Analysis and Forecasting of Economic Structural Change.* Berlin: Springer-Verlag.
Hackl, P. and Westlund, A.H. (1989), Statistical analysis of "Structural Change": An annotated bibliography. *Empirical Economics*, **14**, 167-192.
Krämer, W. (ed.) (1989), *Econometrics of Structural Change.* Heidelberg: Physica-Verlag.
Krelle, W. (ed.) (1989), *The Future of the World Economy. Economic Growth and Structural Change.* Berlin: Springer-Verlag.
Poirier, D.J. (1976), *The Econometrics of Structural Change.* Amsterdam: North-Holland.
Schulze, U. (1987), *Mehrphasenregression: Stabilitätsprüfung, Schätzung, Hypothesenprüfung.* Berlin: Akademie-Verlag.

# Contents

## CHAPTER 1

# Introduction

*Peter Hackl and Anders H. Westlund*

This volume contains 21 papers that are grouped into four parts. These are:

I. Identification of structural change.

II. Model building in the presence of structural change.

III. Forecasting in the presence of structural change.

IV. Economic modeling and the use of empirical data.

It is to be hoped that economists as well as statisticians and econometricians will read this volume. According to our experience, the understanding of the notion "structural change" is quite different for these two groups. Consequently, we think that a few comments about this notion is in order right at the beginning. Then, we will give some introductory remarks to each part of the volume.

## 1.1 The Notion of Structural Change

The "structural change" concept is often used in everyday life without an exact definition. In research, howevever, clarifications and definitions are urgent. The use of the concept "change" presupposes statements about degree of change, which requires measurement. But even in research there is obviously a certain conceptual confusion concerning what is actually meant by structural changes, and as a consequence, an uncertainty of how these phenomena are to be identified, characterized, and modeled.

Structural changes have been of major concern in economics. Different theories on economic development and growth assume that economic relationships and processes are changing over time, and these changes are described and explained basically in a descriptive way, without being statistically estimated and tested. One view of structural changes that is discussed among economists comprises changes in the composition of the output vector of an economic system, or changes in the composition of instrumental as well as exogenous input vectors. The approach is adopted by Krelle in Chapter 17; he discusses the way in which structural change is used to study changes in the sectoral composition of GDP, changes in the commodity composition of demand, changes in labor force, changes in trade relations, etc. This means that the notion of structure is not explicitly defined.

The statistical perspective of structural change is basically related to the model structure. There is no unambiguous definition of the concept "structure", although in relation to the theory of systems, it indicates the relations among variables of a system. A system is, of course, any arbitrarily selected set of variables interacting with each other and with an environment. Active and relevant variables are defined by the purpose of the study and by relevant theories. Thus, the actual "system at work" consists of a finite number of variables. Variables not included will represent the environment. In modeling economic systems the identification requires some knowledge of the relation between the system and the variables of the environment that have a special influence on the system. In the modeling process these variables are characterized as exogenous, while the rest of the environment is represented in the model by random disturbances.

Scientific treatment of the concept of structural change requires a strict definition of the system at work. Theories specify the form of the model equations, and structural form parameters depict the causal structure of the system and the relations to exogenous variables. In this way we relate the concept "change" to a proposition in the model field. In a regression model framework, for example, the change in one or more of the parameters indicate structural changes. Poirier (1976) distinguishes between some kind of general structural variability and structural change, and emphasizes that just considerable and low frequent variability should be associated with structural change (which certainly is consistent with a general nonscientific view).

Important reasons for structural changes in economic systems manifested in parameters shifts are, e.g., personal changes in behavior, the technical progress with corresponding changes in production function parameters, moves from fixed to floating exchange rate regimes. Within macroeconomics the so-called Lucas critique claims for parameter changes. The idea is that parameters of macroeconomic models will be determined by the expectations of economic agents involved concerning future economic policy. If a policy changes, so do the expectations and related parameters.

Apparently, it is extremely difficult to distinguish general misspecification problems from the problem of structural change and its impacts on the modeling process. In order to do that, some identifying theoretical knowledge is necessary (such as *a priori* knowledge of a changing institutional framework). The role of the econometric analysis is now basically to determine the "significance" of the structural change. The modeling process with respect to structural change may continue according to one of at least three lines:

1. To allow for time-varying (or space-varying) parameters.

2. To model the structural change and introduce it into the basic model (if structural

changes are caused by the environment, the interest should be focused on changes in exogenous variables, and on changes in their relation to the system).

3. To redefine (basically widen) the system at work.

Of course, if this is not done correctly, it will introduce misspecifications, with possible consequences such as residual autocorrelation, heteroscedasticity, etc. Thus, any strategy of diagnostic checking, at least among those based on residual analysis, must be influenced by the existing theoretical knowledge and hypotheses about structural change.

Apparently, structural change is a relative concept, and statements about it are restricted to the actual system at work and to the way it is manifested through a specified model. The concept is, however, an internal part of each model-building process, and the role of statistical analysis is "to detect its presence, to find ways to assimilate it in models, and to find methods of statistical inference that are robust to its presence" [see Anderson and Mizon (1989)].

## 1.2 Identification of Structural Change

Changes of the underlying data-generating process vis-à-vis the structure of the model are intrinsic phenomena of modeling in economics. Of course, the stages of model building include diagnostic checking of the model, a crucial part of the process of learning from reality. The identification of changes might be the outcome of diagnostic checks. However, it can hardly be expected that all sorts of driving forces for changing patterns—among them technical progress or other stimulants of economic growth—can be anticipated in a model, not to mention behaviorial changes or effects of political actions. From this and other reasons, the need of tools for the identification or detection of structural change is obvious.

The problem of testing for parameter constancy was tackled for the first time in the late 1950s in the regression model context. A large body of literature on various aspects of this problem has been published since. Methods that are of special interest for economists are those that allow one to test hypotheses of parameter constancy in linear regression relationships, simultaneous equations models, and time series models, corresponding to the most commonly applied types of model. Comprehensive discussions about these methods can be found in Hackl (1980), Chow (1984), Judge *et al.* (1985), Anderson and Mizon (1989), and Dziechciarz (1989) [see also the bibliography by Hackl and Westlund (1989)].

A crucial point for the application of tests and the interpretation of the test outcomes is the robustness of the test statistic with respect to the various assumptions. In recent years, robustness with respect to distributional assumptions has found great interest, and robustified versions of many statistical procedures (mainly estimation procedures) have been suggested. A contribution in the sense of these ideas is Chapter 2 by Marie Hušková, who introduces and discusses CUSUM and MOSUM test procedures for testing the constancy of regression relationships over time that are based on robustified versions of recursive residuals: The (recursive) least squares estimators of the regression coefficients that are used in the definition of traditional recursive residuals are replaced by $M$-estimators. These $M$-tests are introduced under general assumptions. Detailed results based on Monte Carlo estimates allow one to compare the cases where Huber's $\psi$ is used as the score function with the traditional methods.

The main reason for using recursive residuals in the CUSUM and MOSUM technique is the simplicity of their distributional properties as compared with the conventional OLS residuals that are used in other approaches. Walter Krämer and his coauthors discuss in Chapter 3 a CUSUM technique that is based on these OLS residuals. They derive the limiting null distribution for large sample sizes. Monte Carlo comparisons between the recursive residual-based and the OLS residual-based CUSUM procedures indicate that one is not superior to the other. However, the OLS-based version has more power in detecting parameter shifts that occur late in the sample than the recursive residual-based CUSUM test, whose deficiency in this respect is notorious.

A classical procedure for testing the constancy of linear regression models over the sample period consists in comparing the regression coefficients of independent regressions for the respective subsamples. The classical approach [Chow (1960)] assumes equal disturbance variances. For the case when the variances are unequal, Jean-Marie Dufour (Chapter 4) derives exact bounds for the null distribution of a (Wald-type) statistic for testing the equality of the regression coefficients (or any other set of linear restrictions on them). The statistic is that of the test suggested by Ohtani and Kobayashi and generalized by Farebrother; these researchers, however, do not answer the question of how to calculate critical limits for their test statistic. Dufour presents an algorithm that only needs the calculation of the central Fisher distribution.

In an earlier paper, Bernd Schips and Yngve Abrahamsen put testing for structural change into the context of misspecification testing [Abrahamsen and Schips (1989)]. They discussed the use of jackknife versions of the Stone-Geisser prediction test procedure for that purpose and demonstrated a good performance of the prediction tests in a wide range of misspecification cases and a superiority in many situations over other tests. In Chapter 5, Bernd Schips and Yngve Abrahamsen present a Monte Carlo study that compares more general jackknife versions ($p$-step) Stone-Geisser prediction tests with other specification tests that are generally used in econometric model building. The study is based on single equation and simple multi-equation macro models and extends in this repect the above-mentioned earlier paper.

Although many papers in the literature on identification of regression parameter changes deal with the testing problem, little attention has been paid to the issue of inferring about the change point. For the two-phase regression problem Lyle Broemeling and his coauthors derive in Chapter 6 in a Bayesian setup the density of the intersection in closed form starting from a proper joint distribution of the intersection point, the regression coefficients, and the disturbance variance that is assumed to be common to both regimes. Point and interval estimates of the intersection can be obtained by means of numerical integration.

Estimating a change point that is indicated by a test for parameter nonconstancy contrasts to the problem posed in intervention analysis; here, a change in regimes is claimed for a certain point of time, and the question is whether this is true. In Chapter 7, Peter Hackl introduces a sequential testing procedure for this purpose. Different from other authors who discussed this topic, in this paper the sequential character of the procedure makes sure that the user can apply it as a surveillance tool to obtain an affirmative response as soon as possible. The procedure is based on ranks. The null distribution is independent of the distribution of the control variable, another point where this method differs from earlier methods.

The question of robustness of statistical procedures with respect to the underlying assumptions has found interest in much research work done on structural change problems. An example given in this volume is Marie Hušková's chapter. On the other hand, the question might be raised whether certain procedures are robust with respect to parameter nonconstancy. An example for this is Chapter 8 by David Hendry and Adrian Neale. They investigate the robustness of the DF (Dickey-Fuller) and the ADF (Augmented Dickey-Fuller) tests when there is a shift in the intercept of an AR process. Their Monte Carlo study reveals interesting results. Such regime shifts can mimic unit roots in stationary time series; consequently, a unit-root test should be accompanied by a diagnostic test for parameter constancy.

## 1.3 Model Building in the Presence of Structural Change

Model evaluation is, of course, a basic dimension of the model-building process. The relative forecast performance of rival models constitute one natural criterion of evaluation. Model congruence requires the models to be coherent with sample data information and with theory, and to encompass (i.e., perform as well as) alternative models. Encompassing is an essential issue in model building. In Chapter 9, Maozu Lu and Graham Mizon present a class of forecast-encompassing tests with the purpose of comparing models on the basis of a combination of parametric-encompassing and parameter-constancy hypotheses. These tests are feasible for evaluating the forecast performance of large-scale macroeconometric models.

Traditional model building through time series analysis or econometrics often presupposes stationary time series although most series in practice require one or more differentiations to attain stationarity. [A time series is denoted $I(d)$ if its $d$-th differences are stationary.] If two different time series are both $I(1)$ but a linear combination of them is $I(0)$, i.e., stationary, the two series are called cointegrated. Increasing attention has been paid to the cointegration concept in economic model building over the last few years. Cointegration theory and tests have been developed for models with constant parameters [see, e.g., Granger (1986) and Engle and Granger (1987)]. Cointegration might be rejected, however, just because parameters are erroneously assumed to be constant (see Chapter 18 by Teräsvirta in this volume). Cointegration is traditionally a linear concept, and allowing for time-varying parameters (e.g., structural change) makes it much more flexible. Thus, it is important to extend the idea of cointegration to time-varying parameter (TVP) regression. That is done in Chapter 10 by Clive Granger and Hahn Lee. Certain properties of a TVP cointegrated process and of the related estimation procedure are indicated. It is also pointed out by the authors that if a TVP cointegration procedure provides evidence of an equilibrium relationship, but the traditional linear cointegration does not, then this might be an indication of some misspecification of the linear cointegrating model. Incorrectly omitted variables should be looked for.

Disequilibrium modeling and switching regressions are early areas of econometric analyses of structural changes. In Chapter 11, G.S. Maddala discusses the Markov switching model, an example of switching regression models with exogenous switching, and self-selection models. In particular, the literature on these approaches is outlined [see also Maddala (1986)], and compared with the literature on structural change analysis. The paper also shows how switching regression methods can be modified to take into account gradual adjustments and (expectations about) policy changes, constituting one important driving force for economic structural change.

The connection between misspecification phenomena and structural change characteristics in economic modeling has been pointed out above. One strategy of economic modeling that sometimes will take care of misspecification problems is to widen the limits of the economic system, to endogenize and allow for simultaneity. Although simultaneous equations models have attracted numerous econometricians, methodological research related to such models with time-varying parameters is by now of surprisingly limited scope. One of the few exceptions is Chapter 12 by Andrew Harvey and Mariane Streibel that deals with the problem of stochastic trends in simultaneous equations models. Identifiability conditions are verified, and the maximum likelihood estimation of such models is discussed.

Another way of eliminating misspecification errors is to allow for nonlinearities in the model. Nonlinearities arise in many different ways in econometric applications. General nonlinear models are often used in the estimation of demand and production functions. The advent of computer technology now makes it possible for econometricians to estimate rather general nonlinearities [for a survey of nonlinear regression models, with an emphasis on the theory of estimation and hypothesis testing, see Amemiya (1983)]. As it is the case for simultaneous equations models, research on nonlinear regression models with time-varying parameters is very rare. Peter Robinson (in Chapter 13), however, discusses estimation of intrinsically nonlinear regression models, where the parameters change smoothly and nonparametrically over time. A kernel-based analog of nonlinear least squares estimation is suggested and analyzed with respect to its asymptotic properties.

Since early the 1970s [see, e.g., Sarris (1973)], Kalman filtering has been used and evaluated as a procedure for estimating econometric models with time-varying parameters. Several research contributions have been published [see Hackl and Westlund (1989)]. In Chapter 14, Wolfgang Schneider continues this tradition. He specifies a state space model and applies scoring and the EM method for the estimation of the hyperparameters of this model. A descriptive interpretation of Kalman filtering (the so-called flexible least squares approach) is described, and its use as an exploratory data analysis approach to a preliminary descriptive stability analysis of a traditional money-demand function for the Federal Republic of Germany is discussed.

## 1.4 Forecasting in the Presence of Structural Change

Forecasting has always been an essential part of the economic planning process. The general purpose of forecasting is, of course, to provide knowledge about future states of an economic system. As there is often substantial evidence against the assumption of systems stability, the forecasting process should also include judgments about the structural developments of the system. The recent extensive changes in the economy have also aroused doubt about formalized approaches in the forecasting process. Thus, qualitative and non-formalized approaches were considered to be more important when forecasting structurally changing economic systems. As simple extrapolations of historical patterns are not always acceptable, the reduced confidence in formalized approaches is to a certain extent well motivated. A combination of quantitative strategies and qualitative judgments will certainly prove useful when forecasting structurally changing systems. In particular, future structural changes might be identified and characterized (e.g., through scenarios that aim at describing the future structural development in terms that allow for substitution into formalized models). The system states are then conditioned on the structural forecasts within the structural model. It is here essential to search for formalized procedures to combine the qualitative and quantitative strategies.

Forecast combinations are nowadays well-accepted procedures. In the case of structurally changing economic systems a practical difficulty is how to identify the combination weights in the case of structurally changing economic systems. Some methods have been suggested that cope with this problem. One such method consists in adaptively estimating the weight matrix [see, e.g., Diebold and Pauly (1987)]. State space modeling and Kalman filtering will also provide strategies that can explicitly consider the nature of non-stationary weights, although the effectiveness of combining may be seriously undermined. The Bayesian perspective to combining forecasts will provide a formal framework for rational transition between judgmental and data-based estimation of combination weights, and also appears to be of specific relevance, when seeking for the interaction between judgmental and quantitative model-based forecasts. Some properties of a Bayesian combination procedure are explored in Chapter 15 by Francis Diebold. In particular, the asymptotic performance of Bayesian composite forecasts is studied, when none of the forecasting models to be combined corresponds to the true data-generating process. Combining is also discussed by Spyros Makridakis (Chapter 16) as a strategy to be used to improve forecasting of structurally changing economic systems. The main point of this chapter, however, is to show that model selection should be based on the actual out-of-sample forecasting performance. Different forecasting horizons are used to identify not only the preferred forecasting method but also "best" forecasting models. This concept will certainly improve the possibility to cope with structural change problems in economic forecasting.

## 1.5 Economic Modeling and the Use of Empirical Data

It is of great importance to learn from empirical studies about how different test and estimation approaches work. It is also important to implement methodological research in empirical applications. The present volume, thus, gives some empirical studies on economic modeling subject to structural changes. The IIASA-University of Bonn Research Project on Economic Growth and Structural Change is one good example of economic empirical research from which insights are gained with respect to econometric and statistical methodology to deal with structural change problems. This particular research project also points out important areas of methodological research on structural change. Part of this research is indicated in Chapter 17 by Wilhelm Krelle, where the starting point is the argument that long-term economic growth is connected with structural change phenomena. That is demonstrated by a multisectoral world model. As mentioned in Section 1.1, structural changes, however, are not considered in this chapter in the statistical sense, i.e., as parameter changes, but basically as changes in the composition of demand and production of different commodities.

A shorter business-cycle perspective is emphasized in Chapter 18 by Timo Teräsvirta. Some issues, related to the structural change problem within business-cycle analysis and forecasting that are raised involve the question of whether cycles are asymmetric around turning points. The asymmetry problem is analyzed through testing linearity of monthly industrial production time series. Linearity is rejected, but the nonlinearity observed is not evidently related to cycle asymmetry, but more likely to an alternative kind of structural change, viz., to change in seasonal patterns, from constant to variable seasonality. Another important dimension of structural change in dynamic models, among which business-cycle models in general are important examples, concerns changes in the dynamic structure. For example, leading indicator models for business cycle forecasting are often characterized by varying leads. These problems may be analyzed through the use of transfer function models allowing for parameter variability. This approach is adopted by Lennart Claesson and Anders Westlund in Chapter 19. The model parameters are estimated by an iterative recursive instrumental variables approach. The filtering procedure used is based on a state space formulation. Structural changes in state-space models are further considered by Hiroki Tsurumi in Chapter 20; abrupt as well as gradual structural shifts are analyzed through a Bayesian procedure. Empirically, the Bayesian procedure is applied to single equation models of the US dollar/yen exchange rate. Based on monthly data, a structural change in exchange rate regimes is identified for 1985 (a shift probably due to the so-called Plaza Hotel agreement). State space modeling and structural change problems are also considered in Chapter 21 by Masanao Aoki. The resulting impulse response time profiles are used to analyze and verify changes in interaction characteristics of the real GNP processes of the USA, West Germany, and Japan. Changes in the dynamic properties of these interactions are emphasized.

An interesting and, to some extent, more complicated example of empirical use of change point models is given by Pranab Sen in Chapter 22. A methodology is developed for studying structural changes in income distributions. Part of the framework involves stratification of the economic system with respect to income level. Different stratification principles imply overlapping or nonoverlapping strata. Methodologically emphasis is placed on nonparametric and robustness considerations.

## References


Abrahamsen, Y. and Schips, B. (1989), Specification and stability tests versus jackknifing, pp. 37-43 in P. Hackl (ed.), *Statistical Analysis and Economic Structural Change.* Berlin: Springer-Verlag.

Amemiya, T. (1983), Non-linear regression models, pp. 333-389 in Z. Griliches and M.D. Intriligator (eds.), *Handbook of Econometrics, Volume 1.* Amsterdam: North-Holland.

Anderson, G.J. and Mizon, G.E. (1989), What can statistics contribute to the analysis of economic structural change? pp. 3-21 in P. Hackl (ed.), *Statistical Analysis and Economic Structural Change.* Berlin: Springer-Verlag.

Chow, G.C. (1960), Tests of equality between sets of coefficients in two linear regressions. *Econometrica,* **28**, 591-605.

Chow, G.C. (1984), Random and changing coefficient models, chapter 21 in Z. Griliches and M.D. Intriligator (eds.), *Handbook of Econometrics, Volume 2.* Amsterdam: North-Holland.

Diebold, F.X. and Pauly, P. (1987), Structural change and the combination of forecasts. *Journal of Forecasting,* **6**, 21-41.

Dziechciarz, J. (1989), Changing and random coefficient models: A survey, pp. 217-251 in P. Hackl (ed.), *Statistical Analysis and Economic Structural Change.* Berlin: Springer-Verlag.

Engle, R.F. and Granger, C.W.J. (1987), Cointegration and error correction: Representation, estimation and testing. *Econometrica,* **55**, 251-276.

Granger, C.W.J. (1986), Developments in the study of cointegrated economic variables. *Oxford Bulletin of Economics and Statistics,* **48**, 213-228.

Hackl, P. (1980), *Testing the Constancy of Regression Relationships over Time.* Göttingen: Vandenhoeck & Ruprecht.

Hackl, P. and Westlund, A.H. (1989), Statistical analysis of "Structural Change": An annotated bibliography. *Empirical Economics,* **14**, 167-192.

Judge, G.G., Griffiths, W.E., Hill, R.C., Lee, T.C., and Lütkepohl, H. (1985), *The Theory and Practice of Econometrics.* New York: John Wiley.

Maddala, G.S. (1986), Disequilibrium, self-selection and switching models, pp. 1633-1688 in Z. Griliches and M.D. Intrilligator (eds.), *Handbook of Econometrics, Volume 3.* Amsterdam: North-Holland.

Poirier, D.J. (1976), *The Econometrics of Structural Change.* Amsterdam: North-Holland.

Sarris, A. (1973), A Bayesian approach to estimation of time-varying regression coefficients. *Annals of Economic and Social Measurement,* **2**, 501-523.

# Part I

# Identification of Structural Change

CHAPTER 2

# Recursive $M$-Tests for the Change-Point Problem

*Marie Hušková*

## Summary

The chapter concerns robust recursive $M$-tests for testing the constancy of regression relationships over time. More exactly, the robust modifications of the CUSUM and MOSUM procedures, based on robustified recursive residuals and $M$-estimators, are presented. Some results of a simulation study are given; they coincide with those from the theoretical considerations.

## 2.1 Introduction

Let $X_1, \ldots, X_n$ be independent random variables observed at ordered time points $t_1 < \ldots < t_n$; the $X_i$ follow the model

$$X_i = \mathbf{c}_i' \boldsymbol{\theta}_i + e_i, \quad i = 1, \ldots, n,$$

where $\mathbf{c}_i = (c_{i1}, \ldots, c_{ip})'$ are known regression vectors, $\boldsymbol{\theta}_i = (\theta_{i1}, \ldots, \theta_{ip})'$ are unknown (regression) vector parameters, and $e_1, \ldots, e_n$ are independent identically distributed (i.i.d.) random errors with $e_i$ having the distribution function (d.f.) $F$ fulfilling certain regularity conditions (and unknown otherwise).

We are interested in *testing the constancy of the regression relationships over time* formulated as

$$H_0\colon \boldsymbol{\theta}_1 = \ldots = \boldsymbol{\theta}_n = \boldsymbol{\theta}_0 \quad \text{(unknown)}$$

and

$$H_1\text{: there exists } 1 \leq m < n \text{ such that } \boldsymbol{\theta}_1 = \ldots = \boldsymbol{\theta}_m \neq \boldsymbol{\theta}_{m+1} = \ldots = \boldsymbol{\theta}_n.$$

The alternative hypothesis $H_1$ means that the observations $X_i$ follow the regression model with the vector parameter $\boldsymbol{\theta}_1$ till the unknown time point $\tau \in (t_m, t_{m+1}]$ and then they switch to the regression model with a different parameter $\boldsymbol{\theta}_n$. The time point $\tau$ is usually called the *change-point*.

Attention has been paid to the case when the error d.f. $F$ is normal with zero mean and known or unknown variance. The test procedures in this case are closely related to the least squares estimators [see, e.g., Brown *et al.* (1975), Hackl (1980), James *et al.* (1987, 1988)]. It is known that the quality of these estimators is highly sensitive to deviations from normality, to heavy-tailed distributions, and to outlying observations. For this reason, so-called *robust estimators* have been developed, which behave reasonably good and are rather insensitive to deviations from the assumed model. Typical robust estimators are $M$ (maximum likelihood type), $R$ (rank statistics type), and $L$ (linear combinations of order statistics type) estimators. For further information concerning these estimators, see, e.g., the monographs by P. Huber (1981) and Hampel *et al.* (1985).

The same arguments (and results of simulation studies) lead to the need of *robust procedures* for our testing problem $H_0$ vs. $H_1$. These can be based on either of the mentioned type of robust estimators. For the location model $X_i = \theta_i + e_i, i = 1, \ldots, n$, procedures based on both $M$- and $R$-estimators were developed and studied; for detailed information about the literature till 1986, see Hušková and Sen (1989). Recent results can be found in Hušková (1989a), Lombard (1987), and Csörgő and Horváth (1987, 1988).

For the regression model, main attention has been paid to robust procedures based on $M$-estimators for they are more appealing from the computational point of view than those based on $R$-estimators. The basic idea in constructing these $M$-test procedures is to replace the least squares estimators by $M$-estimators (or by certainly modified $M$-estimators) and the residuals by their robust counterparts. Generally, two types of $M$-test procedures have been developed: nonrecursive and recursive.

The $M$-estimator $\hat{\boldsymbol{\theta}}_k(\psi)$ of $\boldsymbol{\theta}_0$ (under $H_0$) generated by the function $\psi$ of $X_1, \ldots, X_k$ is defined as a solution of the equation

$$\sum_{i=1}^{k} \mathbf{c}_i \psi\left(X_i - \mathbf{c}_i' \boldsymbol{\theta}\right) = \mathbf{0}. \tag{2.1}$$

The maximum likelihood test for the normal d.f. $F$ leads to the *nonrecursive $M$-test procedure*, which is based on the $M$-residuals $\psi[X_k - \mathbf{c}_k' \hat{\boldsymbol{\theta}}_n(\psi)]$, $k = p+1, \ldots, n$, where $\hat{\boldsymbol{\theta}}_n(\psi)$ is the $M$-estimator of $\boldsymbol{\theta}_0$ generated by the score function $\psi$ based on the observations $X_1, \ldots, X_n$. These tests together with their asymptotic properties are discussed by Sen (1984) and Hušková (1988,1989a). Results of a simulation study are contained in Antoch and Hušková (1989).

Recursive tests (CUSUM, MOSUM) have been suggested for the normal d.f. $F$ based on the recursive residuals

$$X_i - \mathbf{c}_i'\boldsymbol{\theta}_{i-1}^*, \quad i = p+1, \ldots, n, \tag{2.2}$$

where $\boldsymbol{\theta}_{i-1}^*$ is the least squares estimator of $\boldsymbol{\theta}_0$ based on $X_1, \ldots, X_{i-1}$. [These tests are studied in detail by Hackl (1980)]. *Recursive M-test procedures* are based on the *M-recursive residuals*

$$W_i = \psi[X_i - \mathbf{c}_i'\boldsymbol{\theta}_{i-1}(\psi)], \quad i = p+1, \ldots, n, \tag{2.3}$$

where $\boldsymbol{\theta}_{i-1}(\psi)$ is the $M$-estimator of $\boldsymbol{\theta}_0$ generated by the score function $\psi$ of the observations $X_1, \ldots, X_{i-1}$, or some related estimator (some possibilities are mentioned in Section 2.3). Notice that for $\psi(x) = x$ the classical recursive residuals (2.2) are obtained.

In this chapter I focus on the recursive $M$-test procedure, robust versions of CUSUM and MOSUM tests. In Section 2.2 I describe the test procedures. Estimators $\boldsymbol{\theta}_{i-1}(\psi)$, $p < i \leq n$, are proposed in Section 2.3. The assumptions together with possible choices for $\psi$ are discussed in Section 2.4. In Section 2.5 I present results of a simulation study, while Section 2.6 contains remarks and recommendations.

## 2.2 Recursive $M$-Test Procedures

The CUSUM $M$-tests are based on the statistics

$$W_{k,c} = \left| \sum_{i=p+1}^{k} W_i \right| \hat{\sigma}_k^{-1}, \quad k = k_0, \ldots, n, \tag{2.4}$$

where $W_i$ is defined by (2.3); $k_0$ ($\geq p+1$) is suitably chosen, and $\hat{\sigma}_k^2$ is a consistent estimator of $\int \psi^2(x)dF(x)$ [some suggestions are the equations (2.6)–(2.8)] in Section 2.3.

Critical regions of the CUSUM $M$-tests are of the form

$$\bigcup_{k=k_0}^{n} \{W_{k,c} > w(\alpha, k, n)\},$$

where the critical values $w(\alpha, k, n)$ are chosen so that the asymptotic level is $\alpha$ (or $\leq \alpha$). The $w(\alpha, k, n)$ are not uniquely determined; different (asymptotic) arguments lead to different sets of critical values.

The Bonferroni inequality (together with large deviation results on $W_k$) leads to

$$w_1(\alpha, k, n) = k^{\frac{1}{2}} \Phi^{-1}\left[1 - \frac{\alpha}{2(n-k_0)}\right],$$

*Table 2.1:* Critical values $w_i = w_i(0.05, k, n)$ for the CUSUM $M$-tests.

| $w_i(0.05,k,n)$ | $n = 40$ | $n = 70$ | $n = 100$ |
|---|---|---|---|
| $w_1$ | $3.235\ k^{\frac{1}{2}}$ | $3.39\ k^{\frac{1}{2}}$ | $3.48\ k^{\frac{1}{2}}$ |
| $w_2$ | $3.182\ k^{\frac{1}{2}}$ | $3.2\ k^{\frac{1}{2}}$ | $3.22\ k^{\frac{1}{2}}$ |
| $w_3$ | 14.173 | 18.75 | 22.41 |
| $w_4$ | $4.896+0.316\ k$ | $7.411+0.2316\ k$ | $9.048+0.193\ k$ |

where $\Phi^{-1}$ is the quantile function corresponding to the standard normal distribution; the resulting test is (asymptotically) conservative. The application of modified results of Darling and Erdős (1956) gives

$$\begin{aligned} w_2(\alpha,k,n) &= k^{\frac{1}{2}}\Big[-\log\log(1-\alpha)^{-\frac{1}{2}} + 2\log\log n \\ &\quad + \frac{1}{2}\log\log\log n - \frac{1}{2}\log(4\pi)\Big]\,(2\log\log n)^{-\frac{1}{2}}. \end{aligned}$$

Since

$$\lim_{n\to\infty} \frac{\Phi^{-1}\left(1-n^{-1}\right)}{\sqrt{2\log n}} = 1\,, \tag{2.5}$$

we get

$$\lim_{n\to\infty} \frac{w_1(\alpha,k,n)}{w_2(\alpha,k,n)} = +\infty, \quad k = p+1,\ldots,n, \quad \alpha \in (0,1)\,.$$

Hence for large enough $n$, $w_1(\alpha,k,n) > w_2(\alpha,k,n)$ for $p \le k \le n, \alpha \in (0,1)$. Sen (1984) proposed critical values

$$w_3(\alpha,k,n) = n^{\frac{1}{2}} w_\alpha\,,$$

where $w_\alpha$ is defined by $P[\sup_{t\in[0,1]}\{|W(t)|\} > w_\alpha] = \alpha$ with $\{W(t),\ t \in [0,1]\}$ being the standardized Wiener process. For the classical CUSUM, Brown *et al.* (1975) suggested

$$w_4(\alpha,k,n) = h_\alpha(n-p)^{\frac{1}{2}}(2k+n-3p-3)/(n-p-1)\,,$$

with $h_\alpha$ fulfilling $1-\Phi(3h_\alpha)+\Phi(h_\alpha)e^{-4h_\alpha^2} = \alpha/2$ ($h_{0.01} = 1.143$; $h_{0.05} = 0.948$; $h_{0.1} = 0.85$). Asymptotic results show that these critical values are also appropriate for our situation.

Notice that none of these critical values depends on $\psi$ and that $w_1$ and $w_2$ are parabolic functions of $k$, $w_3$ is constant in $k$, and $w_4$ is linear in $k$. *Table 2.1* contains critical values $w_i(0.05,k,n)$ for $i = 1,\ldots,4$ and $n = 40, 70, 100$. The relation between the critical regions is shown in *Figure 2.1* for $\alpha = 0.05$, $n = 100$, $p = 2$.

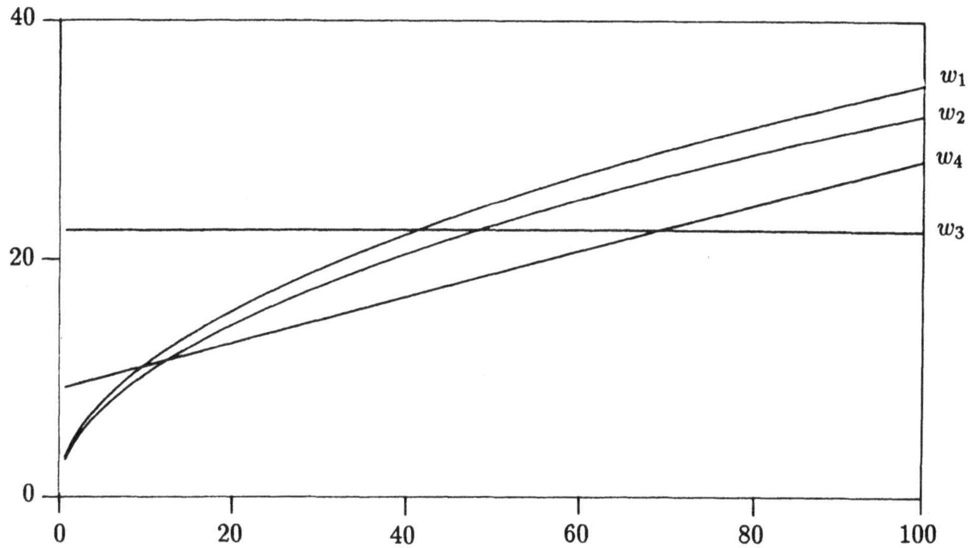


*Figure 2.1:* Critical regions $w_i$, $i = 1, \ldots, 4$, of the CUSUM $M$-tests.

The MOSUM $M$-tests are based on the statistics

$$W_{k,G} = \left| \sum_{i=k-G+1}^{k} W_i \right| \left( \hat{\sigma}_k \sqrt{G} \right)^{-1}, \quad k = k_0, \ldots, n,$$

where $W_i$ ($i = k_0, \ldots, n$) is defined in (2.3), $k_0$ ($\geq k - G + 1$) is suitably chosen, and $\hat{\sigma}_k^2$ is a consistent estimator of $\int \psi^2(x) dF(x)$. The critical regions of the MOSUM $M$-tests are of the form

$$\bigcup_{k=k_0}^{n} \{W_{k,G} > m(\alpha, h, n)\},$$

where $h = G/n$ and the critical values $m(\alpha, h, n)$ are determined so that the asymptotic level is $\alpha$ (or $\leq \alpha$).

The Bonferroni inequality gives critical values

$$m_1(\alpha, h, n) = \Phi^{-1}\left(1 - \frac{\alpha}{2[n - nh - k_0]}\right).$$

The test is asymptotically conservative. A modification of the results of Deheuvels and Révész (1987) leads (for $G/n$ small, see assumption $C_2$ below) to critical values

$$m_2(\alpha, h, n) = \left(2 \log h^{-1}\right)^{\frac{1}{2}} + \left[\log\log h^{-1} - \log \pi - 2 \log\log(1-\alpha)^{-\frac{1}{2}}\right] \left(8 \log h^{-1}\right)^{-\frac{1}{2}}.$$

*Table 2.2:* Critical values $m_1 = m_1(0.05, h, n)$ with $h = G/n$.

| $G$ | $n = 40$ | $n = 70$ | $n = 100$ | $n = 200$ |
|---|---|---|---|---|
| 5 | 3.17 | 3.35 | 3.45 | 3.65 |
| 10 | 3.15 | 3.33 | 3.43 | 3.65 |
| 20 | 3.00 | 3.27 | 3.42 | 3.63 |

*Table 2.2* contains critical values $m_1(\alpha, h, n)$ for $\alpha = 0.05$ and $k_0 = 2$. By direct computation we obtain

$$\begin{aligned} m_2(0.05; 0.05) &= 3.935 \\ m_2(0.05; 0.10) &= 3.780 \\ m_2(0.05; 0.15) &= 3.699 \\ m_2(0.05; 0.20) &= 3.6495. \end{aligned}$$

If $G \to \infty$ and $Gn^{-\beta} \to 0$ as $n \to \infty$ then from (2.5) follows $m_1(\alpha, h, n) > m_2(\alpha, h, n)$ for large enough $n$ and $\alpha \in (0, 1)$.

One should keep in the mind that all presented critical values $w_i(\alpha, k, n)$ and $m_i(\alpha, h, n)$ are based on asymptotic distributions (the convergence could be rather slow); hence $n$ should be large enough. Furthermore, $k_0$ should not be too small because both the estimators $\boldsymbol{\theta}_k(\psi)$ and $\hat{\sigma}_k$ can considerably oscillate for small $k$. Theoretical results are derived by Sen (1984) and Hušková (1989b,c,d).

The CUSUM and the MOSUM $M$-tests described above are not scale invariant (similarly as the usual $M$-estimators), which means that the critical regions corresponding to the observations $(X_1, \ldots, X_n)$ and $(cX_1, \ldots, cX_n)$, $0 < c \neq 1$, are generally different. However, a simple modification (called studentization) of the procedures results in scale invariant procedures. This *studentization* consists in replacing $X_i - \mathbf{c}_i'\boldsymbol{\theta}_{i-1}(\psi)$ and $X_i - \mathbf{c}_i'\boldsymbol{\theta}_k(\psi)$ by $[X_i - \mathbf{c}_i'\boldsymbol{\theta}_{i-1}(\psi)]\, s_k^{-1}$ and $[X_i - \mathbf{c}_i'\boldsymbol{\theta}_k(\psi)]\, s_k^{-1}$, respectively, where $s_k$ is an estimator of a scale functional $\sigma = \sigma(F) > 0$ with the properties

$$s_k[\mathbf{X}(k)] > 0, \quad s_k[a\,\{\mathbf{X}(k) + \mathbf{c}(k)\,b\}] = a\, s_k\,[\mathbf{X}(k)]\,,$$

where $\mathbf{X}(k) = (X_1, \ldots, X_k)'$, $\mathbf{c}(k) = (\mathbf{c}_1, \ldots, \mathbf{c}_k)'$, $b > 0$, $a \in R_1$. The resulting procedures are scale invariant. Since $s_k$ can considerably oscilate for small $k$, $\sigma(F)$ should preferably be estimated independently of $X_1, \ldots, X_n$.

## 2.3 The Estimators $\hat{\sigma}_k$ and $\boldsymbol{\theta}_k(\psi)$

As an estimator of $\int \psi^2(x)\, dF(x)$ we can use either

$$\hat{\sigma}_k^2 = k^{-1} \sum_{i=p+2}^{k} \psi^2[X_i - \mathbf{c}_i'\boldsymbol{\theta}_{i-1}(\psi)]\,, \quad k = k_0, \ldots, n\,, \tag{2.6}$$

or

$$\hat{\sigma}_k^2 = k^{-1}\sum_{i=1}^{k}\psi^2[X_i - \mathbf{c}_i'\boldsymbol{\theta}_k(\psi)], \quad k = k_0, \dots, n, \tag{2.7}$$

which are consistent (for $n \to \infty$) even under the local alternative; however, they can oscillate for small $k$. Intuitively one may expect that the estimator

$$\hat{\sigma}_k^2 = k^{-1}\sum_{i=1}^{k}\psi^2\left[X_i - \mathbf{c}_i'\boldsymbol{\theta}_k(\psi)\right] - \overline{\psi}_k^2, \quad k = k_0, \dots, n, \tag{2.8}$$

with $\overline{\psi}_k = k^{-1}\sum_{i=1}^{k}\psi[X_i - \mathbf{c}_i'\boldsymbol{\theta}_k(\psi)]$ would be better than (2.6) and (2.7) as it suppresses the influence of moderate alternatives.

The estimator $\boldsymbol{\theta}_k(\psi)$ of $\boldsymbol{\theta}_0$ should have the property

$$\max_{k_n \le k \le n}\left\{k^v\left\|\mathbf{C}_k^{\frac{1}{2}}\left[\boldsymbol{\theta}_k(\psi) - \boldsymbol{\theta}_0\right] - \mathbf{C}_k^{-\frac{1}{2}}[\lambda'(0)]^{-1}\sum_{i=1}^{k}\mathbf{c}_i\psi\left(X_i - \mathbf{c}_i'\boldsymbol{\theta}_0\right)\right\|\right\} = O_p(1)$$

as $n \to \infty$ and $k_n \to \infty$ for some $v > 0$; here, $\boldsymbol{\theta}_0$ contains the values of the parameters and $\|\cdot\|$ denotes the Euclidean norm. Reasonable candidates are the stochastic approximation-type estimators $\boldsymbol{\theta}_k(\psi)$ and the recursive $M$-estimators $\tilde{\boldsymbol{\theta}}_k(\psi)$.

The *stochastic approximation-type estimator* is defined by

$$\boldsymbol{\theta}_{k+1}(\psi) = \boldsymbol{\theta}_k(\psi) + \gamma_k^{-1}\mathbf{C}_{k+1}^{-1}\sum_{i=1}^{k+1}\mathbf{c}_i\psi\left[X_i - \mathbf{c}_i'\boldsymbol{\theta}_k(\psi)\right], \quad k = k_n, \dots, n,$$

where $\boldsymbol{\theta}_{k_n}(\psi)$ is an initial estimator based on $X_1, \dots, X_{k_n}$; a suitable choice for $\boldsymbol{\theta}_{k_n}(\psi)$ is the usual $M$-estimator generated by the score function $\psi$. The quantity $\gamma_k$ estimates $\lambda'(0)$, which is the derivative of $\lambda(t) = -\int\psi(x-t)dF(x)$ at $t = 0$. An usual choice is

$$\gamma_k = \begin{cases} \gamma_k^* & \text{if} \quad a_k < \gamma_k^* < a_k^{-1} \\ a_k & \text{if} \quad \gamma_k^* \le a_k \\ a_k^{-1} & \text{if} \quad \gamma_k^* \ge a_k^{-1} \end{cases}$$

for $k = k_n, \dots, n$, where

$$\gamma_k^* = \left(2tk^{\frac{1}{2}}\right)^{-1}\sum_{i=1}^{k}\left\{\psi\left[X_i - \mathbf{c}_i'\boldsymbol{\theta}_k(\psi) + tk^{-\frac{1}{2}}\right] - \psi\left[X_i - \mathbf{c}_i'\boldsymbol{\theta}_k(\psi) - tk^{-\frac{1}{2}}\right]\right\}$$

with $t > 0$ fixed and $a_k \searrow 0$. If the function $\psi$ is smooth, $\gamma_k^*$ can be replaced by

$$\gamma_k^{**} = k^{-1}\sum_{i=1}^{k}\psi'\left[X_i - \mathbf{c}_i'\boldsymbol{\theta}_k(\psi)\right].$$

Asymptotic properties of these estimators are studied in Hušková (1989b).

The *recursive M-estimator* $\tilde{\boldsymbol{\theta}}_k(\psi)$ is defined as follows:

$$\tilde{\boldsymbol{\theta}}_{k+1}(\psi) = \tilde{\boldsymbol{\theta}}_k(\psi) + \tilde{\gamma}_k^{-1}\mathbf{C}_{k+1}^{-1}\mathbf{c}_{k+1}\psi\left[X_{k+1} - \mathbf{c}_{k+1}'\tilde{\boldsymbol{\theta}}_k(\psi)\right], \quad k = k_n, \ldots, n, \tag{2.9}$$

where $\tilde{\boldsymbol{\theta}}_{k_n}(\psi)$ is an initial estimator based on $X_1, \ldots, X_{k_n}$; a suitable choice for $\tilde{\boldsymbol{\theta}}_{k_n}(\psi)$ is the usual $M$-estimator generated by the score function $\psi$. Similarly as in the previous case, $\tilde{\gamma}_k$ is an estimator of $\lambda'(0)$, usually of the form

$$\tilde{\gamma}_k = \begin{cases} \tilde{\gamma}_k^* & \text{if} \quad a_k < \tilde{\gamma}_k^* < a_k^{-1} \\ a_k & \text{if} \quad \tilde{\gamma}_k^* \leq a_k \\ a_k^{-1} & \text{if} \quad \tilde{\gamma}_k^* \geq a_k^{-1} \end{cases}$$

for $k = k_n, \ldots, n$, where

$$\tilde{\gamma}_k^* = (2tk)^{-1}\sum_{i=1}^{k}\left\{\psi\left[X_i - \mathbf{c}_i'\tilde{\boldsymbol{\theta}}_{i-1}(\psi) + ti^{-\frac{1}{2}}\right] - \psi\left[X_i - \mathbf{c}_i'\tilde{\boldsymbol{\theta}}_{i-1}(\psi) - ti^{-\frac{1}{2}}\right]\right\} i^{\frac{1}{2}}$$

with $t > 0$ fixed and $a_k \searrow 0$. If the function $\psi$ is smooth, $\tilde{\gamma}_k^*$ can be replaced by

$$\tilde{\gamma}_k^{**} = k^{-1}\sum_{i=1}^{k}\psi'\left[X_i - \mathbf{c}_i'\tilde{\boldsymbol{\theta}}_k(\psi)\right].$$

Notice that for computation the following relation is useful:

$$\tilde{\gamma}_{k+1}^* = \tilde{\gamma}_k^*\left[k(k+1)\right]^{\frac{1}{2}} + \left[2t(k+1)^{\frac{1}{2}}\right]^{-1}$$
$$\times\left\{\psi\left[X_{k+1} - \mathbf{c}_{k+1}'\tilde{\boldsymbol{\theta}}_k(\psi) + t(k+1)^{-\frac{1}{2}}\right] - \psi\left[X_{k+1} - \mathbf{c}_{k+1}'\tilde{\boldsymbol{\theta}}_k(\psi) - t(k+1)^{-\frac{1}{2}}\right]\right\}$$

The asymptotic properties of these estimators are studied, e.g., in Poljak and Tsypkin (1979).

The recursive $M$-estimators are more appealing from the computational point of view. However, a small simulation study shows that the convergence of the stochastic approximation-type estimators is somewhat faster than that of the recursive ones.

We may also use the usual $M$-estimator generated by the function $\psi$ [see Sen (1984)], which leads to rather long computations (because they are defined implicitely and an iterative procedure has to be applied after every new observation).

## 2.4 Choice of the Score Function $\psi$ and Assumptions

In this section the assumptions on the score function $\psi$, the distribution $F$, and the regression vectors $\mathbf{c}_i$ are formulated and some typical $\psi$ functions are given.

Under the following assumptions, the tests described in Section 2.2 have asymptotic level $\alpha$ (or $\leq \alpha$):

$A_1$: $\psi$ is nondecreasing; there exist positive constants $D_1, D_2$ such that

$$\sup\left\{\int \psi^2(x-a)\,dF(x), |a| \le D_1\right\} < +\infty,$$

$$\int [\psi(x-a) - \psi(x-b)]^2\,dF(x) \le D_2|a-b|^2$$

for $|a| \le D_1$, $|b| \le D_1$.

$A_2$: $\int |\psi(x)|^3 dF(x) < +\infty$.

$A_3$: $\psi$ is bounded.

$B$: The function $\lambda(a) = -\int \psi(x-a)dF(x)$, $a \in R_1$, fulfills $\lambda(0) = 0$; there exists the first derivative $\lambda'(0) > 0$ and

$$|\lambda'(a) - \lambda'(b)| \le D_3|a-b|^r$$

for $|a| \le D_4$, $|b| \le D_4$, some $D_3 > 0$, $D_4 > 0$, and $r > 0$.

$C_1$: The regression vectors $\mathbf{c}_i = (c_{i1}, \ldots, c_{ip})'$, $i = 1, \ldots, n$, fulfill

$$n^{-1}\sum_{i=1}^{[nt]} \mathbf{c}_i\mathbf{c}_i' \to t\mathbf{C} \quad \text{as} \quad n \to \infty \quad \text{for} \quad t \in [0,1]$$

$$\limsup_{n\to\infty} n^{-1}\sum_{i=1}^{n} c_{ij}^4 < +\infty, \quad \limsup_{n\to\infty} \max_{1\le i\le n} \left\{c_{ij}^2 n^{-1}\log^3 n\right\} < +\infty,$$

for $j = 1, \ldots, p$, where $\mathbf{C}$ is a positive definite matrix and $[a]$ denotes the integer part of $a$.

$C_2$: The regression vectors $\mathbf{c}_i = (c_{i1}, \ldots, c_{ip})'$, $i = 1, \ldots, n$, fulfill

$$\limsup_{n\to\infty} \max_{G_n+p<k\le n} G_n^{-1} \sum_{i=k-G_n+1}^{k} |c_{ij}| < +\infty$$

$$G_n/n \to 0, \quad G_n \log^{-3} n \to \infty \qquad \text{as} \quad n \to \infty.$$

For the CUSUM $M$-procedure one needs assumptions $A_1$, $A_2$, $B$, and $C_1$ while assumptions $A_1$, $A_3$, $B$, $C_1$, and $C_2$ are connected with the MOSUM $M$-procedure. Assumptions $A_1$ and $B$ are fulfilled by typical $\psi$ functions [see, e.g., Huber (1981), Hampel *et al.* (1985)] and for a large family of distribution function $F$, e.g.:

- $\psi(x) = x$ (which leads to the classical least squares estimators) and $\int x^2 dF(x) < +\infty$.
- The Huber $\psi$ function $[\psi(x) = \text{sign}x \min(|x|, K),\ x \in R_1]$, and $F$ has a bounded derivative in a neighborhood of $\pm K$.
- $\psi$ is a step function having a finite number of jumps (of finite magnitude), and $F$ has bounded derivatives $f$ and $f'$ in neighborhood of the jump points of $\psi$.

It can easily be shown that also nonmonotone functions $\psi$ can be used; some of them, e.g., the Tukey biweight function $[\psi(x) = 2x\left(1 - x^2\right)$ for $|x| \leq 1$, $\psi(x) = 0$ otherwise$]$ and the Cauchy $\psi$ function $[\psi(x) = 2x(1 + x^2)^{-1},\ x \in R_1]$ are of particular interest.

Assumptions $C_1$ and $C_2$ mean that only a small part of the $c_{ij}$ can be larger and that the design matrix $(\mathbf{c}_1, \ldots, \mathbf{c}_n)'$ is asymptotically, in a certain sense, regular. The assumptions imposed on $G$ require that $G$ is large enough, however small enough with respect to $n$.

## 2.5 A Simulation Study

The following simple linear model was considered:

$$\begin{aligned} X_i &= 5 + 2c_i + e_i, \quad i = 1, \ldots, n/2, \\ X_i &= 5 + 2c_i + e_i + \theta, \quad i = n/2 + 1, \ldots, n, \end{aligned}$$

where the $X_i$ are observations, $c_i = 10(1 + 2i/n)$ for $i = 1, \ldots, n$, and the $e_i$'s are i.i.d. errors distributed according to the d.f. $F$. Simulations were done for sample sizes $n = 40$ and 70, and for shifts $\theta = 0$ and 2. For $F$, the normal distribution $N(0,1)$ with parameters 0 and 1 and the Laplace distribution $L(0,1)$ with parameters 0 and 1 were chosen.

For all of these situations 300 and 100 repetitions of the experiments were done with $n = 40$ and $n = 70$, respectively; the maximum of the MOSUM test statistics

$$\hat{M} = \max_{12(15) \leq k \leq 40(70)} \left\{ \left(\sqrt{G}\hat{\sigma}_k\right)^{-1} \left| \sum_{i=k-G+1}^{k} W_i \right| \right\}$$

and the corresponding index

$$\hat{k} = \arg \max_{12(17) \leq k \leq 40(70)} \left\{ \left(\sqrt{G}\hat{\sigma}_k\right)^{-1} \left| \sum_{i=k-G+1}^{k} W_i \right| \right\}$$

were computed for $G = 10$ and 15 and for $\psi_1(x) = x$, $x \in R_1$, and the Huber function

$$\psi_2(x) = \begin{cases} x & \text{if } |x| \leq 1.7\,s \\ 1.7\,s\,\text{sign}\,x & \text{if } |x| > 1.7\,s\,, \end{cases}$$

where $s$ is an estimator of $2F^{-1}(3/4) - 1$.

*Figures 2.2–2.5* show the histograms of $\hat{k}$ under the null hypothesis $H_0$ ($\theta = 0$) for $n = 40$, $G = 10$, while *Figures 2.6–2.9* present the histograms of $\hat{k}$ under the alternative ($\theta = 2$). *Figures 2.10–2.17* contain the histograms of $\hat{M}$ for the same situations.

It can be seen that the frequencies of $\hat{k}$ under $H_0$ are almost uniformly distributed while under the alternative there is a peak just after the change (i.e., around the 25th observation). There is a considerable difference in the frequencies of $\hat{M}$ between the null and the alternative hypothesis if the error d.f. is $N(0, 0.5)$ and $\psi_1$ is used; the frequencies of $\hat{M}$ correspond to the asymptotic results. However, using $\psi_1$ if the error d.f. is $L(0,1)$, the frequencies of $\hat{M}$ do not behave as one would wish. The results for $\psi_2$ are acceptable in both cases.

The results for $n = 70$ and as well as for the CUSUM procedure are quite similar. *Figure 2.18* shows a typical behavior of

$$W_{k,G} = \left(\sqrt{G}\hat{\sigma}_k\right)^{-1} \left| \sum_{i=k-G+1}^{k} W_i \right|, \quad 12 \leq k \leq 70.$$

*Figure 2.18* indicates that the MOSUM $M$-tests could also be used for the detection of more than one change (if the changes do not follow close to each other). The simulations were done by my colleague J. Antoch; an extensive study will be published elsewhere.

## 2.6 Concluding Remarks

To apply the test procedures described in Section 2.2 one has to choose $\psi$ and the critical value. As for the choice of $\psi$ one should follow the recommendations made for the $M$-estimators: If the error d.f. is normal, choose $\psi(x) = x$, $x \in R_1$; if a nonnormal distribution is suspected (mostly the distribution with some heavier tails such as the Laplace distribution) use the Huber $\psi$ function (or another one). For more advanced recommendations see, e.g., Huber (1981) or Hampel *et al.* (1985).

Since the critical values introduced in Section 2.2 are based on asymptotic distributions (for $n \to \infty$) and the convergence for a nonnormal error d.f. $F$ is quite slow, the mentioned critical values should be used very carefully and probably only for preliminary inference if $n$ is moderate. For the final inference (decision) one should carefully inspect the behavior of either

$$\left(\sqrt{G}\hat{\sigma}_k\right)^{-1} \left| \sum_{i=k-G+1}^{k} W_i \right|, \quad k = k_0, \ldots, n,$$

or

$$W_{k,c}, \quad k = k_0, \ldots, n.$$

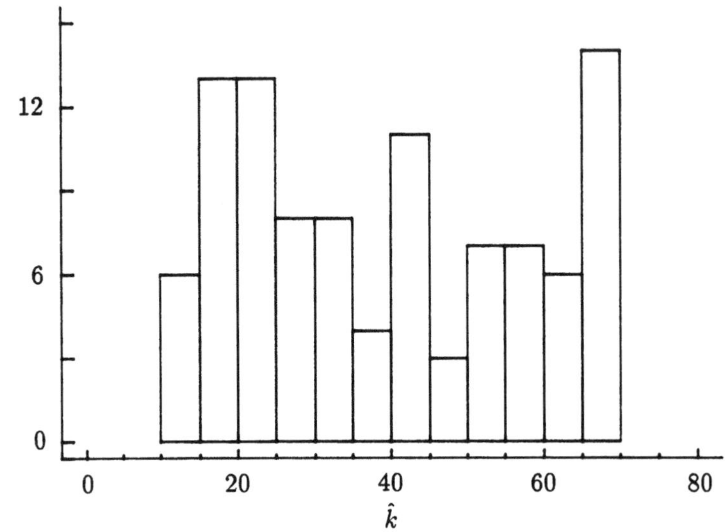


*Figure 2.2:* Histogram of $\hat{k}$ under $H_0$ ($\theta = 0$), $N(0, 0.5)$, Huber's $\psi$.

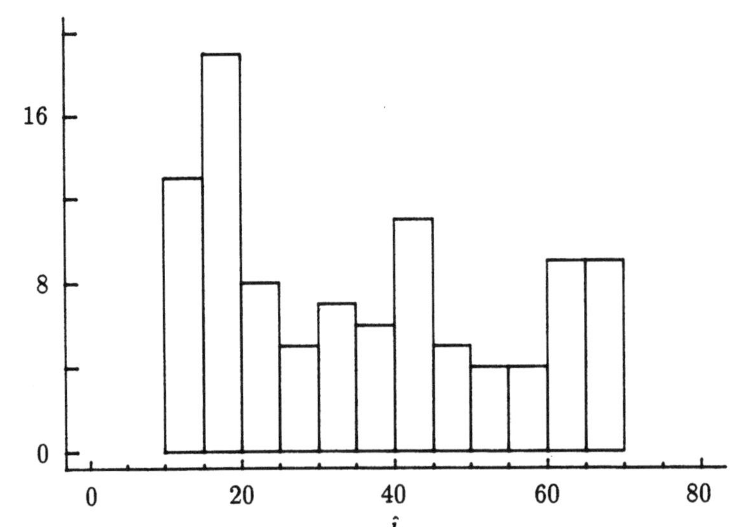


*Figure 2.3:* Histogram of $\hat{k}$ under $H_0$ ($\theta = 0$), $N(0, 0.5)$, $\psi(x) = x$.

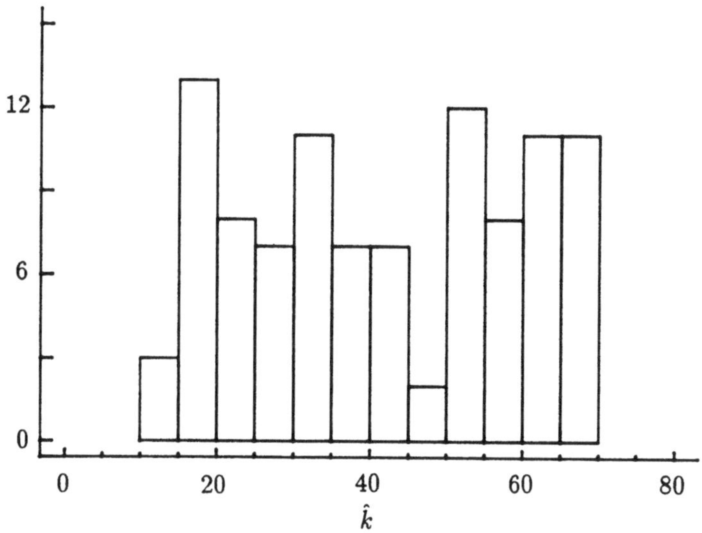


*Figure 2.4:* Histogram of $\hat{k}$ under $H_0$ ($\theta = 0$), $L(0,1)$, Huber's $\psi$.

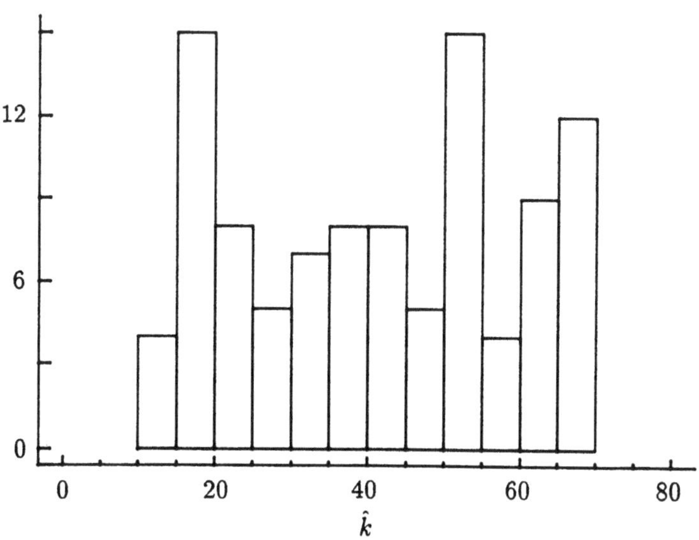


*Figure 2.5:* Histogram of $\hat{k}$ under $H_0$ ($\theta = 0$), $L(0,1)$, $\psi(x) = x$.

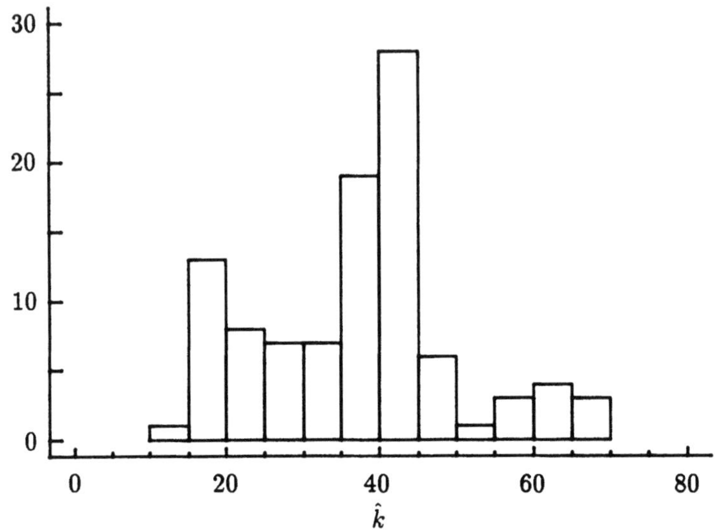


*Figure 2.6:* Histogram of $\hat{k}$ under $\theta = 2$, $N(0, 0.5)$, Huber's $\psi$.

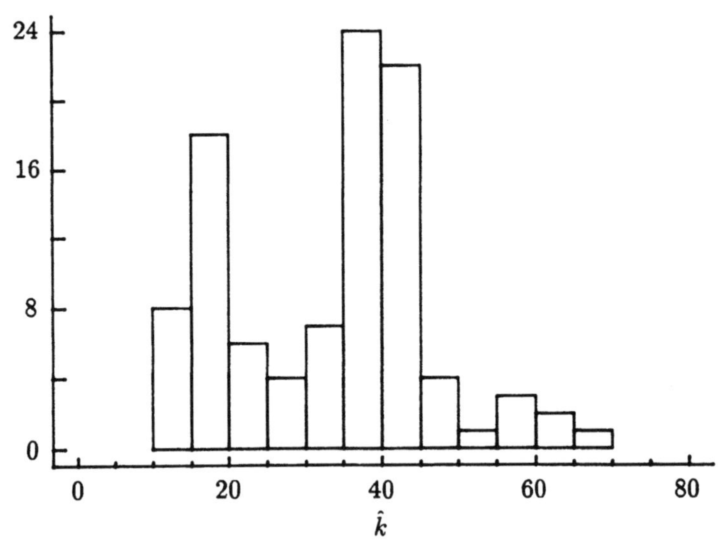


*Figure 2.7:* Histogram of $\hat{k}$ under $\theta = 2$, $N(0, 0.5)$, $\psi(x) = x$.

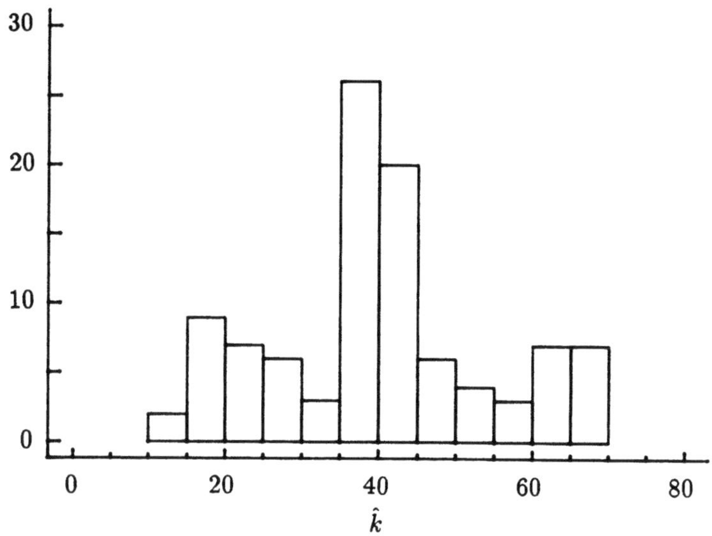


*Figure 2.8:* Histogram of $\hat{k}$ under $\theta = 2$, $L(0,1)$, Huber's $\psi$.

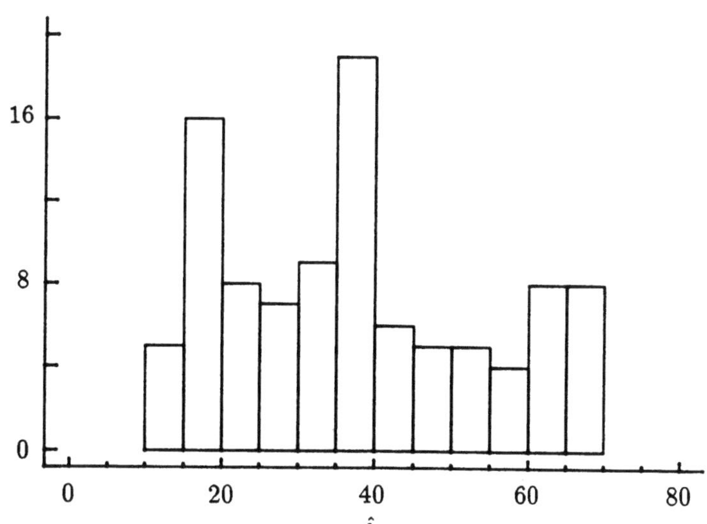


*Figure 2.9:* Histogram of $\hat{k}$ under $\theta = 2$, $L(0,1)$, $\psi(x) = x$.

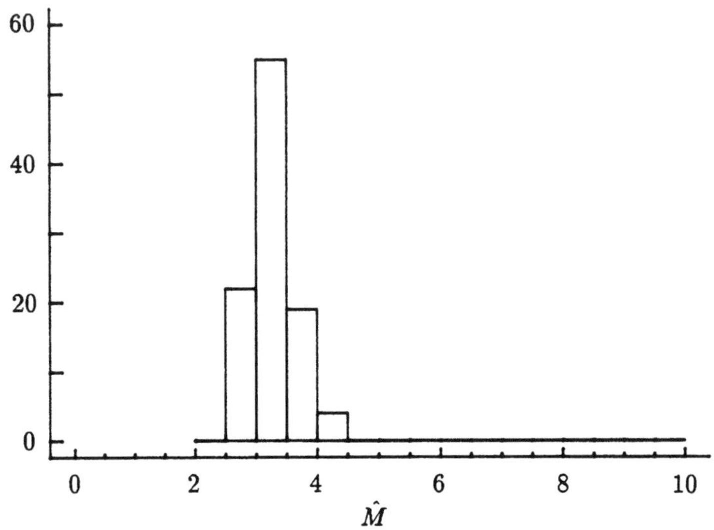


*Figure 2.10:* Histogram of $\hat{M}$ under $H_0$ ($\theta = 0$), $N(0, 0.5)$, Huber's $\psi$.

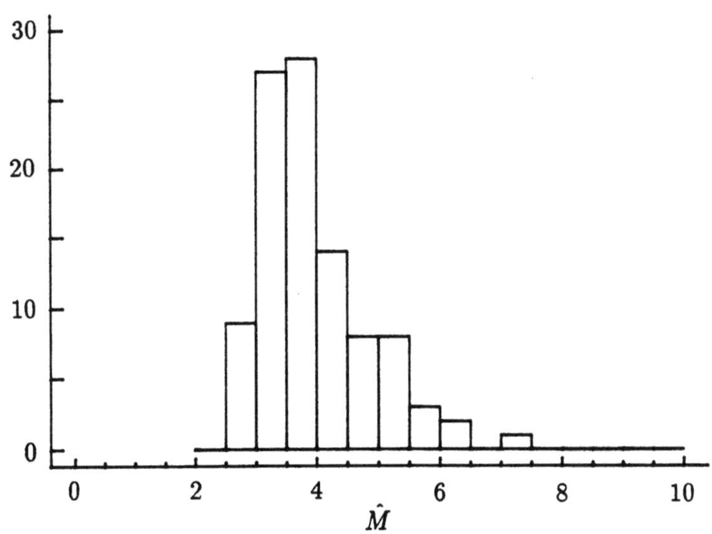


*Figure 2.11:* Histogram of $\hat{M}$ under $H_0$ ($\theta = 0$), $N(0, 0.5)$, $\psi(x) = x$.

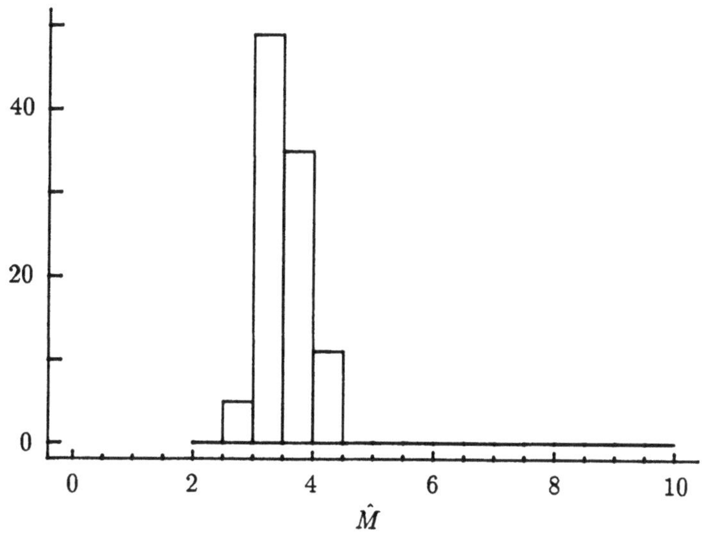


*Figure 2.12:* Histogram of $\hat{M}$ under $H_0$ ($\theta = 0$), $L(0,1)$, Huber's $\psi$.

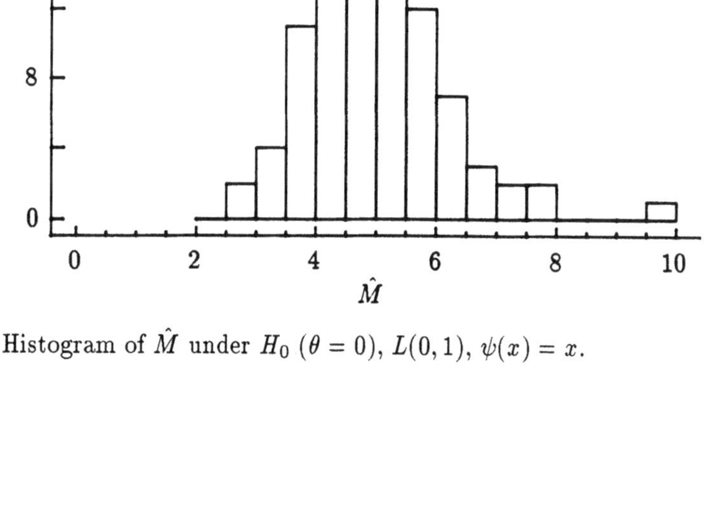


*Figure 2.13:* Histogram of $\hat{M}$ under $H_0$ ($\theta = 0$), $L(0,1)$, $\psi(x) = x$.

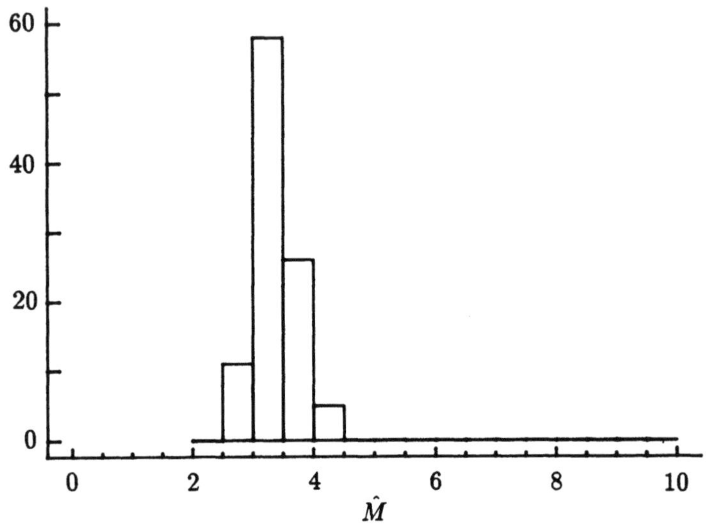


*Figure 2.14:* Histogram of $\hat{M}$ under $\theta = 2$, $N(0, 0.5)$, Huber's $\psi$.

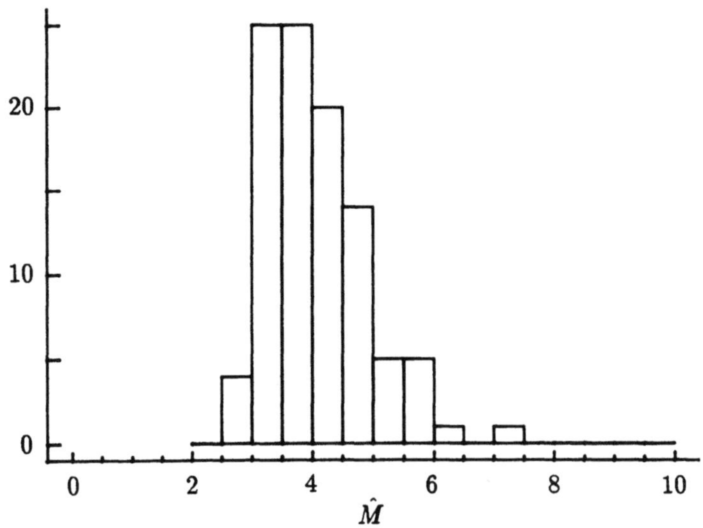


*Figure 2.15:* Histogram of $\hat{M}$ under $\theta = 2$, $N(0, 0.5)$, $\psi(x) = x$.

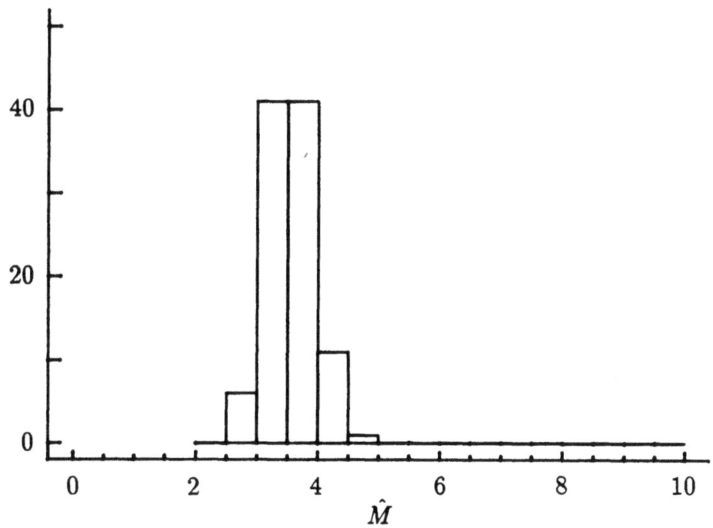


*Figure 2.16:* Histogram of $\hat{M}$ under $\theta = 2$, $L(0,1)$, Huber's $\psi$.

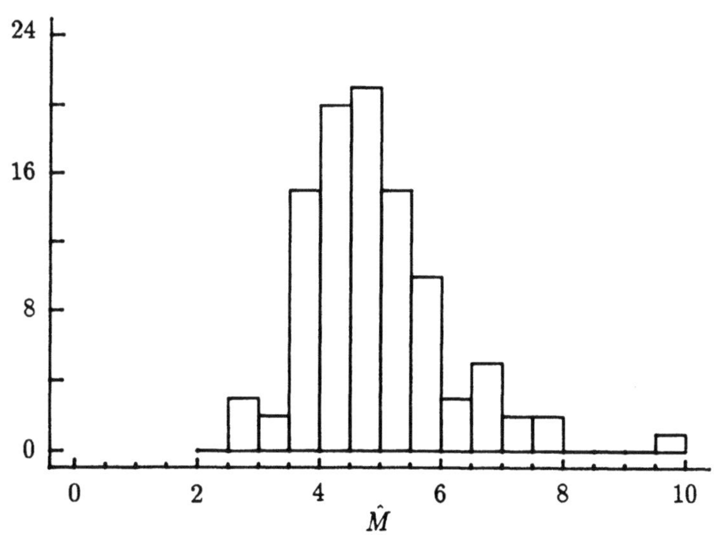


*Figure 2.17:* Histogram of $\hat{M}$ under $\theta = 2$, $L(0,1)$, $\psi(x) = x$.

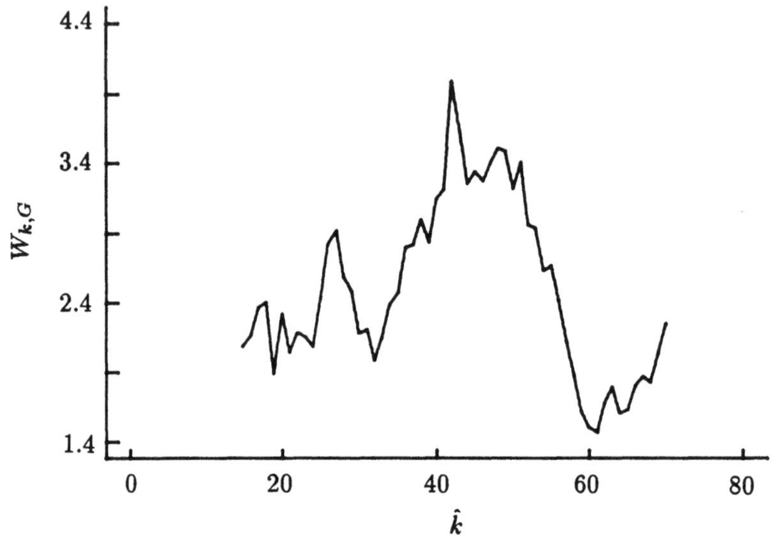


*Figure 2.18:* Statistic $W_{k,G}$ of the MOSUM *M*-test for $k = 12, \ldots, 70$.

## Acknowledgment

I would like to express many thanks to my colleague J. Antoch who kindly has done the simulation study.

## References

Antoch, J. and Hušková, M. (1989), Some *M*-tests for detection of a change in linear models, pp. 123-136 in P. Mandl and M. Hušková (eds.), *Proceedings of the Fourth Prague Symposium on Asymptotic Statistics.*

Brown, R.L., Durbin, J., and Evans, J.M. (1975), Techniques for testing the constancy of regression relationships over time (with discussion). *Journal of the Royal Statistical Society*, **B-37**, 149-192.

Csőrgő, M. and Horváth, L. (1987), Nonparametric tests for the changepoint problem. *Journal of Statistical Planning and Inference*, **17**, 1-9.

Csőrgő, M. and Horváth, L. (1988), Nonparametric method for changepoint problems, pp. 403-425 in P.R. Krishnaiah and C.R. Rao (eds.), *Handbook of Statistics, Volume 7.* New York: John Wiley.

Darling, D.A. and Erdős, P. (1956), A limit theorem for the maximum of normalized sums of independent random variables. *Duke Mathematical Journal*, **23**, 143-155.

Deheuvels, P. and Révész, P. (1987), Weak laws for the increments of Wiener processes, Brownian bridges, empirical processes and partial sums of i.i.d.r.v.'s, pp. 69-88 in M.L. Puri *et al.* (eds.), *Mathematical Statistics and Probability Theory, Volume A.* Dordrecht: D. Reidel.

Hackl, P. (1980), *Testing the Constancy of Regression Models Over Time.* Göttingen: Vandenhoeck and Ruprecht.

Hampel, F.R., Ronchetti, E.M., Rousseeuw, P.J., and Stahel, W.R. (1985), *Robust Statistics. The Approach Based on Influence Functions.* New York: John Wiley.

Huber, P.J. (1981), *Robust Statistics.* New York: John Wiley.

Hušková, M. (1988). Recursive $M$-tests for detection of change. *Sequential Analysis*, **7**, 75-90.

Hušková, M. (1989a), Nonparametric tests for change in regression models at an unknown time point. Paper presented at the international workshop: Theory and Practice in Data Analysis, Berlin, 1988.

Hušková, M. (1989b), Stochastic approximation type estimators in linear models. Submitted.

Hušková, M. (1989c), Some asymptotic results for CUSUM $M$-tests. Submitted.

Hušková, M. (1989d), Some asymptotics for MOSUM $M$-tests. Submitted.

Hušková, M. and Sen, P.K. (1989), Nonparametric tests for shift and change in regression at an unknown time point, pp. 73-87 in P. Hackl (ed.), *Statistical Analysis and Forecasting of Economic Structural Change.* Berlin: Springer-Verlag.

James, B., James, K.L., and Siegmund, D. (1987), Tests for change point. *Biometrika*, **74**, 71-84.

James, B., James, K.L., and Siegmund, D. (1988). Conditional boundary crossing probabilities with applications to change-point problems. *Annals of Probability*, **16**, 825-839.

Lombard, F. (1987), Rank tests for change-point problem. *Biometrika*, **74**, 615-624.

Poljak, B.T. and Tsypkin, Ya.Z. (1979), Adaptive estimation algorithms. *Automation and Remote Control*, **3**, 72-84.

Sen, P.K. (1984), Recursive $M$-tests for the constancy of multivariate regression relationships over time. *Sequential Analysis*, **3**, 191-211.

CHAPTER 3

# Recursive vs. OLS Residuals in the CUSUM Test

*Walter Krämer, Werner Ploberger, and Irene Schlüter*

## Summary

We extend the well-known CUSUM test for the constancy of the coefficients of a linear regression model, which is usually based on recursive residuals, to ordinary least squares (OLS) residuals. We show how to modify the test statistic, derive its limiting distribution under $H_0$, and compare the finite sample power of the two versions of the test via Monte Carlo experiments.

## 3.1 The Model and the Tests

We consider the familiar Brown-Durbin-Evans (1975) CUSUM test for structural change in the linear regression model

$$y_t = x_t'\beta + u_t \qquad t = 1, \ldots, T, \tag{3.1}$$

where $y_t$ is the dependent variable, $x_t = (1, x_{t2}, \ldots, x_{tK})'$ is a $K \times 1$ vector of independent variables (including a constant), such that

$$\operatorname*{plim}_{T\to\infty} \frac{1}{T} \sum_{t=1}^{T} x_t x_t' = R \tag{3.2}$$

for some finite and nonsingular $K \times K$ matrix $R$, the $u_t$'s are i.i.d.$(0, \sigma^2)$ disturbances (not necessarily normal), and $\beta$ is a $K \times 1$ vector of regression coefficients. The null hypothesis under test is that this vector remains constant over time.

The standard CUSUM test is based on recursive residuals (standardized forecast errors)

$$\tilde{u}_t = [y_t - x_t'\hat{\beta}^{(t-1)}]/f_t\,, \tag{3.3}$$

where

$$\hat{\beta}^{(t-1)} = \left(\sum_{i=1}^{t-1} x_i x_i'\right)^{-1} \sum_{i=1}^{t-1} x_i y_i$$

is the OLS estimate for $\beta$ from the first $t-1$ observations ($t = K+1, \ldots, T$; we tacitly assume that $\sum_{i=1}^{K} x_i x_t'$ has rank $K$) and where

$$f_t = \left[1 + x_t' \left(\sum_{i=1}^{t-1} x_i x_i'\right)^{-1} x_t\right]^{1/2} .$$

It rejects the null hypothesis of parameter constancy for large values of

$$\sup_{0 \le z \le 1} \left| \frac{W^{(T)}(z)}{1+2z} \right| , \tag{3.4}$$

where

$$W^{(T)}(z) = \frac{1}{\hat{\sigma}\sqrt{T-K}} \sum_{t=K+1}^{K+z(T-K)} \tilde{u}_t, \qquad 0 \le z \le 1\,, \tag{3.5}$$

is the cumulated sum of the first $z(T-K)$ recursive residuals, and where

$$\hat{\sigma} = \sqrt{\frac{1}{T} \sum_{t=K+1}^{T} \tilde{u}_t^2}\;. \tag{3.6}$$

This chapter is concerned with the OLS-based analogue of (3.5). Let

$$\hat{u}_t^{(T)} = y_t - x_t'\hat{\beta}^{(T)}$$

be the $t$-th OLS residual, where $\hat{\beta}^{(T)} = (\sum_{t=1}^{T} x_t x_t')^{-1} \sum_{t=1}^{T} x_t y_t$ is the OLS estimator for $\beta$ from the full sample and where the superscript $T$ emphasizes the fact that OLS residuals, unlike recursive residuals, change as the sample increases. Following Ploberger and Krämer (1988), we suggest a test based on

$$B^{(T)}(z) = \frac{1}{\hat{\sigma}\sqrt{T}} \sum_{t=1}^{zT} \hat{u}_t^{(T)}, \qquad 0 \le z \le 1\,, \tag{3.7}$$

i.e., the cumulated sum of the first $zT$ OLS residuals. This is very similar to the standard form in (3.5). The only difference is that summation starts at $t = 1$ and that OLS residuals rather than recursive residuals are used. MacNeill (1978) has considered similar tests in the context of very special regressors (constants or polynomial trends). Otherwise, OLS residuals have not been very popular in CUSUM-type specification tests. McCabe and Harrison (1980) provide an OLS-based extension of the CUSUM of squares test, but a similar extension of the CUSUM test has not yet been given. A technical reason is that OLS residuals, unlike recursive residuals, are stochastically dependent and heteroscedastic even under $H_0$. This makes it hard to derive the null distribution of any test. In addition, OLS residuals always sum to zero when the regression contains a constant, whether there is a structural change or not, so the intuition behind the standard CUSUM test (i.e., that the cumulated sum of the recursive residuals will eventually drift off after a structural change) does not apply to the OLS-based version of the test.

On the other hand, the standard CUSUM test is rather weak when structural changes are orthogonal to the $x_t$'s [see Ploberger and Krämer (1989) or Section 3.2 below] or when a structural change occurs late in the sample period (since this leaves the cumulated sums of forecast errors little time to cumulate), whereas a structural change at any time affects all the OLS residuals. Thus there is some prior justification for probing into the usefulness of their cumulated sums.

We suggest to reject the null hypothesis of parameter constancy for large values of

$$\sup_{0 \leq z \leq 1} |B^{(T)}(z)| . \tag{3.8}$$

This differs from the standard case. The standard test statistic (3.4) is equivalent to rejecting the null hypothesis for cumulated sums outside some bounds as in *Figure 3.1*, whereas the statistic (3.8) amounts to rejecting for cumulated sums outside the bounds in *Figure 3.2* (where we have added a typical sample path for illustration).

Similar to the standard test, the exact finite sample null distribution of the statistic (3.8) is not known, i.e., both procedures are only asymptotic tests. Sen (1982) shows that

$$W^{(T)}(z) \xrightarrow{d} W(z) \qquad \text{as } T \to \infty , \tag{3.9}$$

where $\xrightarrow{d}$ denotes convergence in distribution and $W(z)$ is the standard Wiener process. This implies a critical value of 0.945 for the test statistic (3.4), for a significance level of $\alpha = 5\%$. Similarly, Ploberger and Krämer (1988) show that

$$B^{(T)}(z) \xrightarrow{d} B(z) \qquad \text{as } T \to \infty , \tag{3.10}$$

where $B(z)$ is a standard Brownian bridge [or "tied-down Brownian motion"; see Billingsley (1968, p. 64)]. This implies that the test statistic (3.8) has the limiting distribution function

$$F(x) = 1 - 2\sum_{j=1}^{\infty}(-1)^{j+1} \exp(-2j^2x^2) \tag{3.11}$$

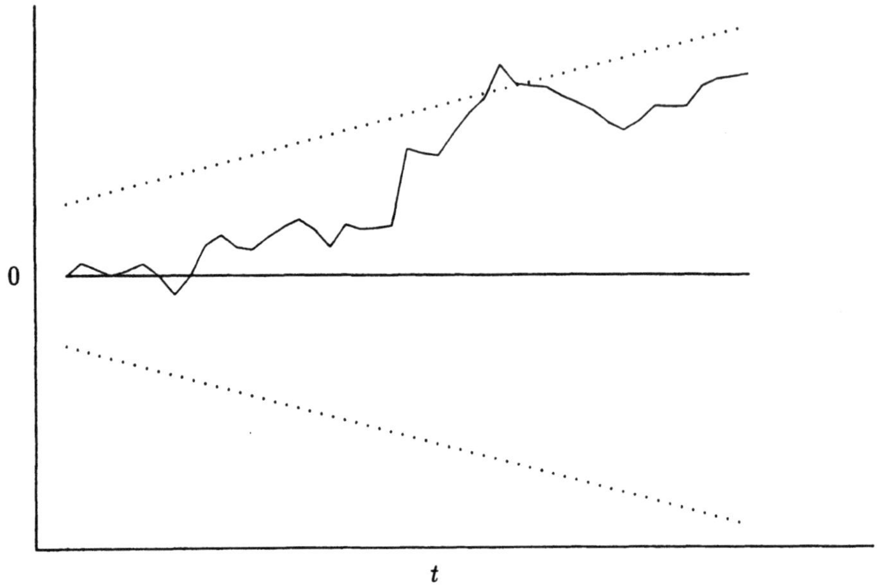


*Figure 3.1:* Sample path and critical lines for standard CUSUM test.

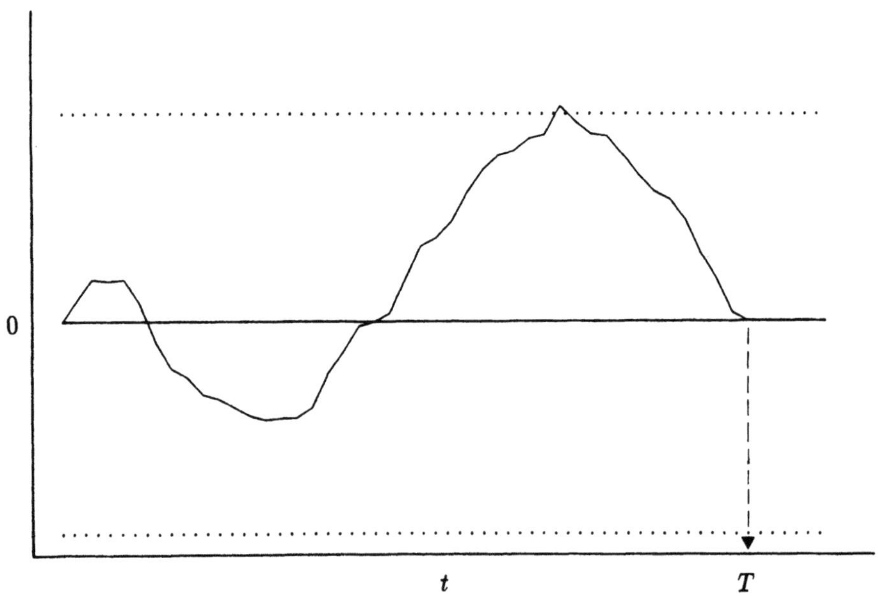


*Figure 3.2:* Sample path and critical lines for OLS-based CUSUM test.

[see Billingsley (1968, p. 85)], which is identical to the asymptotic null distribution of the Ploberger-Krämer-Kontrus (1989) fluctuation test and also to the limiting distribution of yet another variant of the CUSUM test for mean adjusted data [Ploberger and Krämer (1987)]. For $\alpha = 5\%$, the distribution function (3.11) produces a critical value of 1.36 for the OLS-based CUSUM test.

## 3.2 Local Power

Any choice between the OLS-based and the recursive-residual-based versions of the CUSUM test must rest on their respective power to detect a structural change. Little can be said here analytically for finite samples, but there are some local power results. Let the regression coefficients vary according to

$$\beta_{t,T} = \beta + \frac{1}{\sqrt{T}} g(t/T), \tag{3.12}$$

where $g(z)$ is some $K$-dimensional function defined on the [0,1] interval. The relationship (3.12) defines a triangle sequence of regression models (local alternatives), where the intensity of any structural changes is of the order $T^{-1/2}$. It includes a single shift at a fixed quantile of the sample period as a special case [i.e., $g(z) = 0$ $(z < z^*)$ and $g(z) = \Delta\beta$ $(z \geq z^*)$].

The limiting rejection probability [as $T \to \infty$ and as the regression coefficients vary according to (3.12)] is called the local power of the test. In econometrics, this has become the most popular means of discriminating among consistent tests.

The local power of both versions of the CUSUM test depends crucially on the "mean regressor"

$$c \equiv \lim_{T\to\infty} \frac{1}{T} \sum_{t=1}^{T} x_t \,. \tag{3.13}$$

This always exists by virtue of assumption (3.2) and coincides with the first column (or row) of the matrix $R$ defined in (3.2).

Given alternatives (3.12) and given any $g$-function which can be expressed as a uniform limit of step-functions, Ploberger and Krämer (1988) show that

$$W^{(T)}(z) \xrightarrow{d} W(z) + \frac{1}{\sigma}\left[c' \int_0^z g(u)\,du - c' \int_0^z \frac{1}{v} \int_0^v g(w)\,dw\,dv\right]. \tag{3.14}$$

This means that the cumulated sums of the recursive residuals tend in distribution to a Wiener process plus some rather complicated nonstochastic function, which depends on the particular type of structural change and the mean regressor $c$. The point is that this function is identically zero if $c$ is orthogonal to $g(z)$ for all $z$ in $[0, 1]$. In this case,

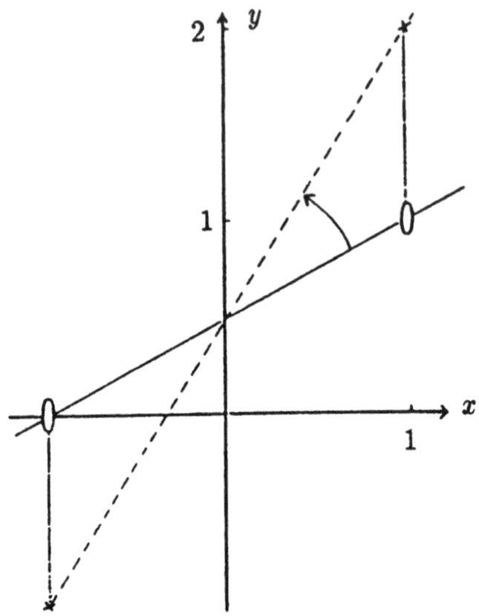


*Figure 3.3:* Mean regressor orthogonal to structural change.

the limiting distribution of the standard CUSUM test is the same as under $H_0$, i.e., the CUSUM test has limiting power equal to its size.

The corresponding expression for the OLS-based version of the CUSUM test is [see Ploberger and Krämer (1988)]

$$B^{(T)}(z) \xrightarrow{d} B(z) + \frac{1}{\sigma}\left[c' \int_0^z g(u)\,du - c'z \int_0^1 g(u)\,du\right] . \tag{3.15}$$

Again, the limiting distribution of the test is the same for $H_0$ and local alternatives if the mean regressor $c$ is orthogonal to any structural change. Both versions of the test therefore share the defect of overlooking certain types of structural shift.

For illustration, consider the regression

$$y_t = \frac{1}{2} + \frac{1}{2}(-1)^t + u_t \tag{3.16}$$

with a one-time increase in slope from 1/2 to 3/2 at some time $T^*$. The true regression lines, both before and after the structural change, are depicted in *Figure 3.3*. The mean regressor in the relationship (3.16) is $c = [1,0]$, whereas the structural shift is $\Delta\beta = [0,1]$. Structural shift and mean regressor are therefore orthogonal.

From *Figure 3.3*, it is intuitively obvious that the standard CUSUM test will never detect this shift, since forecast errors subsequent to the shift will have opposite signs and will cancel each other, giving their cumulated sum little chance to cross the critical lines. The same happens with the OLS-based CUSUM test. The regression line as estimated from the full sample will lie in between the pre-shift and the post-shift true regression lines, with the same intercept, but some intermediate slope. This implies that successive OLS residuals both before and after the shift will tend to have opposite sign, again leaving their cumulated sums no chance to cumulate.

If $g(z)$ is not orthogonal to $c$ for all $z$ in $[0,1]$, the limiting distribution under $H_0$ will differ from the limiting distribution under local alternatives for both tests, and the limiting rejection probability exceeds the size of the test. It can, at least in principle, be computed as a crossing probability for Brownian motion or the Brownian bridge from (3.14) and (3.15), but all our attempts at doing so have failed. The critical lines implied by (3.14) and (3.15) are much more complicated than the simple cases (mostly straight lines), which are discussed by, e.g., Durbin (1971).

## 3.3 Power in Finite Samples

Next we report on a Monte Carlo investigation into the finite sample relevance of the asymptotic results from Section 3.2. To economize on computer time and for ease of comparison with Ploberger and Krämer (1988), we confine ourselves to bivariate models with $u_t \sim N(0,1)$, and a single structural shift. The regressors are either

$$x_t = [1, \sin t]' \quad \text{or} \quad x_t = [1, \mathrm{GNP}_t]',$$

where $\mathrm{GNP}_t$ are (centered) monthly growth rates of West German GNP from 1976 to 1986. In both cases, the mean regressor is given by

$$c = \lim_{T\to\infty} \frac{1}{T} \sum x_t = [0,1]'. \tag{3.17}$$

Both regressors were used for sample sizes $T = 60$ and $T = 120$. Given $T$, we let the shift occur at $T^* = z^*T$, where $z^* = 0.1, 0.3, 0.5, 0.7$, and $0.9$. The shift itself is

$$\Delta\beta = \frac{1}{\sqrt{T}}[\cos\psi, \sin\psi]', \tag{3.18}$$

where $\psi$ is the angle between $\Delta\beta$ and the mean regressor $c$, i.e., the quantity we are primarily interested in considering the local power results from Section 3.2. This angle is given the values 0°, 18°, 36°, 54°, 72°, and 90°, with Section 3.2 predicting a monotone loss of power as $\psi$ increases.

The intensity of the shift is $\|\Delta\beta\| = |b|/\sqrt{T}$, and is likewise varied systematically. For any given combination of $b$, $\psi$, $T$, and $z^*$, $N = 1{,}000$ runs were performed for any given combination of $b$, $\psi$, $T$, and $z^*$ (except when $b = 0$, in which case we performed $N = 10{,}000$ runs to get a more precise figure for the size of the test). We uniformly used a nominal size of $\alpha = 5\%$.

*Table 3.1:* Finite sample power of the CUSUM tests for $x_t = [1, \sin t]'$ and $T = 60$.

| | | Angle $\psi$ | | | | | |
|---|---|---|---|---|---|---|---|
| | $b$ | 0° | 18° | 36° | 54° | 72° | 90° |
| | (a) CUSUM Test (size = 0.0321) | | | | | | |
| $z^* = 0.1$ | 4.8 | .158 | .143 | .112 | .073 | .042 | .038 |
| | 7.2 | .334 | .331 | .222 | .130 | .057 | .033 |
| | 9.6 | .547 | .506 | .373 | .197 | .073 | .027 |
| | 12.0 | .748 | .703 | .542 | .304 | .088 | .024 |
| $z^* = 0.3$ | 4.8 | .220 | .203 | .147 | .084 | .045 | .037 |
| | 7.2 | .472 | .426 | .300 | .160 | .060 | .031 |
| | 9.6 | .739 | .696 | .531 | .266 | .074 | .025 |
| | 12.0 | .922 | .887 | .727 | .396 | .092 | .020 |
| $z^* = 0.5$ | 4.8 | .140 | .130 | .089 | .061 | .046 | .035 |
| | 7.2 | .324 | .295 | .204 | .093 | .048 | .032 |
| | 9.6 | .572 | .530 | .364 | .170 | .046 | .023 |
| | 12.0 | .793 | .737 | .559 | .263 | .053 | .013 |
| $z^* = 0.7$ | 4.8 | .054 | .056 | .049 | .042 | .038 | .041 |
| | 7.2 | .116 | .109 | .065 | .046 | .032 | .035 |
| | 9.6 | .218 | .200 | .128 | .055 | .028 | .024 |
| | 12.0 | .330 | .303 | .221 | .075 | .022 | .010 |
| $z^* = 0.9$ | 4.8 | .036 | .038 | .038 | .040 | .039 | .041 |
| | 7.2 | .034 | .036 | .035 | .036 | .037 | .040 |
| | 9.6 | .031 | .029 | .028 | .031 | .035 | .037 |
| | 12.0 | .025 | .026 | .029 | .025 | .027 | .028 |
| | (b) CUSUM (OLS) Test (size = 0.0334) | | | | | | |
| $z^* = 0.1$ | 4.8 | .057 | .055 | .052 | .043 | .035 | .039 |
| | 7.2 | .103 | .090 | .062 | .048 | .034 | .034 |
| | 9.6 | .164 | .151 | .100 | .059 | .034 | .030 |
| | 12.0 | .290 | .243 | .157 | .066 | .034 | .027 |
| $z^* = 0.3$ | 4.8 | .351 | .322 | .222 | .112 | .052 | .034 |
| | 7.2 | .707 | .663 | .504 | .254 | .074 | .028 |
| | 9.6 | .947 | .925 | .777 | .439 | .105 | .026 |
| | 12.0 | .994 | .988 | .950 | .641 | .159 | .022 |
| $z^* = 0.5$ | 4.8 | .504 | .466 | .348 | .182 | .062 | .033 |
| | 7.2 | .871 | .831 | .684 | .401 | .118 | .029 |
| | 9.6 | .983 | .975 | .919 | .633 | .181 | .021 |
| | 12.0 | .999 | .966 | .985 | .832 | .265 | .016 |
| $z^* = 0.7$ | 4.8 | .363 | .332 | .247 | .124 | .052 | .037 |
| | 7.2 | .720 | .683 | .511 | .284 | .079 | .031 |
| | 9.6 | .942 | .920 | .808 | .476 | .119 | .026 |
| | 12.0 | .992 | .988 | .951 | .696 | .180 | .018 |
| $z^* = 0.9$ | 4.8 | .051 | .051 | .046 | .041 | .037 | .034 |
| | 7.2 | .098 | .087 | .070 | .045 | .040 | .033 |
| | 9.6 | .170 | .145 | .109 | .059 | .038 | .031 |
| | 12.0 | .276 | .242 | .158 | .081 | .036 | .030 |

*Table 3.2:* Finite sample power of the CUSUM tests for $x = [1, \text{GNP}_t]'$ and $T = 60$.

| | | Angle $\psi$ | | | | | |
|---|---|---|---|---|---|---|---|
| | $b$ | 0° | 18° | 36° | 54° | 72° | 90° |
| | (a) CUSUM Test (size = 0.0322) | | | | | | |
| $z^* = 0.1$ | 4.8 | .143 | .133 | .099 | .061 | .040 | .033 |
| | 7.2 | .297 | .265 | .027 | .116 | .047 | .030 |
| | 9.6 | .482 | .445 | .342 | .187 | .065 | .030 |
| | 12.0 | .705 | .643 | .489 | .281 | .096 | .029 |
| $z^* = 0.3$ | 4.8 | .192 | .179 | .137 | .076 | .040 | .031 |
| | 7.2 | .413 | .378 | .274 | .156 | .054 | .035 |
| | 9.6 | .686 | .646 | .482 | .263 | .081 | .029 |
| | 12.0 | .878 | .843 | .693 | .400 | .126 | .028 |
| $z^* = 0.5$ | 4.8 | .125 | .115 | .087 | .052 | .035 | .038 |
| | 7.2 | .285 | .259 | .180 | .103 | .042 | .032 |
| | 9.6 | .527 | .475 | .336 | .172 | .057 | .034 |
| | 12.0 | .755 | .706 | .540 | .269 | .083 | .030 |
| $z^* = 0.7$ | 4.8 | .052 | .048 | .039 | .035 | .031 | .031 |
| | 7.2 | .113 | .106 | .077 | .043 | .032 | .032 |
| | 9.6 | .206 | .190 | .129 | .074 | .035 | .031 |
| | 12.0 | .334 | .299 | .212 | .109 | .038 | .031 |
| $z^* = 0.9$ | 4.8 | .032 | .032 | .030 | .030 | .030 | .030 |
| | 7.2 | .029 | .031 | .032 | .030 | .029 | .029 |
| | 9.6 | .026 | .027 | .028 | .032 | .030 | .031 |
| | 12.0 | .026 | .025 | .026 | .029 | .030 | .034 |
| | (b) CUSUM (OLS) Test (size = 0.0267) | | | | | | |
| $z^* = 0.1$ | 4.8 | .056 | .053 | .047 | .036 | .035 | .038 |
| | 7.2 | .100 | .090 | .067 | .048 | .034 | .037 |
| | 9.6 | .173 | .153 | .115 | .063 | .035 | .040 |
| | 12.0 | .280 | .251 | .178 | .094 | .045 | .035 |
| $z^* = 0.3$ | 4.8 | .282 | .258 | .178 | .100 | .044 | .037 |
| | 7.2 | .648 | .605 | .436 | .217 | .069 | .036 |
| | 9.6 | .912 | .878 | .746 | .405 | .106 | .035 |
| | 12.0 | .990 | .981 | .924 | .635 | .163 | .036 |
| $z^* = 0.5$ | 4.8 | .439 | .403 | .286 | .142 | .054 | .037 |
| | 7.2 | .825 | .763 | .611 | .340 | .083 | .036 |
| | 9.6 | .977 | .965 | .875 | .581 | .157 | .037 |
| | 12.0 | .995 | .992 | .978 | .801 | .260 | .036 |
| $z^* = 0.7$ | 4.8 | .328 | .297 | .210 | .091 | .043 | .034 |
| | 7.2 | .687 | .628 | .483 | .254 | .060 | .035 |
| | 9.6 | .941 | .906 | .782 | .456 | .107 | .033 |
| | 12.0 | .988 | .985 | .948 | .660 | .192 | .030 |
| $z^* = 0.9$ | 4.8 | .048 | .046 | .044 | .040 | .036 | .034 |
| | 7.2 | .080 | .073 | .062 | .043 | .038 | .032 |
| | 9.6 | .147 | .129 | .091 | .059 | .040 | .030 |
| | 12.0 | .262 | .232 | .151 | .077 | .044 | .030 |

*Table 3.1* summarizes the experiments for $T = 60$ and $x_t = [1, \sin t]'$. It shows that power for both versions of the test is indeed decreasing as $\psi$ increases, even dropping below the size as $\psi = 90°$. Power for the OLS-based version is comparable for $z^*$ and $1 - z^*$, whereas the standard version peaks at $z^* = 0.3$, and performs increasingly worse as $z^* \to 1$, again as predicted by the asymptotic results from Section 3.2. Only for very early shifts is the OLS-based version outperformed by its standard counterpart. Since the exact sizes of the tests are almost identical (and in both cases below the nominal size, confirming previous Monte Carlo work), there is no need to correct for any difference here.

Not unexpectedly, power for both tests increase, *ceteris paribus*, as the intensity of the shift increases. An exception is $\psi = 90°$, and $z^* = 0.9$ (standard version), where power *decreases* as $b$ increases. This is due to the particular form (3.6) of our estimator for $\sigma$, which enters the denominator of both test statistics and is likely to blow up, and thus decrease, the chance of crossing the respective critical limits, as $b$ increases. This issue is addressed by Harvey (1975) and Alt and Krämer (1986), but outside the scope of this chapter.

*Table 3.2* gives the analoguous results for $x_t = [1, \text{GNP}_t]'$. It mainly demonstrates that our previous results are no artifact of the particular regressor sequence, confirming what was observed for $x_t = [1, \sin t]'$. The exact size of the OLS-based version is here markedly below the corresponding figures for the standard test, so the superior performance of the former version would even be enhanced by any correction.

*Tables 3.3* and *3.4* extend the sample size to $T = 120$. The main result here again is that nothing much has changed. Exact size is for both tests now closer to the nominal size of $\alpha = 5\%$, as expected. *Ceteris paribus* (i.e., given $z^*$, $\psi$, and $b$), power for $T = 60$ and $T = 120$ does not differ much, indicating that local power (i.e., the limiting power as the intensity $b/\sqrt{T}$ tends to zero) is a reasonable guide for samples as small as 60.

## 3.4 Conclusion

The CUSUM test can equally well be based on OLS residuals. It then reacts also to structural shifts that occur late in the sample, which are likely to go unnoticed by the standard version of the test. No version is uniformly superior to the other.

### Acknowledgments

Research supported by Deutsche Forschungsgemeinschaft (DFG). We are grateful to Grayham Mizon, Wang Liqun, and Marie Hušková for helpful suggestions and advice.

*Table 3.3:* Finite sample power of the CUSUM tests for $x_t = [1, \sin t]'$ and $T = 120$.

| | $b$ | Angle $\psi$ 0° | 18° | 36° | 54° | 72° | 90° |
|---|---|---|---|---|---|---|---|
| | | (a) CUSUM Test (size = 0.037) | | | | | |
| $z^* = 0.1$ | 4.8 | .179 | .165 | .122 | .083 | .052 | .033 |
| | 7.2 | .365 | .328 | .248 | .140 | .066 | .031 |
| | 9.6 | .584 | .528 | .417 | .231 | .090 | .029 |
| | 12.0 | .780 | .746 | .585 | .335 | .106 | .027 |
| $z^* = 0.3$ | 4.8 | .250 | .229 | .161 | .097 | .056 | .028 |
| | 7.2 | .540 | .489 | .348 | .184 | .068 | .027 |
| | 9.6 | .823 | .775 | .603 | .311 | .094 | .023 |
| | 12.0 | .949 | .930 | .819 | .479 | .134 | .020 |
| $z^* = 0.5$ | 4.8 | .168 | .147 | .113 | .071 | .047 | .029 |
| | 7.2 | .411 | .359 | .236 | .117 | .049 | .026 |
| | 9.6 | .684 | .623 | .451 | .207 | .061 | .022 |
| | 12.0 | .890 | .850 | .660 | .324 | .076 | .019 |
| $z^* = 0.7$ | 4.8 | .079 | .067 | .057 | .046 | .035 | .031 |
| | 7.2 | .152 | .132 | .100 | .058 | .036 | .028 |
| | 9.6 | .293 | .253 | .161 | .083 | .037 | .022 |
| | 12.0 | .487 | .434 | .266 | .116 | .039 | .016 |
| $z^* = 0.9$ | 4.8 | .035 | .033 | .032 | .031 | .030 | .030 |
| | 7.2 | .038 | .039 | .036 | .031 | .030 | .028 |
| | 9.6 | .042 | .042 | .037 | .029 | .028 | .027 |
| | 12.0 | .049 | .047 | .040 | .030 | .026 | .024 |
| | | (b) CUSUM (OLS) Test (size = 0.039) | | | | | |
| $z^* = 0.1$ | 4.8 | .081 | .073 | .063 | .048 | .039 | .040 |
| | 7.2 | .131 | .123 | .093 | .067 | .039 | .039 |
| | 9.6 | .225 | .199 | .141 | .086 | .043 | .034 |
| | 12.0 | .390 | .334 | .216 | .114 | .049 | .028 |
| $z^* = 0.3$ | 4.8 | .389 | .348 | .262 | .150 | .066 | .039 |
| | 7.2 | .779 | .734 | .555 | .297 | .103 | .034 |
| | 9.6 | .964 | .947 | .842 | .513 | .147 | .028 |
| | 12.0 | .999 | .996 | .969 | .745 | .204 | .021 |
| $z^* = 0.5$ | 4.8 | .547 | .508 | .372 | .210 | .079 | .041 |
| | 7.2 | .907 | .876 | .733 | .422 | .139 | .037 |
| | 9.6 | .993 | .988 | .940 | .690 | .210 | .028 |
| | 12.0 | 1.000 | .999 | .995 | .876 | .284 | .018 |
| $z^* = 0.7$ | 4.8 | .405 | .360 | .250 | .142 | .057 | .037 |
| | 7.2 | .765 | .707 | .543 | .271 | .095 | .035 |
| | 9.6 | .965 | .947 | .810 | .485 | .124 | .024 |
| | 12.0 | .997 | .994 | .962 | .686 | .178 | .021 |
| $z^* = 0.9$ | 4.8 | .076 | .069 | .059 | .046 | .039 | .039 |
| | 7.2 | .140 | .121 | .092 | .058 | .043 | .037 |
| | 9.6 | .238 | .204 | .148 | .078 | .044 | .035 |
| | 12.0 | .383 | .328 | .215 | .106 | .046 | .033 |

*Table 3.4:* Finite sample power of the CUSUM tests for $x = [1, \mathrm{GNP}_t]'$ and $T = 120$.

| | | Angle $\psi$ | | | | | |
|---|---|---|---|---|---|---|---|
| | $b$ | 0° | 18° | 36° | 54° | 72° | 90° |
| | | (a) CUSUM Test (size = 0.037) | | | | | |
| $z^* = 0.1$ | 4.8 | .167 | .150 | .118 | .081 | .050 | .033 |
| | 7.2 | .354 | .325 | .244 | .141 | .066 | .033 |
| | 9.6 | .583 | .543 | .405 | .234 | .082 | .033 |
| | 12.0 | .773 | .740 | .592 | .341 | .110 | .033 |
| $z^* = 0.3$ | 4.8 | .235 | .210 | .157 | .094 | .060 | .035 |
| | 7.2 | .542 | .500 | .355 | .186 | .072 | .034 |
| | 9.6 | .819 | .774 | .609 | .337 | .100 | .034 |
| | 12.0 | .958 | .930 | .830 | .526 | .140 | .034 |
| $z^* = 0.5$ | 4.8 | .146 | .134 | .096 | .065 | .047 | .035 |
| | 7.2 | .328 | .300 | .207 | .108 | .063 | .035 |
| | 9.6 | .613 | .555 | .395 | .195 | .066 | .035 |
| | 12.0 | .834 | .788 | .627 | .314 | .093 | .035 |
| $z^* = 0.7$ | 4.8 | .067 | .067 | .060 | .054 | .039 | .035 |
| | 7.2 | .135 | .126 | .089 | .062 | .046 | .035 |
| | 9.6 | .242 | .219 | .150 | .084 | .055 | .035 |
| | 12.0 | .415 | .363 | .249 | .132 | .059 | .036 |
| $z^* = 0.9$ | 4.8 | .032 | .033 | .031 | .034 | .034 | .035 |
| | 7.2 | .040 | .039 | .035 | .031 | .034 | .035 |
| | 9.6 | .050 | .048 | .042 | .034 | .033 | .035 |
| | 12.0 | .051 | .051 | .049 | .040 | .031 | .035 |
| | | (b) CUSUM (OLS) Test (size = 0.0281) | | | | | |
| $z^* = 0.1$ | 4.8 | .053 | .050 | .047 | .039 | .035 | .032 |
| | 7.2 | .103 | .096 | .073 | .047 | .034 | .033 |
| | 9.6 | .207 | .180 | .119 | .069 | .039 | .032 |
| | 12.0 | .357 | .312 | .212 | .102 | .044 | .032 |
| $z^* = 0.3$ | 4.8 | .365 | .320 | .240 | .131 | .047 | .032 |
| | 7.2 | .775 | .717 | .535 | .279 | .086 | .031 |
| | 9.6 | .959 | .943 | .845 | .504 | .147 | .032 |
| | 12.0 | .999 | .996 | .965 | .750 | .216 | .032 |
| $z^* = 0.5$ | 4.8 | .447 | .408 | .284 | .155 | .053 | .032 |
| | 7.2 | .835 | .793 | .630 | .340 | .110 | .033 |
| | 9.6 | .979 | .966 | .886 | .598 | .172 | .032 |
| | 12.0 | .999 | .997 | .979 | .815 | .252 | .032 |
| $z^* = 0.7$ | 4.8 | .295 | .262 | .185 | .104 | .043 | .032 |
| | 7.2 | .649 | .605 | .435 | .220 | .072 | .032 |
| | 9.6 | .910 | .873 | .723 | .404 | .110 | .032 |
| | 12.0 | .991 | .983 | .916 | .624 | .166 | .031 |
| $z^* = 0.9$ | 4.8 | .056 | .050 | .040 | .032 | .032 | .032 |
| | 7.2 | .097 | .091 | .070 | .045 | .032 | .032 |
| | 9.6 | .175 | .155 | .116 | .063 | .033 | .032 |
| | 12.0 | .315 | .284 | .177 | .094 | .038 | .032 |

## References

Alt, R. and Krämer, W. (1986), A modification of the CUSUM test. Research Memorandum No. 229. Vienna: Institute for Advanced Studies.

Billingsley, P. (1968), *Convergence of Probability Measures*. New York: John Wiley.

Brown, R.L., Durbin, J., and Evans, J.M. (1975), Techniques for testing the constancy of regression relationships over time. *Journal of the Royal Statistical Society*, **B-37**, 149-163.

Durbin, J. (1971), Boundary-crossing probabilities for the Brownian motion and techniques for computing the power of the Kolmogorov-Smirnov test. *Journal of Applied Probability*, **8**, 431-453.

Harvey, A. (1975), Comment on the paper by Brown, Durbin and Evans. *Journal of the Royal Statistical Society*, **B-37**, 179-180.

Krämer, W., Ploberger, W., and Alt, R. (1988), Testing for structural change in dynamic models. *Econometrica*, **56**, 1355-1369.

McCabe, B.P.M. and Harrison, M.J. (1980), Testing the constancy of regression relationships over time using least squares residuals. *Applied Statistics*, **29**, 142-148.

MacNeill, I.B. (1978), Properties of sequences of partial sums of polynomial regression residuals with applications to tests for change of regression at unknown times. *The Annals of Statistics*, **6**, 422-433.

Ploberger, W. and Krämer, W. (1987), Mean adjustment and the CUSUM test for structural change. *Economics Letters*, **25**, 255-258.

Ploberger, W. and Krämer, W. (1988), The CUSUM test with OLS residuals. Mimeo.

Ploberger, W. and Krämer, W. (1989), The local power of the CUSUM and CUSUM of squares tests. Forthcoming in *Econometric Theory*.

Ploberger, W., Krämer, W., and Kontrus, K. (1989), A new test for structural stability in the linear regression model. *Journal of Econometrics*, **40**, 307-318.

Sen, P.K. (1982), Invariance principles for recursive residuals. *The Annals of Statistics*, **10**, 307-312.

CHAPTER 4

# Kimball's Inequality and Bounds Tests for Comparing Several Regressions under Heteroscedasticity

*Jean-Marie Dufour*

## Summary

This chapter studies the problem of comparing the coefficients of several independent linear regressions with unequal variances. Using an extension of Kimball's inequality, we give simple exact bounds for the null distribution of a general Wald-type statistic for testing any set of linear restrictions linking the coefficients of the regressions. The bounds proposed are based on central Fisher distributions, so that the $p$-values can be obtained by using any program that computes the central Fisher distribution. In particular, the bounds suggested are much easier to compute than earlier bounds proposed by Ohtani and Kobayashi (1986) and Farebrother (1989), especially when more than two regressions are considered.

## 4.1 Introduction

A common problem in econometrics and statistics consists of comparing the coefficients of several independent linear regressions. Such problems are met, in particular, when assessing whether the coefficients of a linear relationship are identical in different subsamples (corresponding, for example, to different subperiods). When the variances of the disturbances are equal, standard analysis-of-covariance methods can be used for this purpose [see Chow (1960) and Dufour (1982)]. On the other hand, when the variances are unequal, it is well known that such procedures are unreliable [see Toyoda (1974), Schmidt and Sickles (1977), and Ohtani and Toyoda (1985a)].

This chapter considers the following model

$$\begin{aligned} y_j &= X_j\beta_j + u_j, \quad j = 1, \ldots, m, & (4.1) \\ u_j &\sim N(0, \sigma_j^2 I_{n_j}), \quad j = 1, \ldots, m, & (4.2) \end{aligned}$$

where $y_j$ is an $n_j \times 1$ vector of observations on a dependent variable, $X_j$ is an $n_j \times k_j$ fixed matrix such that $1 \leq \text{rank}(X_j) = k_j < n_j$, $u_j$ is an $n_j \times 1$ vector of random disturbances, and $u_1, \ldots, u_m$ are independent. I study the problem of testing general linear restrictions of the form

$$H_0: \sum_{j=1}^{m} R_j\beta_j = r\,, \tag{4.3}$$

where $R_j$ is a $q \times k_j$ matrix, $r$ is a $q \times 1$ vector, and the matrix $[R_1, R_2, \ldots, R_m]$ has rank $q$.

For the case of two regressions with equal numbers of coefficients ($m = 2, k_1 = k_2$), finite-sample procedures for testing $H_0$: $\beta_1 - \beta_2 = 0$ were suggested by Jayatissa (1977), Tsurumi (1984), Ohtani and Kobayashi (1986), and Dalal and Mudholkar (1988). Among these, the method of Ohtani and Kobayashi (1986), which is based on bounding the distribution of a Wald-type statistic, is the most generally applicable. The three other methods impose additional restrictions that are inappropriate in many econometric problems. For example, Tsurumi (1984) requires $n_1 = n_2$ (which may lead to dropping a portion of the sample) and paired observations (where different pairings can lead to different values of the test statistic), Dalal and Mudholkar (1988) need $X_1 = cX_2$ (where $c$ is a constant), while Jayatissa's (1977) test calls for sample sizes that are sufficiently large with respect to the number of regressors ($n_i - k_i \geq k_i, i = 1, 2$). Further, the last procedure, which is based on Hotelling's $T^2$ statistic, is computationally burdensome, requires several arbitrary choices, and makes an inefficient use of the sample; for additional discussion, see Watt (1979), Tsurumi (1984), Ohtani and Toyoda (1985a), Honda and Ohtani (1986), and Ali and Silver (1985). Other available procedures are based on approximate distributions (e.g., large-sample approximations) and thus lead to tests whose size may exceed the stated level; see Goldfeld and Quandt (1978), Watt (1979), Honda (1982), Erlat (1984), Rothenberg (1984), Ohtani and Toyoda (1985a, 1985b), Ali and Silver (1985), Honda and Ohtani (1986), Kobayashi (1986), Toyoda and Ohtani (1986), and Conerly and Mansfield (1988). The procedure suggested by Weerahandi (1987) is not a test in the usual sense, and it is not known whether the probability of type I error is smaller than the nominal "level"; on this issue, see also Griffiths and Judge (1989). Bayesian approaches to the problem were suggested by Zellner and Siow (1980) and Tsurumi and Sheflin (1985).

The problem of testing general hypotheses of the form (4.3) with $m$ equations ($m > 2$) and arbitrary numbers of regressors (when $k_1 = \ldots = k_m$ does *not* necessarily hold) has been much less studied, especially from a finite-sample perspective. The main results on this issue are due to Farebrother (1988, 1989) who proposed a generalization of the approach of Ohtani and Kobayashi (1986). [Farebrother also proposed extensions of the approaches of Jayatissa (1977) and Weerahandi (1987). The shortcomings of these two methods are the same as in the two sample case.] Farebrother, however, did not propose an algorithm for computing the bounds, so that the generalized method is not operational. Finding these bounds requires one to evaluate joint probabilities for ratios of non-independent ratios of chi-square variables. Even though Ohtani and Kobayashi (1986) proposed an algorithm applicable when $m = 2$, it appears that their approach would be difficult to apply when $m \geq 3$. Even for $m = 2$, it is fairly complex to use.

In this chapter, I give a result that can simplify considerably the application of Farebrother's generalization of the Ohtani-Kobayashi bounds test as well as the latter itself. For this purpose, we use an extension of Kimball's (1951) inequality to give simple exact bounds for the distribution of the Wald-type statistic considered by these authors. The bounds proposed are based on central Fisher distributions, and $p$-values can be obtained by using any program that computes the central Fisher distribution. Although less tight than the bounds considered by Ohtani-Kobayashi-Farebrother, the bounds suggested can be computed much more easily, irrespective of the number $m$ of regressions involved.

In Section 4.2, the test statistic studied is defined, various notations are introduced, and further details on the problem are given. In Section 4.3, the extension of Kimball's inequality is stated and the simple bounds based on it are derived. Section 4.4 discusses the result obtained and its application.

## 4.2 Test Statistic

To test general hypotheses of the form $\sum_{j=1}^{m} R_j\beta_j = r$, Farebrother (1988, 1989) generalized the approach of Ohtani and Kobayashi (1986). The test statistic considered is the Wald-type statistic

$$W = (R\hat{\beta} - r)' \left( \sum_{j=1}^{m} s_j^2 Q_j \right)^{-1} (R\hat{\beta} - r), \tag{4.4}$$

where $\hat{\beta} = (\hat{\beta}_1', \hat{\beta}_2', \ldots, \hat{\beta}_m')'$, $\hat{\beta}_j' = (X_j'X_j)^{-1}X_j'y_j$, $s_j^2 = \hat{u}_j'\hat{u}_j/\nu_j$, $\nu_j = n_j - k_j$, $\hat{u}_j = y_j - X_j\hat{\beta}_j$, $Q_j = R_j(X_j'X_j)^{-1}R_j'$, and the $q \times k$ matrix $R = [R_1, R_2, \ldots, R_m]$ has rank $q$. This statistic is not identical to the Wald statistic for this problem; the latter would use $\hat{\sigma}_j^2 = \hat{u}_j'\hat{u}_j/n_j$ instead of $s_j^2$. The null hypothesis $H_0$ is rejected when $W$ is large.

Let $S_j = s_j^2/\sigma_j^2$, $j = 1, \ldots, m$, and

$$W_0 = (R\hat{\beta} - r)' \left( \sum_{j=1}^{m} \sigma_j^2 Q_j \right)^{-1} (R\hat{\beta} - r). \tag{4.5}$$

Then the statistic $W$ can be bounded as follows:

$$W_L \leq W \leq W_U, \tag{4.6}$$

where

$$W_L = \min\{W_0/S_1, \ldots, W_0/S_m\}, \tag{4.7}$$
$$W_U = \max\{W_0/S_1, \ldots, W_0/S_m\}. \tag{4.8}$$

$W_0$, $S_1, \ldots, S_m$ are mutually independent with $W_0 \sim \chi^2(q)$ and $\nu_j S_j \sim \chi^2(\nu_j)$, $j = 1, \ldots, m$. However, the ratios $W_0/S_1, \ldots, W_0/S_m$ are not independent. The distribution and tail areas of $W$ can now be bounded either by computing the distributions of $W_L$ and $W_U$ or by bounding the latter in turn.

For the case of two regressions ($m = 2$) with $k_1 = k_2 = q$ and $R_1 = -R_2 = I_q$, Ohtani and Kobayashi (1986) proposed an algorithm based on evaluating joint probabilities of the forms $P[W_0/S_j > c, S_2/S_1 > 1]$ and $P[W_0/S_j > c, S_2/S_1 < 1]$, $j = 1, 2$, from doubly infinite series of gamma functions. Using this method, critical bounds for $W$ were evaluated for $\alpha = 0.05$ and a limited number of values of $q$ and sample sizes ($q = 2, \ldots, 6$; $n_1, n_2 = 12, 13, \ldots, 20, 30, 40, 60$). It is easy to see that this algorithm can also deal with cases where $k_1 = k_2 = q$ and $R_1 = -R_2 = I_q$ do not hold (provided $m = 2$). However, when more than two regressions are considered ($m \geq 3$), it appears that the approach proposed by Ohtani and Kobayashi for computing the distributions of $W_L$ and $W_U$ would be much more difficult to apply. It is complex to implement for even $m = 2$.

In Section 4.3 simple bounds for $P[W \geq c]$ are given which are based on central Fisher distributions and are applicable irrespective of the values of $m$, $q$, $k_1, \ldots, k_m$, without special restrictions on $R_1, \ldots, R_m$.

## 4.3 Kimball Bounds

To derive bounds for $P[W \geq c]$, we will use the following lemma, which slightly extends earlier inequalities given by Kimball (1951), Khatri (1967), and Tong (1980, p. 14).

**Lemma 4.1** *Let $Y$ be a real random variable and let $f_i(y)$, $i = 1, \ldots, p$, be Borel-measurable real-valued functions of $y \in R$ such that $E[|f_i(Y)|] < \infty$, $i = 1, \ldots, p$, and $E[|\prod_{i=1}^{N} f_i(Y)|] < \infty$, $N = 1, \ldots, p$. If the functions $f_i(y)$ are nonnegative and monotonic in the same direction (either nondecreasing or nonincreasing), then*

$$E\left[\prod_{i=1}^{p} f_i(Y)\right] \geq \prod_{i=1}^{p} E\left[f_i(Y)\right]. \tag{4.9}$$

*Proof:* The result is trivial for $p = 1$. Let $p \geq 2$. Since the functions $f_1(y), \ldots, f_p(y)$ are nonnegative and monotonic in the same direction, the functions $\prod_{i=1}^{N} f_i(y)$, $N = 1, \ldots, p$, are also nonnegative and monotonic in the same direction as $f_i(y)$, $i = 1, \ldots, p$. Using *Lemma 2.2.1* of Tong (1980, p. 13), we get (for N = 2, ..., p)

$$E\left[\prod_{i=1}^{N} f_i(Y)\right] = E\left\{\left[\prod_{i=1}^{N-1} f_i(Y)\right] f_N(Y)\right\} \geq E\left[\prod_{i=1}^{N-1} f_i(Y)\right] E\left[f_N(Y)\right]. \tag{4.10}$$

The inequality (4.9) then follows by applying (4.10) successively in $E[\prod_{i=1}^{p} f_i(Y)]$. □

In particular, it is clear that the conditions $E[|f_i(Y)|] < \infty$, $i = 1, \ldots, p$, and $E[|\prod_{i=1}^{N} f_i(Y)|] < \infty$, $N = 1, \ldots, p$, hold when the functions $f_1(Y), \ldots, f_p(Y)$ are bounded. We can now prove our basic result.

**Theorem 4.1** *Under the assumptions (4.1), (4.2), and (4.3),*

$$\prod_{j=1}^{m} P[F(q,\nu_j) \geq x] \leq P[W/q \geq x] \leq 1 - \prod_{j=1}^{m} P[F(q,\nu_j) < x] \tag{4.11}$$

*for all $x$, where $W$ is the Wald statistic in (4.4), $\nu_j = n_j - k_j$, and $F(q,\nu_j)$ follows a Fisher distribution with $(q,\nu_j)$ degrees of freedom.*

*Proof:* From (4.6), we have $W_L \leq W \leq W_U$. For any $x$,

$$\begin{aligned} P[W_L/q \geq x] &= P[\min\{W_0/S_1, \ldots, W_0/S_m\} \geq qx] \\ &= P[W_0/S_j \geq qx, j = 1, \ldots, m] \\ &= P[1/S_j \geq qx/W_0, j = 1, \ldots, m] \\ &= E_0\{P[1/S_j \geq qx/W_0, j = 1, \ldots, m|W_0]\} \\ &= E_0\{\prod_{j=1}^{m} P[1/S_j \geq qx/W_0|W_0]\}, \end{aligned} \tag{4.12}$$

where $E_0$ is the expected value with respect to the distribution of $W_0$ and the last identity follows by noting that $W_0, S_1, \ldots, S_m$ are mutually independent. Since the functions $f_j(Y) = P[1/S_j \geq qx/W_0|W_0]$, $j = 1, \ldots, m$, are nonnegative, bounded, and monotonically decreasing in $x$, we can use *Lemma 4.1*, and we find

$$\begin{aligned} P[W_L/q \geq x] &\geq \prod_{j=1}^{m} E_0\{P[1/S_j \geq qx/W_0|W_0]\} \\ &= \prod_{j=1}^{m} P[1/S_j \geq qx/W_0] \\ &= \prod_{j=1}^{m} P[(W_0/q)/S_j \geq x] \\ &= \prod_{j=1}^{m} P[F(q,\nu_j) \geq x] \end{aligned} \tag{4.13}$$

for all $x$, where $(W_0/q)/S_j = (W_0/q)/(s_j^2/\sigma_j^2)$ follows a Fisher distribution with $(q,\nu_j)$ degrees of freedom. Similarly, for $W_U/q$ we have

$$\begin{aligned} P[W_U/q < x] &= P[\max\{W_0/S_1, \ldots, W_0/S_m\} < qx] \\ &= P[W_0/S_j < qx, j = 1, \ldots, m] \end{aligned}$$

$$
\begin{aligned}
&= E_0\{P[1/S_j < qx/W_0, j = 1, \ldots, m|W_0]\} \\
&= E_0\{\prod_{j=1}^{m} P[1/S_j < qx/W_0|W_0]\} \\
&\geq \prod_{j=1}^{m} E_0\{P[1/S_j < qx/W_0|W_0]\} \\
&= \prod_{j=1}^{m} P[F(q, \nu_j) < x] \qquad (4.14)
\end{aligned}
$$

for any $x$, where the fact that the functions $g_j(x) = P[1/S_j < qx/W_0|W_0]$ are nonnegative, bounded, and monotonically increasing in $x$ has been used. Thus, for any $x$,

$$
\begin{aligned}
\prod_{j=1}^{m} P[F(q, \nu_j) \geq x] &\leq P[W_L/q \geq x] \leq P[W/q \geq x] \\
&\leq P[W_U/q \geq x] \leq 1 - \prod_{j=1}^{m} P[F(q, \nu_j) < x]. \qquad \square
\end{aligned}
$$

## 4.4 Discussion

From the proof of *Theorem 4.1*, it is clear that

$$
\prod_{j=1}^{m} P[F(q, \nu_j) \geq x] \leq P[W_L/q \geq x] \leq P[W_U/q \geq x] \leq 1 - \prod_{j=1}^{m} P[F(q, \nu_j) < x]
$$

so that the bounds in (4.11) are less tight than those based on computing $P[W_L/q \geq x]$ and $P[W_U/q \geq x]$, as done by Ohtani and Kobayashi (1986). On the other hand, they have the merit of being much easier to compute. They are based on the central Fisher distributions with $(q, \nu_j)$, $j = 1, \ldots, m$, degrees of freedom. The simplest way to use them in practice is to find $p$-values for $W$ with any program that computes the Fisher distribution (such programs are widely available). Letting $G(x; q, \nu_j) = P[F(q, \nu_j) < x]$ be the cumulative distribution function of $F(q, \nu_j)$, we see that the critical region $1 - \prod_{j=1}^{m} G(W/q; q, \nu_j) \leq \alpha$ is conservative at level $\alpha$, while the critical region $\prod_{j=1}^{m}\{1 - G(W/q; q, \nu_j)\} \leq \alpha$ is liberal at level $\alpha$. One can thus use the following bounds test:

$$
\begin{aligned}
&\text{reject } H_0 \text{ if } 1 - \prod_{j=1}^{m} G(W/q; q, \nu_j) \leq \alpha, \\
&\text{accept } H_0 \text{ if } \prod_{j=1}^{m}\{1 - G(W/q; q, \nu_j)\} > \alpha, \\
&\text{consider the test inconclusive, otherwise.} \qquad (4.15)
\end{aligned}
$$

For further discussion of bounds procedures, see Dufour (1989, 1990).

Clearly, it is also possible to build tables to perform tests of given levels (say, $\alpha = 0.10, 0.05$). This requires finding $F_L(\alpha)$ and $F_U(\alpha)$ such that

$$\prod_{j=1}^{m} P[F(q,\nu_j) \geq F_L(\alpha)] = \alpha, \quad \prod_{j=1}^{m} P[F(q,\nu_j) < F_U(\alpha)] = 1 - \alpha. \tag{4.16}$$

The rejection region $W/q > F_U(\alpha)$ and the acceptance region $W/q \leq F_L(\alpha)$ are equivalent to the corresponding regions in (4.15). However, the critical bounds $F_L(\alpha)$ and $F_U(\alpha)$ depend on $m$, $q$, $\nu_1, \ldots, \nu_m$, and tables of such critical values may rapidly become very large. In this respect, if $\alpha_0 = 1 - (1-\alpha)^{1/m}$, $0 < \alpha < 1$, $P[F(q,\nu_j) \geq F(\alpha_0; q, \nu_j)] = \alpha_0$ and

$$G_U(\alpha) = \max\{F(\alpha_0; q, \nu_j), j = 1, \ldots, m\}. \tag{4.17}$$

It is useful to observe that

$$\begin{aligned} P[W/q \geq G_U(\alpha)] &\leq 1 - \prod_{j=1}^{m} P[F(q,\nu_j) < G_U(\alpha)] \\ &\leq 1 - \prod_{j=1}^{m} P[F(q,\nu_j) < F(\alpha_0; q, \nu_j)] = \alpha. \end{aligned} \tag{4.18}$$

Similarly, if $\alpha_0' = \alpha^{1/m}$ and

$$G_L(\alpha) = \min\{F(\alpha_0'; q, \nu_j), j = 1, \ldots, m\}, \tag{4.19}$$

we have

$$\begin{aligned} P[W/q \geq G_L(\alpha)] &\geq \prod_{j=1}^{m} P[F(q,\nu_j) \geq G_L(\alpha)] \\ &\geq \prod_{j=1}^{m} P[F(q,\nu_j) \geq F(\alpha_0'; q, \nu_j)] = \alpha. \end{aligned} \tag{4.20}$$

Each critical bound, $G_U(\alpha)$ and $G_L(\alpha)$, requires one to find at most $m$ different quantiles of central Fisher distributions (only one when $\nu_1 = \ldots = \nu_m$). On the other hand, it is clear that $G_L(\alpha) \leq F_L(\alpha) \leq F_U(\alpha) \leq G_U(\alpha)$ so that the rejection region $W/q \geq G_U(\alpha)$ is more conservative than the region $W/q \geq F_U(\alpha)$, and the acceptance region $W/q < G_L(\alpha)$ is smaller than the region $W/q < F_L(\alpha)$.

## Acknowledgments

This work was supported by the Centre de recherche et développement en économique (C.R.D.E.) of the Université de Montréal, the Natural Sciences and Engineering Research Council of Canada, and the Fonds FCAR (Government of Québec). The author thanks Adolf Buse, Noxy Dastoor, Richard Farebrother, Eric Ghysels, and Quang Vuong for useful comments.

## References


Ali, M.M. and Silver, J. (1985), Tests for equality between sets of coefficients in two linear regressions under heteroscedasticity. *Journal of the American Statistical Association*, **80**, 730-735.

Chow, G.C. (1960), Tests of equality between sets of coefficients in two linear regressions. *Econometrica*, **28**, 591-603.

Conerly, M.D. and Mansfield, E.R. (1988), An approximate test for comparing heteroskedastic regression models. *Journal of the American Statistical Association*, **83**, 811-817.

Dalal, S.R. and Mudholkar, G.S. (1988), A conservative test and confidence region for comparing heteroscedastic regressions. *Biometrika*, **75**, 149-152.

Dufour, J.-M. (1982), Generalized Chow tests for structural change: A coordinate-free approach. *International Economic Review*, **23**, 565-575.

Dufour, J.-M. (1989), Non-linear hypotheses, inequality restrictions, and non-nested hypotheses: Exact simultaneous tests in linear regressions. *Econometrica*, **57**, 335-355.

Dufour, J.-M. (1990), Exact tests and confidence sets in linear regressions with autocorrelated errors. *Econometrica*, **58**, 475-494.

Erlat, H. (1984), Testing for structural change under heteroscedasticity: A note. *The Manchester School of Economic and Social Studies*, **52**, 380-390.

Farebrother, R.W. (1988), Testing linear restrictions with unequal variances (problem 88.2.1). *Econometric Theory*, **4**, 349.

Farebrother, R.W. (1989), Testing linear restrictions with unequal variances (solution). *Econometric Theory*, **5**, 324-326.

Goldfeld, S.M. and Quandt, R.E. (1978), Asymptotic tests for the constancy of regressions in the heteroscedastic case. Research memorandum 229, Econometric Research Program, Princeton University.

Griffiths, W. and Judge, G. (1989), Testing and estimating location vectors under heteroscedasticity. Working Paper, Econometrics Department, University of New England, Armidale, Australia.

Honda, Y. (1982), On tests of equality between sets of coefficients in two linear regressions when disturbance variances are unequal. *The Manchester School of Economic and Social Studies*, **50**, 116-125.

Honda, Y. and Ohtani, K. (1986), Modified Wald tests in tests of equality between sets of coefficients in two linear regressions under heteroscedasticity. *The Manchester School of Economic and Social Studies*, **54**, 208-218.

Jayatissa, W.A. (1977), Tests of equality between sets of coefficients in two linear regressions when disturbance variances are unequal. *Econometrica*, **45**, 1291-1292.

Khatri, C.G. (1967), On certain inequalities for normal distributions and their applications to simultaneous confidence bounds. *Annals of Mathematical Statistics*, **38**, 1853-1867.

Kimball, A.W. (1951), On dependent tests of significance. *Annals of Mathematical Statistics*, **22**, 600-602.

Kobayashi, M. (1986), A bounds test for equality between sets of coefficients in two linear regressions when disturbance variances are unequal. *Journal of the American Statistical Association*, **81**, 510-513.

Ohtani, K. and Kobayashi, M. (1986), A bounds test for equality between sets of coefficients in two linear regression models under heteroscedasticity. *Econometric Theory*, **2**, 220-231.

Ohtani, K. and Toyoda, T. (1985a), Small sample properties of tests of equality between sets of coefficients in two linear regressions under heteroscedasticity. *International Economic Review*, **26**, 37-43.

Ohtani, K. and Toyoda, T. (1985b), A Monte Carlo study of the Wald, LM, and LR tests in a heteroscedastic linear model. *Communications in Statistics, Simulation and Computation*, **14**, 735-746.

Rothenberg, T.J. (1984), Hypothesis testing in linear models when the error covariance matrix is nonscalar. *Econometrica*, **52**, 827-842.

Schmidt, P. and Sickles, R. (1977), Some further evidence on the use of the Chow test under heteroscedasticity. *Econometrica*, **45**, 1293-1298.

Tong, Y.L. (1980). *Probability inequalities in multivariate distributions.* New York: Academic Press.

Toyoda, T. (1974), The use of the Chow test under heteroscedasticity. *Econometrica*, **42**, 601-608.

Toyoda, T. and Ohtani, K. (1986), Testing equality between sets of coefficients after a preliminary test for equality of disturbance variances in two linear regressions. *Journal of Econometrics*, **31**, 67-80.

Tsurumi, H. (1984), On Jayatissa's test of constancy of regressions under heteroscedasticity. *Economic Studies Quarterly*, **34**, 57-62.

Tsurumi, H. and Sheflin, N. (1985), Some tests for the constancy of regressions under heteroscedasticity. *Journal of Econometrics*, **27**, 221-234.

Watt, P.A. (1979), Tests of equality between sets of coefficients in two linear regressions when disturbance variances are unequal: Some small sample properties. *The Manchester School of Economic and Social Studies*, **47**, 391-396.

Weerahandi, S. (1987), Testing regression equality with unequal variances. *Econometrica*, **55**, 1211-1215.

Zellner, A. and Siow, A. (1980), Bayesian posterior odds ratios for several frequently encountered hypotheses in regression analysis, in J. Bernardo *et al.* (eds.), *Bayesian Statistics.* Valencia: University of Valencia Press.

## CHAPTER 5

# A Further Look at Model Evaluation

*Bernd Schips and Yngve Abrahamsen*


## Summary

In this chapter we compare the specification tests generally used in econometric model building with recently developed $p$-step Stone-Geisser prediction tests using jackknife procedures. The comparisons are based on a Monte Carlo study using different structural forms. The structural forms are partly misspecified. Three sets of simulations are presented: A variety of single equation models, a simple multi-equation macro model, and Klein's Model I using US data. The jackknife based statistics enable a more critical evaluation of the out-of-sample performance of the estimated multi-equation models.


## 5.1 Introduction

The basic idea of this study is quite simple: The methods and techniques of classical inferential statistics in econometrics are necessary but not sufficient. The classical approach is inadequate in practice because we do not know the form of the underlying mechanism producing the economic data. Hence, we have not only errors in variables but also errors in equations.

The traditional econometric methods for estimating simultaneous equation models require the identification of the structural parameters. The asymptotically justified estimation methods can be used only when the identification problem is really solved, because identification is a necessary condition for the existence of estimators with the desired asymptotic properties like consistency. However, the identification problem cannot be solved by the more or less pragmatic use of exclusion restrictions. The basic hypothesis that the considered model is true is the critical and important point. Econometric models are only approximations, they are never true.

The identification of dynamic macroeconomic models is difficult. The reason for this is that the identifying restrictions are pragmatically adjusted to avoid obvious conflicts with the data, so that these restrictions can only be regarded as simplifications, not as *a priori* knowledge imposed on the data. Although this is a weakness in econometric model building, it does not follow that macroeconomic models are of no value in the preparation of economic policy decisions. (The well-known rational expectations critique is based on the fact that statistical models are likely to become unreliable for conditions far outside the historically "normal" range experienced in the sample data.)

The main point is that only a careful statistical modeling of the historical structure can be used to make conditional forecasts appropriate for economic policy analysis. Statistical models and estimation methods should be used which do not strictly require identifying parameters of behavior which are invariant to unprecendented changes.

The concepts of unbiasedness, consistency, and efficiency utterly lose their meaning by the time an applied econometrician starts to work [Leamer (1983)]. Often the relevant economic theory does not yield precise information regarding functional forms of relationships, lag structures, and other elements involved in a stochastic specification of a model. Therefore, a "trial and error" process of specification search is used. However, the usual specification search invalidates the classical inferential procedures. This unhappy state of affairs may be remedied by appropriate sequential procedures, but sequential specification tests have yet to be developed.

The statistical basis of econometric methods might give the impression that "applied" econometrics also possesses the robustness of a science. One tends to forget that the raw material for such applications are economic data and the hypotheses of economic theory. Any practitioner of empirical economics soon realizes that there is nothing more changeable than the quality of economic data and economic models. Therefore, in a soft science like economics, the element of judgment is of paramount importance if empirically estimated models are to be of practical use. Some tools which are helpful for guiding one's judgment are discussed below.

## 5.2 Predictive Model Evaluation

From the viewpoint of "classical" statistical inference the procedure and presentation of model testing are straightforward. As soon as the distribution of the random variables is determined, the computation of parameter estimators and test statistics with the desired properties is possible. Tests of hypotheses are conditional on the assumed distribution of the random variables of the model.

Disregarding the technical problems in the process of deducing estimators and tests (which may be considerable), it seems as if efficient methods of model testing would result from the application of the classical inference statistics. Yet one problem of classical statistical inference is that statements about the model's structure are dependent upon the distributive assumptions. These distributive assumptions cannot be verified themselves. Tests concerning the stochastic specification of an econometric model are always applied in connection with the implied economic hypotheses and are limited to the verification of specific distributive assumptions within a certain set of distributions. The fundamental problem with currently available specification tests is that only certain aspects of model specification or only particular competing hypotheses can be examined. Other aspects of the hypotheses are assumed to be correct and remain untested. The process of specifying an econometric model therefore includes the decision about which assumptions will remain untested. To avoid this conditional testing, we seek to evaluate the entire model specification.

The first step in this direction was the development of resampling methods like the jackknife or the bootstrap. The second step was the formulation of jackknife-based statistics that shed light upon the predictive quality of a model. These statistics were suggested by Stone (1974) and Geisser (1974) and are jackknife versions of the $Q^2$ statistic originally introduced by Ball (1963):

$$Q^2 = 1 - \frac{\sum_t (y_t - y_t^*)^2}{\sum_t (y_t - \bar{y})^2},$$

where $y_t$ is the observed value, $y_t^*$ is the predicted or estimated value, and $\bar{y}$ is the mean of the observed values. Bergström and Wold (1983) used this $Q^2$ and related statistics to test the forecast properties of different estimators for general models. The Stone-Geisser statistic is described in Bergström and Wold (1983) as follows: With $T$ observations, a model is estimated $T$ times using $(T - 1)$ observations, one observation being deleted each time. (Observation 1 is deleted for the first estimation, observation 2 for the second estimation, etc.) Using these estimates, forecasts are produced for the deleted observations and the $Q^2$ statistic is computed using these forecasts. For an early reference to the jackknife, see Tukey (1958).

The jackknife and the bootstrap are nonparametric methods for assessing the errors in a statistical estimation problem [Miller (1974)]. In the course of computing the Stone-Geisser statistics, a jackknife-based estimate of the variance is obtained as a by-product.

Jackknifing is especially appealing in situations where no explicit formulas are available for variances or the computations of variances are too complicated. However, it is known that jackknife variance estimates are biased upward [see Efron and Gong (1983)]. This may be a disadvantage compared with estimates obtained through bootstrapping, but to be on the safe side is an advantage not only in econometric modeling.

We recommend using the jackknife $Q^2$ statistics as measures that shed light upon the predictive quality of a model [see Abrahamsen (1986) and Abrahamsen and Schips (1989)]. We interpret poor predictive quality as revealed by these statistics as an indication of model misspecification. First the usefulness of the Stone-Geisser statistics shall be examined by using single-equation models that are *a priori* specified incorrectly in different ways. For these models there exists a large number of "standard" specification tests. The model misspecifications examined are listed in *Table 5.1* [see also Abrahamsen and Schips (1989, p. 40)].

*Table 5.1:* Types of misspecification (omitted variable, incorrect functional form).

| | Data-generating model [a] | Estimation model [b] |
|---|---|---|
| F1 | $C_t = u_t$ | $C_t = b_1 + b_2Y_t + v_t$ |
| F2 | $C_t = 10 + 0.5Y_t + u_t$ | $C_t = b_1 + b_2Y_t + v_t$ |
| F3 | $C_t = 10 + 0.2Y_t + 0.375C_{t-1} + u_t$ | $C_t = b_1 + b_2Y_t + b_3C_{t-1} + v_t$ |
| F4 | $C_t = 10 + 0.3Y_t + 0.15Y_{t-1} + u_t$ | $C_t = b_1 + b_2Y_t + v_t$ |
| F5 | $C_t = 10 + 0.2Y_t + 0.375C_{t-1} + u_t$ | $C_t = b_1 + b_2Y_t + v_t$ |
| F6 | $C_t = 10 + 0.2Y_t + 0.375C_{t-1} + u_t$ | $C_t = b_1 + b_2C_{t-1} + v_t$ |
| F7 | $C_t = 10 + 0.3Y_t + 0.0015Y_t^2 + u_t$ | $C_t = b_1 + b_2Y_t + v_t$ |
| F8 | $C_t = 10 + 0.7Y_t^{0.9} + u_t$ | $C_t = b_1 + b_2Y_t + v_t$ |

[a] $u_t \sim N(0,1)$; the variances of the error terms correspond to the estimated standard deviation of residuals in econometric modeling.
[b] $v_t$ is assumed to be distributed as $N(0, \sigma_v^2)$.

In the Monte Carlo simulations of the single-equation models, i.i.d. normal random variables were used as the error terms.

Abrahamsen (1986) presented a comparison of specification tests using jackknife $Q^2$ statistics to reveal incorrect specification of single equation models. For the analysis Abrahamsen (1986) performed Monte Carlo simulations. Monte Carlo simulation is a natural tool to obtain knowledge of the finite sample properties of the estimation and to test procedures for linear and nonlinear models. In his paper Abrahamsen applied—besides the usual significant tests—modified jackknife versions of the original $Q^2$ statistic suggested by Stone (1974) and Geisser (1974). Instead of a comparison of the predictive quality of the values for $y_t^*$ (these are the predictions for the observations which have been omitted in the process of jackknifing) with the predictive quality of the mean value $\bar{y}$ of the observations, Abrahamsen (1986) made comparisons with other "naive" predictors. First, he compared the values for $y_t^*$ with $y_{t-1}$, the predictor corresponding to a random-walk model:

$$Q^2(RW) = 1 - \frac{\sum_t (y_t - y_t^*)^2}{\sum_t (y_t - y_{t-1})^2} .$$

Second, he compared $y_t^*$ with the prediction that extrapolates from the previous change $y_{t-1} - y_{t-2}$:

$$Q^2(LC) = 1 - \frac{\sum_t (y_t - y_t^*)^2}{\sum_t (y_t - 2y_{t-1} + y_{t-2})^2} .$$

*Table 5.2* lists the criteria used for detecting misspecification in the first set of Monte Carlo experiments. For the "classical" test statistics, which have known finite sample or asymptotic distributions, misspecification was determined by a 5% significance level test. For the other statistics a critical value of 0.8 was used. This value was chosen after informal

*Table 5.2:* Criteria for detecting misspecification (the significance level of all tests is 0.05).

| | Criteria |
|---|---|
| K1 | $t$-test $H_0$: $b_i = 0$ |
| K2 | JK $t$-test $H_0$: $b_i = 0$ |
| K3 | Durbin Watson test or Durbin's $h$-test |
| K4 | $R^2 < 0.8$ |
| K5 | Bartlett's $M$ specification error test (BAMSET) |
| K6 | regression specification error test (RESET) |
| K7 | $R^2(SG) < 0.8$ |
| K8 | $R^2(RW) < 0.8$ |
| K9 | $R^2(LC) < 0.8$ |

inspection of $Q^2$ when estimating the "correct" specification: For most models, a critical value of 0.8 gives an acceptably small probability of Type I error. If one explanatory variable is the lagged dependent, the OLS estimators are consistent and asymptotically efficient. The usual tests are asymptotically justified. But even when the disturbances are independent (as in these simulations), the situation is not favorable from the viewpoint of a practitioner; the small sample properties of the familar estimators and tests are not satisfactory [Abrahamsen and Schips (1989)]. In this case, the $Q^2$ measures $Q^2(RW)$ and $Q^2(LC)$ reflect the bias of the estimators. The question remains: How large must a sample be to allow inference based on the asymptotic behavior of an estimator?

Abrahamsen and Schips (1989) also analyzed the problem of the possible instability of the model's structure (i.e., a shift in the value of the coefficients for a subset of the sample period). The $Q^2$ statistics normally tend to increase according to the number of observations in a correctly specified model. If a $Q^2$ statistic for a sub-period exceeds the corresponding measure for the complete period, we get a clear indication of instability in the model's structure. The detection criterion for the $Q^2$ statistic is accordingly simple: There is misspecification if the $Q^2$ statistic for the whole estimation period did not exceed the $Q^2$ statistic for both sub-periods [Abrahamsen and Schips (1989, p. 41)].

The frequencies of detected structural change suggest that the standard tests (Chow tests, sequence of Chow tests, and CUSUM) are less sensitive to this particular form of misspecification than the jackknife $Q^2$ statistics. We gather from this and from the first set of Monte Carlo experiments that at least nothing opposes the use of jackknife methods in connection with prediction-oriented measures of model evaluation. At first glance, these $Q^2$ statistics seem to be more likely to detect model misspecification than the frequently used tests originating from classical inference statistics. One interesting result is the comparison of the "traditional" $t$-test and the jackknife $t$-test. The simulations show that the power and decision outcome of both tests are roughly the same. This conclusion holds if we increase the number $p$ of deleted observations in the jackknifing procedure. *Table 5.3* shows the frequency of misspecification detection using jackknife estimates with $p \geq 1$.

*Table 5.3:* Evaluation of the Monte Carlo simulations: Number of indicated misspecifications out of 100 simulations.

| Criteria | | Model F1 | Model F2 |
|---|---|---|---|
| K1 | | 94 | 0 |
| K2 | $p=1$ | 93 | 0 |
| | $p=2$ | 93 | 0 |
| | $p=3$ | 95 | 0 |
| K3 | | 6 | 6 |
| K4 | | 100 | 0 |
| K5 | | 17 | 17 |
| K6 | | 10 | 10 |
| K7 | $p=1$ | 100 | 0 |
| | $p=2$ | 100 | 0 |
| | $p=3$ | 100 | 0 |
| K8 | $p=1$ | 100 | 62 |
| | $p=2$ | 100 | 38 |
| | $p=3$ | 96 | 35 |
| K9 | $p=1$ | 39 | 12 |
| | $p=2$ | 53 | 22 |
| | $p=3$ | 45 | 17 |

An econometrician might object to our choice of criteria (*Table 5.2*). Abrahamsen (1986) used only two specification tests—the RESET and BAMSET test—out of the whole battery of specification tests. [The justification for the combined application of RESET and BAMSET was given by Ramsey and Gilbert (1972).] There exist some other well-known specification tests, such as the Hausman test or the Lagrange multiplier test. These tests, however, require knowledge about the specific type of misspecification, and in practice we rarely have this information. In fact, this information exists only for simulation experiments. Sometimes the model builder has ideas about alternative specifications. *Table 5.4* gives an impression of the usefulness of the $Q^2$ statistics compared with the Lagrange multiplier test under ideal conditions (the alternative hypothesis is the "correct" specification used for the simulation).

If the errors are normal i.i.d., the familiar $t$-tests have the correct size and possess optimal power properties. In practice, however, we are rarely sure about the proper specification of the errors. Looking at the detection frequencies in *Table 5.5*, we see that the jackknife $Q^2$ statistics are useful not only for the "simple" case where only the error is misspecified, but also for the more complicated case where variables are omitted as well. However, the detection frequencies for the jackknifed $t$-test (second row) give cause for concern regarding the jackknife variance estimators when errors are autocorrelated.

When the number $p$ of deleted observations in the jackknife procedure is greater than one, one can obtain $Q^2$ statistics appropriate for evaluating the quality of the $k$-th

*Table 5.4:* Evaluation of the Monte Carlo simulations: Number of indicated misspecifications out of 100 simulations.

| Criteria | Data-generating/estimation model | | | |
|---|---|---|---|---|
| | F4 | F5 | F6 | F7 |
| K1 | 0 | 0 | 100 | 49 |
| K2, $p = 3$ | 0 | 0 | 100 | 76 |
| K3 | 3 | 35 | $(4^a)/2^b$ | 7 |
| K4 | 0 | 0 | 0 | 0 |
| K5 | 12 | 10 | 16 | 33 |
| K6 | 6 | 11 | 5 | 1 |
| *LM* test[c] | 53 | 48 | 100 | 31 |
| K7, $p = 3$ | 0 | 0 | 4 | 0 |
| K8, $p = 3$ | 79 | 94 | 100 | 26 |
| K9, $p = 3$ | 52 | 73 | 99 | 14 |

[a] Durbin Watson test
[b] Durbin's $h$-test
[c] $H_0$: estimation model; $H_1$: correct specification.

*Table 5.5:* Evaluation of the Monte Carlo simulations: Number of indicated misspecifications out of 100 simulations. Data-generating models with autocorrelated errors[a].

| Criteria | Data-generating/estimation model | | | |
|---|---|---|---|---|
| | F2A | F3A | F5A | F6A |
| K1 | 0 | 25 | 100 | 100 |
| K2, $p = 3$ | 0 | 52 | 100 | 100 |
| K3 | 20 | $(4^b)/26^c$ | 71 | $(5^b)/3^c$ |
| K4 | 0 | 0 | 0 | 0 |
| K5 | 19 | 21 | 13 | 12 |
| K6 | 10 | 15 | 27 | 4 |
| *LM* test[d] | – | – | 78 | 100 |
| K7, $p = 3$ | 0 | 0 | 0 | 0 |
| K8, $p = 3$ | 67 | 100 | 100 | 100 |
| K9, $p = 3$ | 29 | 100 | 97 | 100 |

[a] $u_t = \rho u_{t-1} + \varepsilon_t$; $\rho = 0.3, \varepsilon_t \sim N(0,1)$.
[b] Durbin Watson test
[c] Durbin's $h$-test
[d] $H_0$: estimation model; $H_1$: correct specification.

prediction within the $p$ *adjacent* omitted periods, $k = 1, \ldots, p$. First we redefine the $Q^2(SG)$ measures in a manner that gives predictions and comparisons on the same information base. For $p = 1$, we omit the observation $t$ when calculating the "alternative prediction", e.g., the mean. This implies that we now have to build $T$ different means for the observation set. The above-defined $Q^2$ then becomes

$$Q^2(SG) = 1 - \frac{\sum_{t=1}^{T}(y_t - y_t^*)^2}{\sum_{t=1}^{T}(y_t - \frac{1}{T-1}\sum_{j\neq t} y_j)^2} = 1 - \frac{\sum_{t=1}^{T}(y_t - y_t^*)^2}{\sum_{t=1}^{T}(y_t - \bar{y}_t)^2}.$$

For $p > 1$ we define the $Q^2$ for predictions done for the $k$-th of the $p$ adjacent omitted observations in the following manner ($N$ is the number of estimations, which equals $T/p$, rounded to the nearest integer, if necessary):

$$Q^2_{k,p,N}(SG) = 1 - \frac{\sum_{i=0}^{N-1}(y_{k+ip} - y^*_{k+ip})^2}{\sum_{i=0}^{N-1}(y_{k+ip} - \frac{1}{N-1}\sum_{j\neq i} y_{k+jp})^2} = 1 - \frac{\sum_{i=0}^{N-1}(y_{k+ip} - y^*_{k+ip})^2}{\sum_{i=0}^{N-1}(y_{k+ip} - \bar{y}_i)^2}.$$

With $k = p = 1$, we have $T = N$ and the formula is identical with the previous $Q^2(SG)$ definition. For $p > 1$ we may compute $p$ different $Q^2$ measures, indexed by $k$ in the formula. We may reformulate the $Q^2(RW)$ and $Q^2(LC)$ in the same manner:

$$\begin{aligned} Q^2_{k,p,N}(RW) &= 1 - \frac{\sum_{i=0}^{N-1}(y_{k+ip} - y^*_{k+ip})^2}{\sum_{i=0}^{N-1}(y_{k+ip} - y_{ip})^2} \\ Q^2_{k,p,N}(LC) &= 1 - \frac{\sum_{i=0}^{N-1}(y_{k+ip} - y^*_{k+ip})^2}{\sum_{i=0}^{N-1}(y_{k+ip} - (k+1)y_{ip} + ky_{ip-1})^2}. \end{aligned}$$

The results for forecasts up to three steps ahead ($p = 3$, $k = 1,2,3$) are given in *Table 5.6*. An interesting finding is that in well-specified models the $Q^2$ statistics have the following pattern: With increasing $k$ the number of $Q^2$ statistics less than 0.8 decreases fast (i.e., the model forecasts appear to improve). The range of the average percent error and the maximum percent error for different $k$ are roughly the same. Only in cases of misspecification do the percent errors increase with $k$.

## 5.3 Evaluation of Multi-Equation Models

We now turn to the evaluation process of multi-equation models. Evaluation of this kind of model is normally limited to verifying the economic interpretation of the estimated parameters and the multipliers based on these parameters and to analyzing the statistics concerning the *ex post* (and in some cases the *ex ante*) predictive qualities of the estimated structure.

However, this evaluation is problematic [Chong and Hendry (1986)]. These problems become obvious when we consider forecasting models. In particular, conventional evaluation procedures which inspect prediction errors over the *ex post* period fail to detect misclassification of variables as exogenous. By construction, the jackknife estimates and the $Q^2$ measures provide valuable information about this misspecification.

*Table 5.6:* Analysis of $k$-step dynamic forecast simulations. Evaluation of the Monte Carlo simulations: Number of indicated misspecification out of 100 simulations.

| Criteria | | Data-generating/estimation model | | | |
|---|---|---|---|---|---|
| | | F3 | F3A | F6 | F6A |
| K1 | $p = 3$ | 54 | 25 | 100 | 100 |
| K2 | | 51 | 52 | 100 | 100 |
| K3 | | $(0^a)/7^b$ | $(4^a)/26^b$ | $(4^a)/2^b$ | $(5^a)/3^b$ |
| K4 | | 0 | 0 | 0 | 0 |
| K5 | | 17 | 21 | 16 | 12 |
| K6 | | 12 | 15 | 51 | 4 |
| K7 | $p = 1$ | 0 | 0 | 4 | 2 |
| | $p = 2$ | 0 | 1 | 49 | 55 |
| | $p = 3$ | 0 | 3 | 80 | 13 |
| K8 | $p = 1$ | 89 | 97 | 100 | 100 |
| | $p = 2$ | 50 | 67 | 100 | 100 |
| | $p = 3$ | 25 | 43 | 100 | 100 |
| K9 | $p = 1$ | 67 | 81 | 99 | 99 |
| | $p = 2$ | 23 | 39 | 90 | 93 |
| | $p = 3$ | 6 | 13 | 76 | 84 |

[a] Durbin Watson test.
[b] Durbin's $h$-test.

To illustrate the use of the jackknife for dynamic models, we use a simple multi-equation macroeconomic model, whose structure and assumed "true" coefficients are given in *Table 5.7.* This model is assumed to have normal i.i.d. errors, with no cross-equation correlation. The various estimation models are described in *Table 5.8.* Note that model *M1* is correctly specified; models *M2*, *M3*, and *M4* have one misspecified equation, and *M5* combines all of these misspecifications. *Tables 5.9* and *5.10* give the frequencies of detected misspecifications for the different estimation models using the 0.8 cutoff. Examining the results more closely it is found that the different $Q^2$ measures easily detect a misspecification of the models and even determine the grossly misspecified equations in *M2, M3, M4*, and *M5.*

An interesting point is the behavior of the average percent errors and the maximum percent errors for different $k$. (A similar pattern was evident in *Table 5.6.*) *Table 5.11* shows the percent errors. The pattern of the percent errors in the evaluation of multi-equation models is not easy to analyze. The $Q^2$ measures are more sensitive compared to the percent errors.

Finally, we examine the well-known model "Klein I" [Klein (1950)]. This model has been used extensively as an example of the properties of econometric estimators. The specification (*Table 5.12*) follows Bergström and Wold (1983).

Using US data for the period 1921–1941, the estimated coefficients of the three behavioral equations are listed in *Table 5.13.* The estimates from OLS, TSLS, FIML, and FP

*Table 5.7:* Data-generating model.

| | Equation |
|---|---|
| (1) | $C_t = 1.1433 + 0.1997Y_t + 0.6942C_{t-1} + u(1)_t$ |
| (2) | $I_t = -93.4437 - 0.0453(Y_t - Y_{t-1}) + 0.0381A_t + u(2)_t$ |
| (3) | $X_t = -15.8380 + 0.0132YF_t + 0.0891PX_t + u(3)_t$ |
| (4) | $M_t = -1.3594 + 0.0389Y_t + 1.2566X_t - 0.0787PM_t + u(4)_t$ |
| (5) | $Y_t \equiv C_t + I_t + X_t - M_t + R_t$ |

Endogenous variables: $C$ (private consumption), $I$ (gross investment), $Y$ (gross national product), $X$ (export), $M$ (import).
Exogenous variables: $A$ (labor supply), $YF$ (gross national product of important foreign countries), $PX$ (price index of exported goods), $PM$ (price index of imported goods), $R$ (other GNP components).

*Table 5.8:* Estimation models.

| Model | | Equation |
|---|---|---|
| *M1* | (1) | $C_t = b_{11} + b_{12}Y_t + b_{13}C_{t-1} + v(1)_t$ |
| | (2) | $I_t = b_{21} + b_{22}(Y_t - Y_{t-1}) + b_{23}A_t + v(2)_t$ |
| | (3) | $X_t = b_{31} + b_{32}YF_t + b_{33}PX_t + v(3)_t$ |
| | (4) | $M_t = b_{41} + b_{42}Y_t + b_{43}X_t + b_{44}PM_t + v(4)_t$ |
| | (5) | $Y_t \equiv C_t + I_t + X_t - M_t + R_t$ |
| *M2* | (1) | $C_t = b_{11} + b_{12}Y_t + v(1)_t$ |
| *M3* | (2) | $I_t = b_{21} + b_{22}Y_t + b_{23}A_t + v(2)_t$ |
| *M4* | (4) | $M_t = b_{41} + b_{42}Y_t + b_{43}PM_t + v(4)_t$ |
| *M5* | (1) | $C_t = b_{11} + b_{12}Y_t + v(1)_t$ |
| | (2) | $I_t = b_{21} + b_{22}Y_t + b_{23}A_t + v(2)_t$ |
| | (3) | $X_t = b_{31} + b_{32}YF_t + b_{33}PX_t + v(3)_t$ |
| | (4) | $M_t = b_{41} + b_{42}Y_t + b_{43}PM_t + v(4)_t$ |
| | (5) | $Y_t \equiv C_t + I_t + X_t - M_t + R_t$ |

*Table 5.9:* Interdependent models (*M1, M3, M4*, and *M5*[a]). Estimation method JK-TSLS, $p = 3$. Evaluation of the Monte Carlo simulations: Number of indicated misspecifications out of 100 simulations, misspecificated equations in italics.

| Criteria | Equation | *M1* | *M3* | *M4* | *M5* |
|---|---|---|---|---|---|
| K7 | (1) | 0 | 0 | 0 | *100* |
| $p = 1$ | (2) | 62 | *62* | 96 | *100* |
| | (3) | 0 | 0 | 0 | 0 |
| | (4) | 0 | 0 | *0* | *0* |
| | (5) | 0 | 0 | 1 | 46 |
| K7 | (1) | 0 | 0 | 0 | *97* |
| $p = 2$ | (2) | 62 | *86* | 86 | *89* |
| | (3) | 0 | 0 | 0 | 0 |
| | (4) | 0 | 0 | *0* | *0* |
| | (5) | 0 | 0 | 0 | 46 |
| K7 | (1) | 2 | 1 | 2 | *97* |
| $p = 3$ | (2) | 39 | *72* | 83 | *83* |
| | (3) | 0 | 0 | 0 | 0 |
| | (4) | 0 | 0 | *0* | *0* |
| | (5) | 0 | 0 | 6 | 51 |
| K8 | (1) | 35 | 32 | 85 | *100* |
| $p = 1$ | (2) | 7 | *81* | 49 | *90* |
| | (3) | 0 | 0 | 0 | 0 |
| | (4) | 0 | 0 | *100* | *100* |
| | (5) | 2 | 0 | 100 | 100 |
| K8 | (1) | 11 | 9 | 62 | *100* |
| $p = 2$ | (2) | 7 | *13* | 8 | *25* |
| | (3) | 0 | 0 | 0 | 0 |
| | (4) | 0 | 0 | *100* | *99* |
| | (5) | 0 | 0 | 94 | 100 |
| K8 | (1) | 2 | 1 | 8 | *100* |
| $p = 3$ | (2) | 5 | *15* | 27 | *34* |
| | (3) | 0 | 0 | 0 | 0 |
| | (4) | 0 | 0 | *3* | *7* |
| | (5) | 0 | 0 | 23 | 100 |
| K9 | (1) | 51 | 50 | 85 | *100* |
| $p = 1$ | (2) | 4 | *29* | 18 | *40* |
| | (3) | 22 | 22 | 22 | 22 |
| | (4) | 12 | 14 | *100* | *100* |
| | (5) | 0 | 0 | 89 | 100 |
| K9 | (1) | 14 | 12 | 53 | *100* |
| $p = 2$ | (2) | 1 | *2* | 2 | *5* |
| | (3) | 0 | 0 | 0 | 0 |
| | (4) | 0 | 0 | *100* | *100* |
| | (5) | 0 | 0 | 17 | 58 |
| K9 | (1) | 2 | 1 | 8 | *99* |
| $p = 3$ | (2) | 0 | *0* | 0 | *1* |
| | (3) | 0 | 0 | 0 | 0 |
| | (4) | 0 | 1 | *89* | *84* |
| | (5) | 0 | 0 | 0 | 8 |

[a] Result for model *M2* are not reported because estimated equations yield an explosive simultaneous system. This is due to not imposing the adding up constraint ($Y \equiv C + I + \ldots$) during estimation when we use a limited information method like TSLS.

*Table 5.10:* Interdependent models (*M1-M5*), estimation method JK-FP, $p = 3$. Evaluation of the Monte Carlo simulations: Number of indicated misspecifications out of 100 simulations, misspecificated equations in italics.

| Criteria | Equation | *M1* | *M2* | *M3* | *M4* | *M5* |
|---|---|---|---|---|---|---|
| K7 | (1) | 0 | *100* | 0 | 0 | *100* |
| $p = 1$ | (2) | 66 | 86 | *100* | 91 | *100* |
| | (3) | 0 | 0 | 0 | 0 | 0 |
| | (4) | 0 | 0 | 0 | *0* | *0* |
| | (5) | 0 | 100 | 0 | 1 | 91 |
| K7 | (1) | 0 | *100* | 0 | 1 | *100* |
| $p = 2$ | (2) | 62 | 90 | *80* | 79 | *90* |
| | (3) | 0 | 0 | 0 | 0 | 0 |
| | (4) | 0 | 0 | 0 | *0* | *0* |
| | (5) | 0 | 100 | 0 | 1 | 76 |
| K7 | (1) | 2 | *100* | 3 | 2 | *100* |
| $p = 3$ | (2) | 39 | 57 | *69* | 73 | *80* |
| | (3) | 0 | 0 | 0 | 0 | 0 |
| | (4) | 0 | 0 | 0 | *0* | *0* |
| | (5) | 0 | 100 | 0 | 0 | 69 |
| K8 | (1) | 40 | *100* | 39 | 79 | *100* |
| $p = 1$ | (2) | 9 | 19 | *81* | 41 | *87* |
| | (3) | 0 | 0 | 0 | 0 | 0 |
| | (4) | 0 | 35 | 0 | *100* | *100* |
| | (5) | 2 | 100 | 1 | 100 | 100 |
| K8 | (1) | 12 | *100* | 12 | 0 | *100* |
| $p = 2$ | (2) | 7 | 11 | *15* | 9 | *21* |
| | (3) | 0 | 0 | 0 | 0 | 0 |
| | (4) | 0 | 0 | 0 | *76* | *97* |
| | (5) | 1 | 100 | 0 | 81 | 100 |
| K8 | (1) | 2 | *100* | 2 | 19 | *100* |
| $p = 3$ | (2) | 5 | 9 | *15* | 18 | *31* |
| | (3) | 0 | 0 | 0 | 0 | 0 |
| | (4) | 0 | 0 | 0 | *8* | *36* |
| | (5) | 0 | 100 | 0 | 25 | 100 |
| K9 | (1) | 50 | *100* | 54 | 80 | *100* |
| $p = 1$ | (2) | 4 | 8 | *28* | 15 | *32* |
| | (3) | 22 | 22 | 22 | 22 | 22 |
| | (4) | 14 | 81 | 19 | *100* | *100* |
| | (5) | 0 | 100 | 0 | 66 | 86 |
| K9 | (1) | 13 | *100* | 12 | 52 | *100* |
| $p = 2$ | (2) | 1 | 1 | *2* | 2 | *3* |
| | (3) | 0 | 0 | 0 | 0 | 0 |
| | (4) | 0 | 12 | 0 | *99* | *100* |
| | (5) | 0 | 97 | 0 | 8 | 46 |
| K9 | (1) | 1 | *100* | 2 | 17 | *100* |
| $p = 3$ | (2) | 1 | 0 | *0* | 0 | *1* |
| | (3) | 0 | 0 | 0 | 0 | 0 |
| | (4) | 0 | 1 | 0 | *93* | *97* |
| | (5) | 0 | 30 | 0 | 0 | 11 |

*Table 5.11:* Means of the average percent error (APE) and the maximum percent error (MPE) in 100 Monte Carlo simulations of interdependent Models (M1–M5); estimation method: JK-FP.

| | | *M1* | | *M2* | | *M3* | | *M4* | | *M5* | |
|---|---|---|---|---|---|---|---|---|---|---|---|
| | | APE | MPE | APE | MPE | APE | MPE | APE | MPE | APE | MPE |
| $k = 1$ | (1) | 1.4 | 4.1 | 15.3 | 65.2 | 1.4 | 4.3 | 1.9 | 6.1 | 11.0 | 50.0 |
| | (2) | 3.3 | 7.7 | 4.0 | 9.3 | 5.9 | 15.7 | 4.3 | 10.4 | 6.2 | 16.6 |
| | (3) | 0.8 | 1.9 | 0.8 | 1.9 | 0.8 | 1.9 | 0.8 | 1.9 | 0.8 | 1.9 |
| | (4) | 1.5 | 4.7 | 4.5 | 20.2 | 1.6 | 5.3 | 8.2 | 34.1 | 12.4 | 53.6 |
| | (5) | 1.3 | 3.5 | 11.0 | 42.2 | 1.4 | 3.7 | 3.0 | 9.0 | 5.6 | 22.5 |
| $k = 2$ | (1) | 1.6 | 4.6 | 10.3 | 35.1 | 1.7 | 5.0 | 2.3 | 6.7 | 8.1 | 26.8 |
| | (2) | 3.0 | 6.8 | 4.0 | 9.9 | 3.7 | 7.9 | 3.4 | 7.7 | 3.8 | 8.6 |
| | (3) | 0.7 | 1.6 | 0.7 | 1.6 | 0.7 | 1.6 | 0.7 | 1.6 | 0.7 | 1.6 |
| | (4) | 1.3 | 3.5 | 2.9 | 9.1 | 1.3 | 3.6 | 5.3 | 13.1 | 8.9 | 38.3 |
| | (5) | 1.3 | 3.6 | 7.6 | 22.7 | 1.4 | 3.7 | 2.9 | 6.8 | 4.6 | 12.0 |
| $k = 3$ | (1) | 1.6 | 4.3 | 7.3 | 21.5 | 1.6 | 4.5 | 2.3 | 5.9 | 6.3 | 14.9 |
| | (2) | 3.1 | 6.8 | 3.6 | 7.8 | 3.7 | 7.8 | 3.7 | 7.8 | 4.2 | 9.1 |
| | (3) | 0.7 | 1.6 | 0.7 | 1.6 | 0.7 | 1.6 | 0.7 | 1.6 | 0.7 | 1.6 |
| | (4) | 1.1 | 2.9 | 2.2 | 5.3 | 1.1 | 2.9 | 4.3 | 10.6 | 7.2 | 26.1 |
| | (5) | 1.3 | 3.3 | 5.5 | 14.9 | 1.3 | 3.4 | 2.5 | 5.2 | 4.0 | 7.3 |

only hint at the misspecification of the model. The estimates of limited and full information methods differ. The OLS errors in equation (1) may be autocorrelated, and the coefficient of the variable ($\Pi_{t-1}$) is not different from zero ($\alpha = 0.05$). In equation (2) and (3) the intercept is not significant ($\alpha = 0.05$). The $R^2$ and the *ex post* prediction errors are acceptable. The Chow tests and a sequence of Chow tests do not indicate structural changes. The same holds for the CUSUM test. The specification tests of the Ramsey type do not give any hints of a misspecification. These results are not surprising. The original author did a good job.

This impression changes after looking at the jackknife estimates (a sequence of independent jackknife estimates) and the $Q^2$-measures (*Table 5.14*). These give some indication of misspecification in the model. First, the coefficient estimates show a large variation over the different subsamples (each sample period deletes three adjacent observations) and the estimates are transformed in the following manner:

$$\beta_{-i} = T\hat{\beta} - (T - p)\hat{\beta}_i, \quad i = 1, \ldots, N.$$

Second, the jackknife-based $t$-tests show that at least two of the three equations are not well specified (*Table 5.15*). These results may look surprising at first. However, the $t$-test based upon a jackknifed system estimator, e.g., the fixed-point estimator, do not neglect the interdependencies in the model.

Third, the $Q^2$ measures (*Table 5.16*) display the patterns described earlier (see discussion of *Table 5.6*): The $Q^2$ measures decrease with $k$, and the percent errors increase in

*Table 5.12:* Model Klein I.

| | Equation |
|---|---|
| 1 | $C_t = \beta_{11} + \beta_{12}\Pi_t + \beta_{13}\Pi_{t-1} + \beta_{14}W_t$ |
| 2 | $I_t = \beta_{21} + \beta_{22}\Pi_t + \beta_{23}\Pi_{t-1} + \beta_{24}K_{t-1}$ |
| 3 | $W_t = \beta_{31} + \beta_{32}t + \beta_{33}YUTMW2_t + \beta_{34}YUTMW2_{t-1}$ |
| 4 | $T_t \equiv YUT_t - Y_t$ |
| 5 | $Y_t \equiv C_t + I_t + G_t - T_t$ |
| 6 | $\Pi_t \equiv Y_t - W1_t - W2_t$ |
| 7 | $W_t \equiv W1_t + W2_t$ |
| 8 | $K_t \equiv K_{t-1} + I_t$ |
| 9 | $YUTMW2_t \equiv Y_t + T_t - W2_t$ |

*Table 5.13:* Estimated coefficients, model Klein I: FP = fixed point, JK = jackknife.

| | Estimation method | | | | | | |
|---|---|---|---|---|---|---|---|
| | OLS | TSLS | FP | FIML | JK-OLS | JK-TSLS | JK-FP |
| | | | | | $p = 3$ | $p = 3$ | $p = 3$ |
| $C$: (1) | | | | | | | |
| $\beta_{11}$ | 16.2366 | 16.5548 | 16.5038 | 17.5800 | 16.6892 | 15.3195 | 16.1799 |
| $\beta_{12}$ | 0.1929 | 0.0173 | 0.0183 | −0.1357 | 0.1833 | 0.4437 | 0.1802 |
| $\beta_{13}$ | 0.0899 | 0.2162 | 0.2588 | 0.3454 | 0.1214 | −0.0584 | 0.1676 |
| $\beta_{14}$ | 0.7962 | 0.8102 | 0.7942 | 0.7968 | 0.7717 | 0.7527 | 0.7641 |
| $I$: (2) | | | | | | | |
| $\beta_{21}$ | 10.1258 | 20.2783 | 22.0898 | 26.8377 | 12.7327 | 30.3629 | 17.7977 |
| $\beta_{22}$ | 0.4796 | 0.1502 | 0.0929 | −0.6591 | 0.3736 | 0.1950 | 0.4410 |
| $\beta_{23}$ | 0.3330 | 0.6159 | 0.6621 | 1.0123 | 0.3992 | 0.5498 | 0.3443 |
| $\beta_{24}$ | −0.1118 | −0.1578 | −0.1658 | −0.1547 | −0.1203 | −0.2094 | −0.1486 |
| $W1$: (3) | | | | | | | |
| $\beta_{31}$ | 1.4970 | 1.5003 | 2.3823 | 5.2456 | 2.5448 | 2.4969 | 2.3644 |
| $\beta_{32}$ | 0.1302 | 0.1304 | 0.1607 | 0.2861 | 0.1381 | 0.1301 | 0.1847 |
| $\beta_{33}$ | 0.4395 | 0.4389 | 0.4013 | 0.2314 | 0.4134 | 0.4419 | 0.4280 |
| $\beta_{34}$ | 0.1461 | 0.1467 | 0.1704 | 0.2763 | 0.1559 | 0.1267 | 0.1434 |

*Table 5.14:* Estimated coefficients for sub-periods, model Klein I. Estimation method: JK-FP, $p = 3$.

| | | Sample period 1 | 2 | 3 | 4 | 5 | 6 | 7 |
|---|---|---|---|---|---|---|---|---|
| *C*: (1) | $\beta_{11}$ | 7.5131 | 16.3683 | 16.5854 | 8.0321 | 22.2734 | 17.7774 | 24.7101 |
| | $\beta_{12}$ | −0.5189 | −0.2263 | 0.2903 | 2.3773 | 0.1270 | −0.3190 | −0.4689 |
| | $\beta_{13}$ | 0.7175 | 0.4871 | 0.4266 | −1.2193 | −0.1730 | 0.4079 | 0.5261 |
| | $\beta_{14}$ | 1.0181 | 0.8096 | 0.6369 | 0.5473 | 0.7985 | 0.8543 | 0.6843 |
| *I*: (2) | $\beta_{21}$ | −18.5131 | 34.8404 | −2.1168 | 65.8627 | −3.7890 | 51.1875 | −3.0524 |
| | $\beta_{22}$ | 0.5147 | −0.5179 | 0.1160 | 2.0196 | 0.6453 | −0.5838 | 0.8329 |
| | $\beta_{23}$ | 0.3620 | 1.2419 | 0.9636 | −0.7811 | −0.6110 | 1.1386 | 0.0961 |
| | $\beta_{24}$ | 0.0167 | −0.2253 | −0.0647 | −0.4461 | −0.0281 | −0.2956 | −0.0530 |
| *W1*: (3) | $\beta_{21}$ | 4.1085 | 3.6298 | −1.0226 | 5.1956 | 5.2585 | 2.5654 | −3.1842 |
| | $\beta_{32}$ | 0.2343 | 0.2368 | 0.0669 | 0.2472 | 0.2063 | 0.0892 | 0.2118 |
| | $\beta_{33}$ | 0.3185 | 0.4251 | 0.1133 | 0.6880 | 0.2758 | 0.4677 | 0.7075 |
| | $\beta_{34}$ | 0.2292 | 0.1201 | 0.5376 | −0.1862 | 0.2529 | 0.0928 | −0.0429 |

*Table 5.15:* Jackknife-based $t$-tests, model Klein I. Estimation method: JK-FP, $p = 3$.

| | Dependent variable: *C* | | | |
|---|---|---|---|---|
| Coefficient | $\beta_{11}$ | $\beta_{12}$ | $\beta_{13}$ | $\beta_{14}$ |
| $t$-value | 6.5774 | 0.4703 | 0.6612 | 12.9859 |
| | Dependent variable: *I* | | | |
| Coefficient | $\beta_{21}$ | $\beta_{22}$ | $\beta_{23}$ | $\beta_{24}$ |
| $t$-value | 1.4509 | 1.3671 | 1.1072 | −2.2046 |
| | Dependent variable: *W1* | | | |
| Coefficient | $\beta_{21}$ | $\beta_{22}$ | $\beta_{23}$ | $\beta_{24}$ |
| $t$-value | 1.9254 | 6.5593 | 5.2274 | 1.6395 |

*Table 5.16:* Prediction quality of the model Klein I. Estimation method: JK-FP, $p = 3$.

| Dependent variable | $C$ | $I$ | $\Pi$ | $K$ | $W$ | $W1$ | $Y$ | $Y2$ |
|---|---|---|---|---|---|---|---|---|
| $Q^2(SG)$ | | | | | | | | |
| $k = 1$ | 0.7479 | 0.4343 | 0.1573 | 0.9561 | 0.8630 | 0.7946 | 0.6621 | 0.6341 |
| $k = 2$ | 0.4020 | 0.2199 | 0.1502 | 0.7183 | 0.6060 | 0.3906 | 0.3025 | 0.2808 |
| $k = 3$ | 0.3682 | 0.3935 | 0.4003 | 0.1158 | 0.6137 | 0.4742 | 0.4870 | 0.4944 |
| $Q^2(RW)$ | | | | | | | | |
| $k = 1$ | −0.0925 | 0.2302 | 0.0887 | 0.3994 | 0.1981 | −0.0331 | 0.0629 | −0.2028 |
| $k = 2$ | −0.0796 | 0.4993 | 0.4479 | 0.1620 | 0.1330 | 0.0853 | 0.2555 | 0.1345 |
| $k = 3$ | 0.2798 | 0.4524 | 0.5936 | 0.1733 | 0.4612 | 0.4261 | 0.5118 | 0.4966 |
| $Q^2(LC)$ | | | | | | | | |
| $k = 1$ | 0.2522 | 0.7451 | 0.5169 | 0.2302 | 0.6624 | 0.6898 | 0.5691 | 0.4641 |
| $k = 2$ | 0.4833 | 0.8632 | 0.7667 | 0.3947 | 0.6798 | 0.7138 | 0.7164 | 0.6915 |
| $k = 3$ | 0.6964 | 0.8861 | 0.8499 | 0.4799 | 0.8029 | 0.8241 | 0.8281 | 0.8237 |
| APE | | | | | | | | |
| $k = 1$ | 6.9838 | 268.9733 | 21.5341 | 1.1490 | 7.0968 | 8.0566 | 11.1553 | 10.6882 |
| $k = 2$ | 9.9443 | 95.6556 | 24.0454 | 2.4712 | 10.9698 | 12.4910 | 14.2873 | 13.9305 |
| $k = 3$ | 9.8606 | 106.1961 | 33.4684 | 3.4701 | 11.6398 | 13.4668 | 16.1759 | 15.5123 |
| MPE | | | | | | | | |
| $k = 1$ | 10.7931 | 1202.4204 | 37.7318 | 1.6790 | 14.6196 | 16.1676 | 17.3324 | 17.4093 |
| $k = 2$ | 17.6180 | 165.5607 | 63.7750 | 4.1319 | 18.6419 | 21.2356 | 28.7902 | 27.3345 |
| $k = 3$ | 31.1117 | 262.9963 | 139.0147 | 7.9188 | 35.1095 | 41.5261 | 52.7205 | 49.1503 |

$Y2 \equiv YUTMW2$.

most cases with $k$. This leads one to suspect that the model is misspecified and is not "predictive" in the sense of Wold.

## 5.4 Conclusions

In our opinion, the tools described in this chapter are useful for the applied econometrician. These tools provide several advantages over the traditional approach. The tools are easy to describe and can be applied to complicated model structures.

All estimation methods can be used to construct jackknife estimates. The jackknife procedure provides estimates for the variances of the estimators used. As a by-product, calculations of different versions of the $Q^2$ statistic can be made. The computed $k$ step *ex post* forecast errors provide useful information on the operational properties of estimated forecasting models. The analyses of dynamic *ex post* forecasts using the jackknifed estimators and the percent errors give us a kind of confidence interval for the parameters of the endogenous variables of a model. These confidence intervals reflect the different sources of uncertainty in the model-building process.

## References


Abrahamsen, Y. (1986), Jackknifed fixpoint estimators for linear and nonlinear econometric models. Paper presented at the International Conference Macromodels'86, Szczyrk, Poland.

Abrahamsen, Y. and Schips, B. (1989), Specification and stability tests versus jackknifing. Some illustrative examples, pp. 37-43 in P. Hackl (ed.), *Statistical Analysis and Economic Structural Change*. Berlin: Springer-Verlag.

Ball, R.J. (1963), The significance of simultaneous methods of parameter estimates in econometric models. *Applied Statistics*, **12**, 14-25.

Bergström, R. and Wold, H. (1983), *Fixpoint estimation in theory and practice*. Göttingen: Vandenhoeck & Ruprecht.

Chong, Y.Y. and Hendry, D.F. (1986), Econometric evaluation of linear macro-economic models. Applied Economics Discussion Paper No. 10, Institute of Economics and Statistics, University of Oxford.

Efron, B. and Gong, G. (1983), A leisurely look at the bootstrap, the jackknife, and cross validation. *The American Statistician*, **37**, 36-48.

Geisser, S. (1974), A predictive approach to the random effect model. *Biometrika*, **61**, 101-107.

Klein, L.R. (1950), *Economic fluctuations in the United States 1921-1941*. New York: John Wiley.

Leamer, E.E. (1983), Let's take the con out of econometrics. *American Economic Review*, **33**, 31-43.

Miller, R.G. (1974), The jackknife: A review. *Biometrika*, **61**, 1-15.

Ramsey, J.B. and Gilbert, R. (1972), A Monte Carlo study of some small sample properties of tests for specification errors. *Journal of the American Statistical Association*, **67**, 180-186.

Stone, M. (1974), Cross-validatory choice and assessment of statistical predictions. *Journal of the Royal Statistical Society*, **B-36**, 111-133.

Tukey, J.W. (1958), Bias and confidence in not-quit large samples. *Annals of Mathematical Statistics*, **29**, 614.

Wold, H. (1969), Econometrics as pioneering in nonexperimental model building. *Econometrica*, **37**, 369-381.

CHAPTER 6

# Bayesian Inferences about the Intersection of Two Regressions

*Lyle D. Broemeling, Peyton J. Cook, and Juanita H. Chin Choy*

## Summary

Previous Bayesian studies of structural change in linear models have focused on the shift point or on the transition function of the model. In this study we take a Bayesian approach for making inferences about the intersection between two regression lines. Using a proper prior density for all of the parameters, the marginal posterior density of the intersection is derived. This density can be expressed in closed form, but it is not a standard density. With numerical integration, the density can be normalized and point and interval estimators computed.

## 6.1 Introduction

Bayesian inferences for the parameters of a two-phase regression are well known and are given in Broemeling (1985) and Broemeling and Tsurumi (1987). These studies, among others, used the shift point or the transition function to model the change from one regime to the other, and Bayesian inferences were focused on these parameters. For example, Tsurumi (1980) examined the US demand for Japanese passenger cars by using the transition function approach, while Chin Choy and Broemeling (1980) analyzed an example of Quandt (1958) by finding the marginal posterior distribution of the shift point.

Very little has appeared from a Bayesian perspective [see, e.g., Holbert (1973) and Chin Choy (1977)] in estimating the intersection in a two-phase regression problem, but the problem has been studied by many, beginning with Hinkley (1969,1971). The problem continues to attract attention and has been generalized to the intersection between stochastic processes. A recent reference is Rosen (1986).

In this chapter, we return to the two-phase regression model with one change, from one simple linear regression to another. All the parameters are assumed unknown, where the prior information is expressed with a proper prior density, and the change is represented by a shift point. Since the abscissa of the intersection point is a ratio of regression coefficients, and the coefficients have a multivariate $t$-distribution, by transforming to the intersection (using the fact that the expected value of the absolute value of a $t$-random variable is known), the marginal posterior density of the intersection is expressed in a closed form. An example that employs numerical integration illustrates the Bayesian approach to making inferences about the intersection.

## 6.2 The Two-Phase Regression Model

Suppose that $y_1, y_2, \ldots, y_n$ is a sequence of independent random variables, where

$$y_i = \begin{cases} \alpha_1 + \beta_1 x_i + e_i, & i = 1, 2, \ldots, m, \\ \alpha_2 + \beta_2 x_i + e_i, & i = m+1, \ldots, n, \end{cases} \tag{6.1}$$

and the $e_i \sim N(0, \tau^{-1})$, $i = 1, 2, \ldots, n$, and $m = 1, 2, \ldots, n-1$. The $y_i$ are the known values of some regressor and the unknown parameters are $m$, $\theta = (\alpha_1, \beta_1, \alpha_2, \beta_2)' \in R^4$, and $\tau > 0$. Thus, one change will occur and the parameter of interest is the abscissa of the intersection

$$\gamma = (\alpha_2 - \alpha_1)/(\beta_1 - \beta_2). \tag{6.2}$$

The most general case is considered, and a proper distribution will express our prior information.

Before finding the posterior distribution of $\gamma$, we must know the posterior distribution of $\theta$, when the prior distribution of $(m, \theta, \tau)$ is given by

$$g(m, \theta, \tau) = g_0(m)\, g_0(\theta, \tau), \tag{6.3}$$

where $g_0(m) = (n-1)^{-1}$, $m = 1, 2, \ldots, n-1$, and $g_0(\theta, \tau)$ is a normal-gamma density

$$g_0(\theta, \tau) \propto \tau^{(a+2)-1} \exp\left\{-\frac{\tau}{2}\left[2b + (\theta - \theta_\mu)' Q (\theta - \theta_\mu)\right]\right\}, \quad \theta \in R^4, \quad \tau > 0,$$

and $a$, $b$, $\theta_\mu$, and $Q$ are hyperparameters that must be specified by the user. We assume that $m$ is independent of $\theta$ and $\tau$.

Under these assumptions, it can be shown [see Broemeling (1985, Chapter 7)] that the posterior density of $\theta$ is a mixture of multivariate $t$-densities. Thus, we have

$$g(\theta|y) = \sum_{m=1}^{n-1} g(m|y)\, t\left[\theta; 4, 2a^*, \theta^*(m), P(m)\right], \quad \theta \in R^4, \tag{6.4}$$

where $y = (y_1, y_2, \ldots, y_n)$, $g(m|y)$ is the marginal p.m.f. of the shift point and $t[\cdot]$ denotes a four-dimensional $t$-density for $\theta$ with $2a^*$ degrees of freedom, location vector $\theta^*(m)$, and precision matrix $P(m)$. The parameters of the above density are given in detail in Chin Choy and Broemeling (1980) and will not be repeated here.

We now transform from $\theta$ to $\gamma$ and show that the marginal posterior density of $\gamma$ can be expressed in closed form.

## 6.3 The Posterior Distribution of the Intersection

The transformation from $\theta$ to $\gamma$ is done in two stages. First, consider the transformation from $\theta$ to $w = (w_1, w_2)$, where

$$w = T\theta \quad \text{and} \quad T = \begin{bmatrix} -1 & 0 & 1 & 0 \\ 0 & 1 & 0 & -1 \end{bmatrix}, \tag{6.5}$$

thus $w_1 = \alpha_2 - \alpha_1$ and $w_2 = \beta_1 - \beta_2$. The second transformation is from $w$ to $(\gamma_1, \gamma_2)$, where $\gamma_1 = w_1/w_2$ and $\gamma_2 = w_2$. Since the distribution of $\theta$ is a mixture of four-dimensional multivariate $t$-densities, the distribution of $w$ is a mixture (with the same mixing distribution) of two-dimensional $t$-densities. This is because $w$ is a linear function of $\theta$ [see DeGroot (1970)]. The posterior density of $w$ is

$$g(w|y) = \sum_{m=1}^{n-1} g(m|y)\, t[w; 2, 2a^*, w(m), V(m)], \quad w \in R^2, \tag{6.6}$$

where $w(m) = T\theta^*(m)$ and $V(m) = [TP^{-1}(m)T']^{-1}$. At the second stage of the transformation process from $w$ to $(\gamma_1, \gamma_2)$, the marginal posterior density is

$$g(\gamma_1, \gamma_2|y) = \sum_{m=1}^{n-1} g(m|y)\, k(m)\, l(\gamma_2, \gamma_1, m), \quad \gamma_i \in R, \tag{6.7}$$

where

$$\begin{aligned}
k(m) &= |V(m)|^{1/2}\Gamma(a^*+1)/2a^*\pi\Gamma(a^*) \\
l(\gamma_2, \gamma_1, m) &= \left\{1 + (1/2a^*)\left[A(\gamma_1, m)\gamma_2^2 - 2B(\gamma_1, m)\gamma_2 + c(m)\right]\right\}^{-(a^*+1)} \\
A(\gamma_1, m) &= (\gamma_1, 1)\, V(m)\, (\gamma_1, 1)' \\
B(\gamma_1, m) &= (\gamma_1, 1)\, V(m)\, T\,\theta^*(m) \\
c(m) &= [T\,\theta^*(m)]'\, V(m)\, [T\,\theta^*(m)]\,.
\end{aligned}$$

This form (6.7) of the density can be simplified by completing the square on $\gamma_2$ in $l(\gamma_2, \gamma_1, m)$, which results in

$$g(\gamma_1, \gamma_2, m) = \sum_{m=1}^{n-1} K(m)\, [G(\gamma_1, m)]^{-(a^*+1)}\, |\gamma_2|\, [H(\gamma_2, m)]^{-(2a^*+2)/2}\,, \tag{6.8}$$

where

$$\begin{aligned} H(\gamma_2, m) &= 1 + [s(\gamma_1, m)/(2a^* + 1)]\,[\gamma_2 - \gamma_2^*(\gamma_1, m)]^2 \\ G(\gamma_1, m) &= 1 + (1/2a^*)\,[c(m) - B^2(\gamma_1, m)/A(\gamma_1, m)] \\ \gamma_2^*(\gamma_1, m) &= B(\gamma_1, m)/A(\gamma_1, m) \\ s(\gamma_1, m) &= (2a^* + 1)\,A(\gamma_1, m)/\,[2a^* G(\gamma_1, m)]\,. \end{aligned}$$

The variable $\gamma_2$ can now be eliminated by integrating (6.8) with respect to $\gamma_2$. This is possible if we know the expectation of the absolute value of a $t$-random variable. Note that $H$ as a function of $\gamma_2$ is the kernel of a $t$-density with $2a^* + 1$ degrees of freedom, location $\gamma_2^*(\gamma_1, m)$, and precision $s(\gamma_1, m)$. In the Appendix, the expectation of the absolute value of a $t$-random variable is derived. When this is used to eliminate $\gamma_2$ from (6.8), one finds that the marginal density of $\gamma_1$ is

$$\begin{aligned} g(\gamma_1|y) = {} & \frac{\Gamma\,[(2a^* + 1)/2]}{\Gamma(a^*)(2\pi a^*)^{1/2}} \\ & \times \sum_{m=1}^{n-1} |V(m)|^{1/2} g(m|y)\, A^{-1/2}(\gamma_1, m)\,[G(\gamma_1, m)]^{-(a^*+1/2)}\, E|\gamma_2|, \quad \gamma_1 \in R\,, \end{aligned} \tag{6.9}$$

and $E|\gamma_2|$ denotes the expectation of the absolute value of $\gamma_2$ with respect to a $t$-distribution with $2a^* + 1$ degrees of freedom, location $\gamma_2^*(\gamma_1, m)$, and precision $s(\gamma_1, m)$. Using the result in Appendix A, we have

$$\begin{aligned} E|\gamma_2| = {} & \frac{(2a^* + 1)^{1/2}\Gamma(a^* + 1)}{a^* \sqrt{\pi}\,\Gamma\,[(2a^* + 1)/2]} \\ & \times s^{-1/2}(\gamma_1, m)\,\{1 + [1/(2a^* + 1)]\, s(\gamma_1, m)\, \gamma_2^*(\gamma_1, m)\}^{-a^*} \\ & + \gamma_2^*(\gamma_1, m)\left\{2\Phi_{(2a^*+1)}\left[\gamma_2^*(\gamma_1, m)\, s^{1/2}(\gamma_1, m)\right] - 1\right\}, \end{aligned} \tag{6.10}$$

where $\Phi_{(d)}$ denotes the cdf of a $t$-random variable with $d$ degrees of freedom. Therefore (6.9) and (6.10) complete the specification of the posterior distribution of the intersection between two linear regressions. Although (6.9) is rather involved, it is relatively easy to compute the median, mode, and credible intervals for $\gamma_1$.

If one knows $m$, where the change occurs, then the appropriate density is the conditional density of $\gamma_1$ given $m$, which is

$$\begin{aligned} & g(\gamma_1|m, y) \\ & = \frac{\Gamma\,[(2a^* + 1)/2]}{(2\pi a^*)^{1/2}\Gamma(a^*)} |V(m)|^{1/2} A^{-1/2}(\gamma_1, m)\,[G(\gamma_1, m)]^{-(a^*+1/2)}\, E|\gamma_2|\,, \end{aligned} \tag{6.11}$$

and this is easier to work with numerically. We will use densities (6.10) and (6.11) to illustrate the Bayesian methodology in Section 6.4.

## 6.4 An Example

Data from Pool and Borchgrevink (1964) proved a good example to illustrate Bayesian inferences for the intersection in two-phase regression. The data are shown in Appendix B, where the independent variable $x$ is Warfarin and the dependent variable is blood factor VII production. Hinkley (1971), Holbert (1973), and Chin Choy (1977) use this data set to illustrate the techniques that they have developed.

For purposes of illustration, the parameters of the prior distribution (6.3) are given the values

$$\begin{aligned} \phi_\mu &= (0, 0.2, 0.95, 0)' \\ Q &= I_4 \\ a &= 2 \\ b &= 0.0017, \quad \text{i.e., } E[\tau^{-1}] = 0.0017. \end{aligned}$$

These estimates were obtained by Hinkley. From this information, the mass function of $m$ is calculated and shown in Appendix B. The location estimators of $m$ are the mode = 6 and the mean = 6.13. Thus, we are confident that the shift point is at $m = 6$. That is, the first six observations follow the first regression, while the remaining nine observations follow the second regression. *Figure 6.1* gives the graphs of the marginal and conditional (given $m = 6$) posterior densities of $\gamma$. These were graphed from formulas (6.9) and (6.11), respectively. In addition, point and interval estimates of $\gamma$ were computed by numerical integration, and are listed in *Table 6.1.*

*Table 6.1:* Point and interval estimates of $\gamma_1$.

| | $g(\gamma_1 \mid y)$ | $g(\gamma_1 \mid 6, y)$ |
|---|---|---|
| Point estimates | | |
| mode | 4.81 | 4.79 |
| median | 4.81 | 4.80 |
| HPD regions | | |
| 90 % | (4.55, 5.07) | (4.56, 5.05) |
| 95 % | (4.49, 5.13) | (4.50, 5.11) |
| 99 % | (4.37, 5.26) | (4.41, 5.23) |

From examining the graphs and the computed estimates, we see how well the conditional density approximates the marginal density. The intersection $\gamma_1$ is the ratio of correlated $t$-random variables, thus the mean and variance do not exist and were not computed.

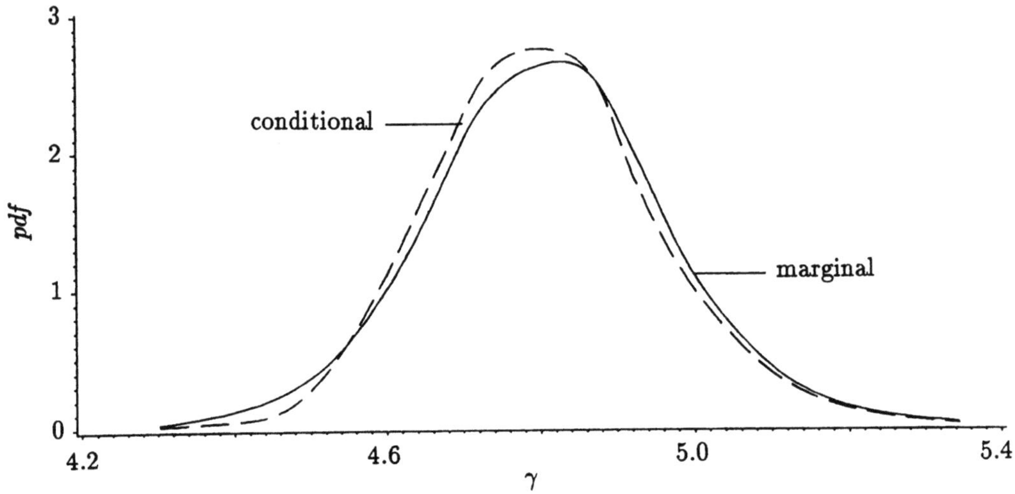


*Figure 6.1:* Marginal and conditional posterior *pdf* of $\gamma$

## 6.5 Comments and Summary

We have shown that it is possible to make Bayesian inferences about the intersection between two simple linear regressions when one uses a shift-point representation of change and a normal-gamma density to express prior information. The marginal posterior density of the intersection can be derived in closed form. The density is easily graphed, and the point and interval estimates are computed. If one is not so confident of prior information, is it possible to employ a vague improper prior density for the parameters? The answer is yes, but certain adjustments have to be made. For example, suppose we let

$$g(m,\theta,\tau) = g_0(m)\, g_0(\theta,\tau) \tag{6.12}$$

with $g_0(\theta,\tau) = \tau^{-1}$, $\tau > 0$, $\theta \in R^4$, where $g_0(m)$ is as before. Then one may show that the posterior mass function of the shift-point $m$ does not exist at $m = 1$ or at $m = n - 1$. However, if one knows that the change cannot occur at these points, *a priori*, one could use

$$g_0(m) = (n-3)^{-1}, \qquad m = 2,3,\ldots,n-2, \tag{6.13}$$

and together with (6.12), derive the posterior density of $\gamma_1$. This would be similar to the result derived when using a proper prior density.

## References

Broemeling, L.D. (1985), *Bayesian Analysis of Linear Models.* New York: Marcel Dekker.

Broemeling, L.D. and Tsurumi, H. (1987), *Econometrics and Structural Change.* New York: Marcel Dekker.

Chin Choy, J. (1977), A Bayesian analysis of a changing linear model. Doctoral Dissertation, Oklahoma State University, Stillwater, Oklahoma.

Chin Choy, J. and Broemeling, L.D. (1980), Some Bayesian inferences for a changing linear model. *Technometrics*, **22**, 71-78.

DeGroot, M.H. (1970), *Optimal Statistical Decisions.* New York: McGraw-Hill.

Hinkley, D.V. (1969), Inference about the intersection in two-phase regression. *Biometrika*, **56**, 495-504.

Hinkley, D.V. (1971), Inference in two-phase regression. *Journal of the American Statistical Association*, **66**, 736-793.

Holbert, D. (1973), A Bayesian analysis of shifting sequences with applications to two-phase regression. Doctoral dissertation, Oklahoma State University, Stillwater, Oklahoma.

Pool, J. and Borchgrevink, C.F. (1964), Comparison of rat liver response to coumarin administered *in vivo* versus *in vitro*. *American Journal of Physiology*, **206**, 229-238.

Quandt, R.E. (1958), The estimation of the parameters of a linear regression system obeying two separate regimes. *Journal of the American Statistical Association*, **53**, 873-880.

Rosen, J. (1986), Tanaka's formula and renormalization for intersections of planar Brownian motion. *Annals of Probability*, **14**, 1245-1251.

Tsurumi, H. (1980), A Bayesian estimation of structural shifts by gradual switching regressions with an application to the U.S. gasoline market, pp. 213-240 in A. Zellner (ed.), *Bayesian Analysis in Econometrics and Statistics in Honor of Harold Jeffreys.* Amsterdam: North-Holland.

# Appendix A

Suppose $X$ is distributed as a general $t$-distribution with $n$ degrees of freedom, location $\mu$, and precision $\tau$ ($n > 0$, $\mu \in R$, and $\tau > 0$), then

$$E|X| = \frac{2\sqrt{n}\,\Gamma[(n+1)/2]}{(n-1)\sqrt{\pi\tau}\,\Gamma(n/2)}(1+\tau\mu^2/n)^{-(n-1)/2} + \mu\left[2\Phi_n(\mu\sqrt{\tau}) - 1\right] ,$$

where $\Phi_n(x)$ is the cdf of student's $t$-distribution with $n$ degrees of freedom. This result can be verified by integration.

# Appendix B

Data from Pool and Borchgrevink (1964) and the posterior distribution of $m$:

| $i$ | $x_i$ | $y_i$ | $m$ | $g(m\|y)$ |
|---|---|---|---|---|
| 1 | 2. | .370483 | 1 | 0. |
| 2 | 2.52288 | .53797 | 2 | 0. |
| 3 | 3. | .607684 | 3 | 0.00001 |
| 4 | 3.52288 | .723323 | 4 | 0.00053 |
| 5 | 4. | .761856 | 5 | 0.19744 |
| 6 | 4.52288 | .892063 | 6 | 0.48151 |
| 7 | 5. | .956707 | 7 | 0.31535 |
| 8 | 5.52288 | .90349 | 8 | 0.00513 |
| 9 | 6. | .898609 | 9 | 0.0002 |
| 10 | 6.52288 | .953850 | 10 | 0. |
| 11 | 7. | .990834 | 11 | 0. |
| 12 | 7.52288 | .890291 | 12 | 0. |
| 13 | 8. | .990779 | 13 | 0. |
| 14 | 8.52288 | 1.050865 | 14 | 0. |
| 15 | 9. | .982785 | | |

CHAPTER 7

# Nonparametric Sequential Surveillance of Intervention Effects in Time Series

*Peter Hackl*

## Summary

A nonparametric process procedure is proposed as a tool for continuous monitoring to detect intervention effects in a time series. The procedure uses exponentially weighted moving averages of the ranks of the post-intervention one-step-ahead forecast errors. The forecasts are based on the pre-intervention model, and the errors are ranked with respect to residuals in the pre-intervention period. The method is illustrated with the Los Angeles oxidant data previously analyzed by Box and Tiao (1975).

## 7.1 Introduction

Box and Tiao (1975) provide a method to model intervention effects in time series. The question that is answered by intervention analysis is whether or not an intervention has changed the level of a time series. Box and Tiao use difference equation (ARIMA) models to represent the dynamic characteristics of both the noise and the interventions effects.

In general, intervention analysis faces two difficulties.

- The pattern of the intervention effect is often difficult to determine *a priori*; consequently, the need to specify the intervention dynamics leads to arbitrariness in the model choice. This fact, together with the weak discriminatory power of the available model selection criteria (Box-Pierce statistic, Ljung-Box statistic), can affect the conclusions.

- Intervention analysis must be seen as a problem of sequential analysis. Although the timing of the intervention is typically known, the number of observations that contain information about this intervention effect increases over time and is not fixed. In many applications, we want to reach a decision as soon as possible.

Process control techniques are ideally suited to address these two concerns. The means of inferring whether the intervention can be assumed effective are diagnostic checks of the differences between the observations and the one-step-ahead forecasts obtained from the model for the pre-intervention time. Correspondingly, Abraham (1987) suggests to apply CUSUM charts to the one-step-ahead forecast errors in the post-intervention period. This possibility is already mentioned by Box and Tiao (1976). It is certainly a promising topic to investigate the appropriateness and the relative merits of other sequential process control methods.

Recently, interest in process control methods that are to a large extent independent of distributional assumptions led to the development of nonparametric process control procedures (Bakir and Reynolds, 1979; Bhattacharyya and Frierson, 1981; McDonald, 1986; Hackl and Ledolter, 1989). Such methods are comparable, in terms of average run length, to the standard process control procedures, such as Shewhart and CUSUM techniques. In addition, they have the adventage of being independent of distributional assumptions. In this chapter I apply such a nonparametric process control procedure in the context of intervention analysis. This procedure is based on the ranks of the one-step-ahead forecast errors for the post-intervention data. The forecast errors are ranked among the historic one-step-ahead forecast errors (that is the residuals) of the pre-intervention period.

Section 7.2 presents a modification of a process control procedure that is based on exponentially weighted moving averages of the ranked one-step-ahead forecast errors. The application of this procedure to intervention analysis is discussed in Section 7.3. Section 7.4 analyzes the Los Angeles oxidant data given by Box and Tiao (1975). Concluding remarks are given in Section 7.5.

## 7.2 The Rank-Based Sequential Control Procedure

Hackl and Ledolter (1989) derived a nonparametric process control procedure that is based on sequential ranks. The sequential rank of an observation is defined as its rank among the $g$ most recent observations. The control procedure is based on exponentially weighted moving averages of these ranks. The procedure is outlier-resistent and performs well if one is concerned about changing process levels.

The use of sequential ranks makes sure that at each control point the actual observation is compared with the most recent past of the process. In the intervention analysis situation, however, information about the effectiveness of the intervention is obtained by comparing post-intervention with pre-intervention data. A corresponding modification of the process control procedures is presented in this section.

The situation can be formalized as follows. Assume that $g-1$ independent and identically distributed random variables $X_i$, $i = 1, \ldots, g-1$, are available as the ranking basis. The probability distribution function of $X_i$ is denoted by $F$. We want to test the null hypothesis that independent random variables $Y_t$, $t = 1, 2, \ldots$, follow the same distribution $F$, against a change-in-location alternative. We define the rank of $Y_t$ among the $g-1$ $X_i$'s as

$$R_t^* = 1 + \sum_{i=1}^{g-1} I_{[Y_t > X_i]} \ , \quad t = 1, 2, \ldots \ , \tag{7.1}$$

where the indicator function $I_{[Y_t > X_i]} = 1$ if $Y_t > X_i$, and 0 otherwise. The parameter $g$ is denoted as the *ranking size.* The *standardized rank* $R_t$ is defined as

$$R_t = \frac{2}{g}\left(R_t^* - \frac{g+1}{2})\right) \ , \quad t = 1, 2, \ldots \ . \tag{7.2}$$

The $R_t$'s are, for all $t$, uniformly and independently distributed on the $g$ points $\{\frac{1}{g} - 1, \frac{3}{g} - 1, \ldots, 1 - \frac{1}{g}\}$, with expectation zero and variance $\mathrm{Var}\{R_t\} = (g^2 - 1)/(3g^2)$.

In the process control literature it is well established that control procedures that take past observations into account are more powerful in detecting small and moderate changes in the location parameter than procedures that are based on only the most recent observation. For this reason we use exponentially weighted moving averages of standardized ranks as test statistics. Other possible choices are the cumulative sum and the moving sum statistics. The exponentially weighted moving averages of standardized ranks are defined recursively as

$$T_t = (1 - \lambda)T_{t-1} + \lambda R_t \ , \quad t = 1, 2, \ldots, \tag{7.3}$$

where $T_0$ is the *starting value*, usually set to equal zero, and $0 < \lambda \leq 1$ is the *smoothing parameter.* If we want to test against increases in location, one considers a change as effective if the series $\{T_t\}$ crosses a suitably chosen *critical limit* $h$ (that is, $T_t > h$ for any of the control times $t \geq 1$). The choice of a common $h$ for all $t$ is justified by the fact that $\mathrm{Var}\{T_t\}$ is, except for small $t$, independent of $t$. Analogously, $T_t < -h$ and $|T_t| > h$ for any $t \geq 1$ are critical regions of the test against a decrease and a two-sided alternative, respectively.

The performance of a process control technique is usually evaluated on its *average run length* (ARL), which is the average number of observations that are needed to exceed the critical limit for the first time. The parameters are chosen so that the average run length is large if the process is under control, and small in the out-of-control case. Similarly, we want our procedure to have a large average run length under the null hypothesis, but a small one if a change in location is effective. This must be achieved through a suitable choice of the parameters in our procedure: the ranking size $g$, the smoothing parameter $\lambda$, and the critical limit $h$. Crowder (1987) gives an integral equation for the average run length of the

*Table 7.1:* Average run lengths for the exponentially weighted moving averages of independent continuous random variables obtained from Crowder's (1987) integral equation for three different smoothing parameters $\lambda$ and several critical values $h$.

| $h$ | $\lambda = .9$ | $\lambda = .8$ | $\lambda = .7$ | $\lambda = .6$ | $\lambda = .5$ |
|---|---|---|---|---|---|
| 0.85 | 34.83 | 89.62 | 254.94 | 920.97 | 5065.23 |
| 0.80 | 18.97 | 44.21 | 98.27 | 274.70 | 1056.91 |
| 0.75 | 12.63 | 25.84 | 50.80 | 116.49 | 343.14 |
| 0.70 | 9.45 | 16.88 | 31.08 | 60.93 | 149.49 |

exponentially weighted moving average procedure that is based on independent continuous random variables. Hackl and Ledolter (1989) show that Crowder's approach can be applied to derive approximate average run lengths for exponentially weighted moving averages of sequential ranks. The approximation neglects the discreteness and the small correlations among the ranks. The approximate average run lengths are in good agreement with the respective Monte Carlo estimates for $g$ larger than 30 and most values of $h$. The same approximation can be expected to give an even better agreement for exponentially weighted moving averages of the ranks in (7.1) and (7.2); in this case, consecutive ranks are independent and the approximation neglects only the discreteness of the ranks. These approximate average run lengths now depend only on $\lambda$ and $h$. Hence, for moderate to large values of $g$, we can use the solutions of Crowder's integral equation to select the parameters $\lambda$ and $h$. A typical choice of parameter values is $\lambda = .7$ and $h = .85$, resulting in a theoretical average run length of 255. The run length for other parameter values are shown in *Table 7.1*. The average run length under the alternative when there is a change of level depends, besides on the parameters $\lambda$ and $h$ of the control procedure, on the characteristics of the series $\{Y_t\}$. General conclusions about average run lengths are, even in an approximate form, difficult to obtain.

## 7.3 Sequential Surveillance of Intervention Effects

The intervention analysis model assumes that the observations $X_t = Z_t + N_t$ are the sums of two components: A noise component $N_t$ and an intervention component $Z_t$ that affects the level of the series. The noise component $N_t$ follows the ARIMA model

$$\phi(B)N_t = \theta(B)a_t\,, \tag{7.4}$$

where $B$ is the backshift operator ($BX_t = X_{t-1}$) and where $\{a_t\}$ is a sequence of independent and identically distributed random variables with zero mean and variance $\sigma_a^2$. Typically, $\phi(B) = \phi_1(B)\phi_2(B^s)(1-B)^d(1-B^s)^D$ and $\theta(B) = \theta_1(B)\theta_2(B^s)$, where $\phi_1$, $\phi_2$, $\theta_1$, and $\theta_2$ are polynomials of order $p_1$, $p_2$, $q_1$, and $q_2$, respectively. This representation allows for seasonality (with seasonal period $s$) and nonstationarity. The roots of $\phi(B)$ and $\theta(B)$ lie outside the unit circle, and $\phi(B)$ and $\theta(B)$ have no common roots.

Intervention effects $Z_t$ are modeled by applying filters to indicator sequences

$$Z_t \equiv \delta^{-1}(B)\omega(B)I_t^{(\tau)}, \tag{7.5}$$

where $\omega(B)$ and $\delta(B)$ are polynomials of order $r$ and $s$, respectively. The indicator variable $I_t^{(\tau)}$ can represent a step input, that is $I_t^{(\tau)} = 0$ if $t < \tau$ and $I_t^{(\tau)} = 1$ if $t \geq \tau$; or a pulse input, that is $I_t^{(\tau)} = 1$ if $t = \tau$ and $I_t^{(\tau)} = 0$ otherwise; here, $\tau$ denotes the *intervention time*. The dynamic model

$$X_t = \delta^{-1}(B)\omega(B)I_t^{(\tau)} + \phi^{-1}(B)\theta(B)a_t \tag{7.6}$$

is a very general model that allows for a variety of intervention effects as well as for correlation among the errors [see Box and Tiao (1975)].

If the intervention effect is ignored and if the time series model alone is applied to $X_t$ then the error series $\{a_t^*\}$ is given by

$$\begin{aligned} a_t^* &= \theta^{-1}(B)\phi(B)X_t \\ &= a_t + \theta^{-1}(B)\phi(B)\omega(B)\delta^{-1}(B)I_t^{(\tau)} \\ &= a_t + \mu_t \,. \end{aligned} \tag{7.7}$$

The error series $\{a_t^*\}$ reflects the intervention in the sense that the levels of $a_t^*$ are for $t \geq \tau$ not necessarily zero. The correlation structure, on the other hand, is not affected by the intervention. As it is obvious from (7.7), the intervention effect on the error series depends both on the ARIMA model (7.4) and on the intervention model $\omega(B)\delta^{-1}(B)I_t^{(\tau)}$. Under the assumptions implied on the roots of the polynomials it can be shown that $\mu_t \to 0$ for increasing $t$, for both the step and the pulse input.

*Figure 7.1* shows the levels $\mu_t = E\{a_t^*\}$ as a function of $t$ for the $(0,0,1) \times (0,1,1)_{12}$ ARIMA model discussed in Section 7.4. There we find that $\phi(B) = 1 - B^{12}$, $\theta(B) = (1 + .435B)(1 - .7577B^{12})$, $\omega(B) = \omega_0 = 1$, $\delta(B) = 1$, and $I_t^{(\tau)}$ represents a step input with $\tau = 61$. A suspected intervention effect can be assessed on the basis of the errors $\{a_t^*, t \geq \tau\}$. This can be done in a sequential manner, starting at the suspected intervention time $\tau$, and by calculating a test statistic for each observation time $\tau, \tau+1, \ldots$. For that purpose we use the sequential procedure that was introduced in the last section. Let $\hat{\phi}_\tau(B)$ and $\hat{\theta}_\tau(B)$ be the polynomials that are estimated on the basis of the pre-intervention data $X_i, i < \tau$. The quantities that have to be ranked are the post-intervention one-step-ahead forecast errors

$$f_t = X_t - \hat{X}_{t-1}(1)\,, \quad t = \tau, \tau+1, \ldots, \tag{7.8}$$

where $\hat{X}_{t-1}(1)$ is the predicted value for $X_t$ that uses all observations $X_i, i \leq t-1$, that are available at time $t-1$. The predictions are obtained from the pre-intervention model and use polynomials $\hat{\phi}_\tau(B)$ and $\hat{\theta}_\tau(B)$. The forecast errors are ranked among the residuals (the historic one-step-ahead forecast errors) $f_t = X_t - \hat{X}_{t-1}(1), t = 1, \ldots, \tau - 1$, from the

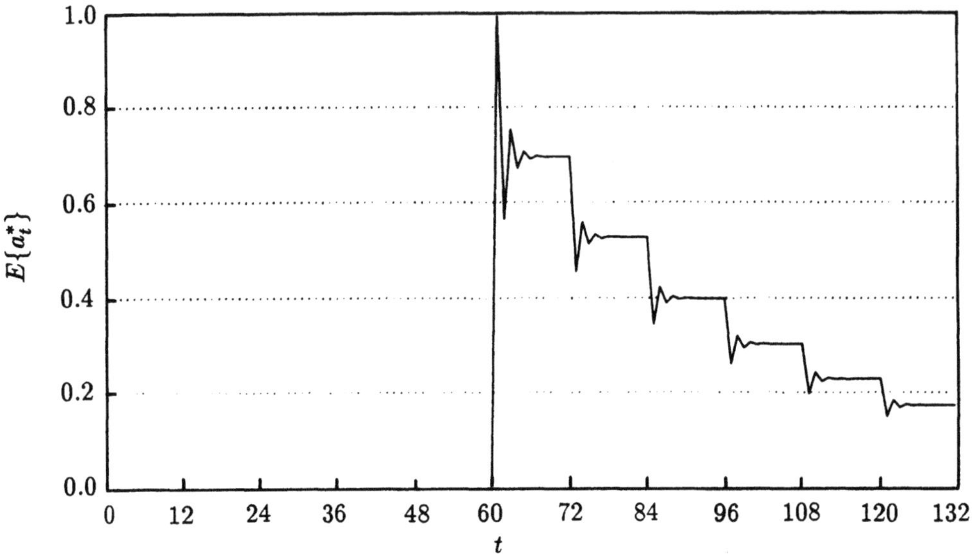


*Figure 7.1:* Expectations $E\{a_t^*\}$ as a function of $t$ for the $(0,0,1) \times (0,1,1)_{12}$ ARIMA model with $\phi(B) = 1 - B^{12}$, $\theta(B) = (1 + .435B)(1 - .7577B^{12})$, $\omega(B) = \omega_0 = 1$, $\delta(B) = 1$, and $I_t^{(\tau)}$ represents a step input with $\tau = 61$.

pre-intervention model. The ranking size $g$ now is $\tau - 1 - p$, where $p$ is the order of the polynomial $\phi(B)$. We denote the standardized rank of $f_t$, $t = \tau, \tau + 1, \ldots$, among $f_i$, $i = 1, \ldots, \tau - 1$, by $R_t$.

If the polynomials $\phi(B)$ and $\theta(B)$ were known, the residuals $f_t$ and the one-step-ahead forecast errors $f_t$ would coincide with the respective errors $a_t^*$; this implies that they would be identical and independently distributed in the no-intervention case, and they would be independent with expectations $\mu_t$ given in (7.7) in the intervention case [Box and Jenkins (1970)].

The exponentially weighted moving averages $T_t$ of the ranks $R_t$

$$T_t = (1 - \lambda)T_{t-1} + \lambda R_t \ , \quad t = \tau, \tau + 1, \ldots \ , \tag{7.9}$$

can be used as a diagnostic whether the intervention has had an effect: The intervention is considered to have changed the process level if $T_t > h$ ($T_t < h$, $|T_t| > h$) for any $t \geq \tau$, where $h$ is the critical limit. The choice of the parameters $g$, $\lambda$, and $h$ determines the performance of the test procedure: Having monthly data, an average run length in the no-intervention case of 60 to 120 may be appropriate. If $g$ is 30 or more we can use the approximate average run length that is derived from Crowder's (1987) integral equation to choose suitable values for $\lambda$ and $h$. This procedure can be applied in a sequential way starting at time $\tau$ and checking $T_t$ in all subsequent times $t = \tau + 1, \ldots$, whenever a new observation becomes available.

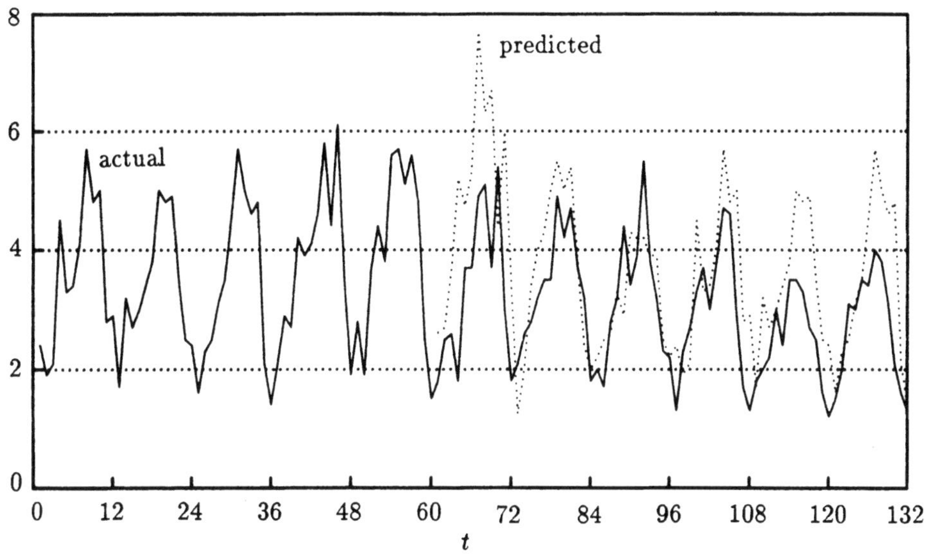


*Figure 7.2:* Monthly averages of the oxidant level in downtown Los Angeles from January 1955 ($t = 1$) to December 1965 and one-step-ahead predictions from January 1960 ($t = 61$) to December 1965 obtained by extrapolating a $(0,0,1) \times (0,1,1)_{12}$ ARIMA model fitted to the data January 1955 to December 1960.

## 7.4 The Los Angeles Oxidant Data

*Figure 7.2* shows the monthly averages of the oxidant level in downtown Los Angeles from January 1955 to December 1965. The events of interest are the opening of the Golden State Freeway and a new law (Rule 63) that reduces the allowable proportion of reactive hydrocarbons in locally sold gasoline. Both events became effective in January 1960. Using the Box-Jenkins method, Box and Tiao (1975) identify a $(0,0,1) \times (0,1,1)_{12}$ ARIMA model for the noise component of the series. Fitting this model

$$(1 - B^{12})X_t = (1 - \theta_1 B)(1 - \theta_2 B^{12})a_t \tag{7.10}$$

to the data of the pre-intervention phase, January 1955 to December 1959 (60 observations), leads to the parameter estimates and standard deviations (in parentheses) given below:

$$\hat{\theta}_1 = \underset{(0.2188)}{-0.4350}\ , \quad \hat{\theta}_2 = \underset{(0.1585)}{0.7577}\ .$$

Diagnostic checks, such as the residual autocorrelations and the Ljung-Box statistic, indicate that the model in (7.10) gives an inadequate description of the time series.

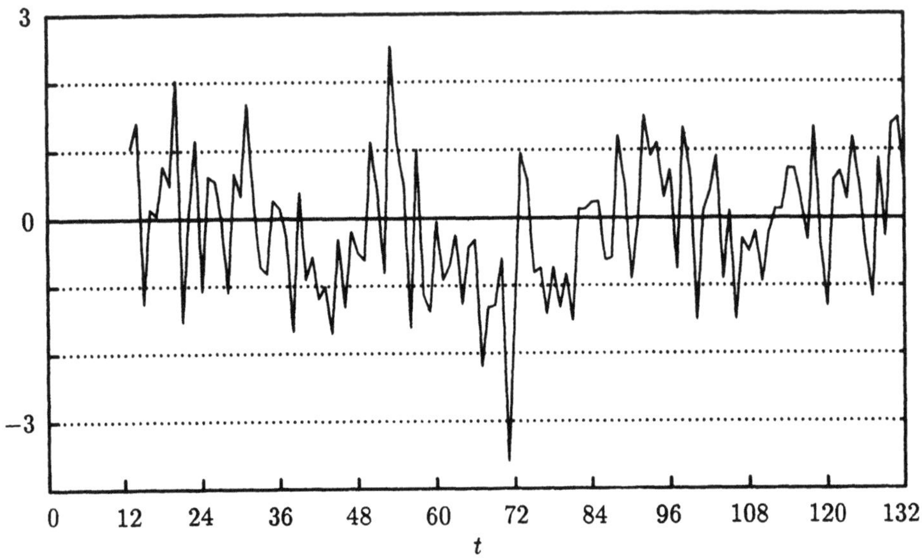


*Figure 7.3:* Residuals from January 1956 ($t = 13$) to December 1959 ($t = 60$) and one-step-ahead forecast errors. Predictions are obtained from model (7.10), with parameter estimates based on the data from January 1955 to December 1959.

The potential intervention starts with 1960. The plots of the post-intervention data (*Figure 7.2*) and of the one-step-ahead forecast errors (*Figure 7.3*) indicate that the level of the series may have changed. In the first two years after the intervention (observations 61 through 84), nearly all one-step-ahead predictions from (7.10) are greater than the actual observations. This confirms the prior expectation that the interventions that become effective in January 1960 (observation 61) have decreased the oxidant level. *Figure 7.3* shows the residuals for $t = 13, \ldots, 60$ and the one-step-ahead forecast errors from January 1960 to December 1966 that are calculated from model (7.10) with parameter estimates based on the pre-intervention data. For applying the diagnostic procedure introduced in Section 7.3 the ranking basis is the $g = 48$ residuals of the pre-intervention phase. The exponentially weighted moving averages $\{T_t\}$ of the ranks of the one-step-ahead forecast errors with ranking size $g = 48$ and smoothing parameter $\lambda = .7$ are given in *Figure 7.4*; $T_{60}$ was set equal zero. The range of interest is the time immediately following the time of the suspected intervention, $\tau = 61$. The trend of $T_t$, $t = 61, \ldots, 71$, is negative, reaching a minimum at $t = 71$: $T_t = -.808$ and $-.856$ for $\lambda = .7$ and $.8$, respectively. For the case that there has been no intervention effect, it can be shown that the average run lengths of the exponentially weighted moving averages of ranks with parameters ($\lambda = .7$, $h = -.808$) and ($\lambda = .8$, $h = -.856$) are 111.8 and 99.5, respectively. The fact that such limits are crossed indicates that it is very likely that the intervention has had an effect on the process.

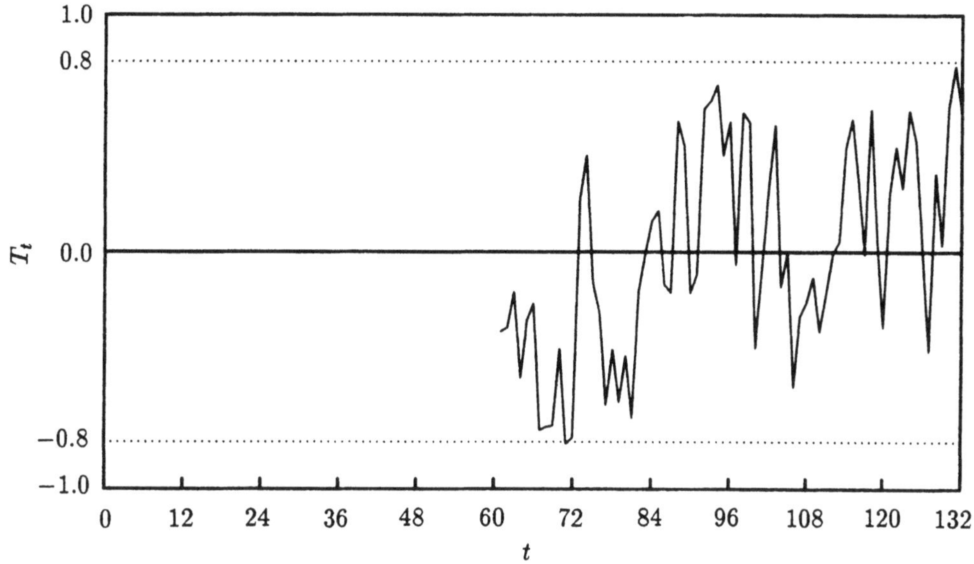


*Figure 7.4:* Exponentially weighted moving averages of the ranks of the post-intervention forecast errors, ranked among the pre-intervention residuals ($g = 48$, $\lambda = .7$); $T_{60}$ was set equal zero.

## 7.5 Concluding Remarks

The sequential surveillance procedure that is proposed in this chapter is related to process control methods. Process control methods are well suited for intervention analysis since in many situations it is difficult to specify the intervention model. Furthermore, intervention analysis, like process control, has to account for the increasing number of observations.

The application of the method requires the calculation of the ranks and the exponentially weighted moving averages of the post-intervention one-step-ahead forecast errors that are obtained from the pre-intervention model. For moderate and large ranking sizes one can obtain approximate average run lengths that are in good agreement with the exact ones. The nonparametric nature of the procedure proposed makes it independent of distributional assumptions.

### References

Abraham, B. (1987), Application of intervention analysis to a road fatality series in Ontario. *Journal of Forecasting*, **6**, 211-219.

Bakir, S.T. and Reynolds, M.R., Jr. (1979), A nonparametric procedure for process control based on within-group ranking. *Technometrics*, **21**, 175-183.

Bhattacharyya, P.K. and Frierson, D., Jr. (1981), A nonparametric control chart for detecting small disorders. *The Annals of Statistics*, **9**, 544-554.

Box, G.E.P. and Jenkins, G.M. (1970), *Time Series Analysis, Forecasting, and Control.* San Francisco: Holden Day.

Box, G.E.P. and Tiao, G.C. (1975), Intervention analysis with applications to economic and environmental problems. *Journal of the American Statistical Association*, **70**, 70-79.

Box, G.E.P. and Tiao, G.C. (1976), Comparison of forecast and actuality. *Applied Statistics*, **25**, 195-200.

Crowder, S.V. (1987), A simple method for studying run-length distributions of exponentially weighted moving average charts. *Technometrics*, **29**, 401-407.

Hackl, P. and Ledolter, J. (1989), A new nonparametric quality control technique. Technical Report No. 160, Iowa City: The University of Iowa.

Lucas, J.M. and Crosier, R.B. (1982), Robust CUSUM: A robustness study for CUSUM quality control schemes. *Communications in Statistics, Theory and Methods*, **11**, 2669-2687.

McDonald, D. (1986), A CUSUM Procedure based on sequential ranks. Technical Report, University of Ottawa.

CHAPTER 8

# A Monte Carlo Study of the Effects of Structural Breaks on Tests for Unit Roots

*David F. Hendry and Adrian J. Neale*

## Summary

The effects of a shift in the intercept of an autoregressive process on the rejection frequencies of standard tests for unit roots are investigated using Monte Carlo methods. Such tests lose power compared with the equivalent parameter values when no breaks occur. $F$-tests for structural breaks fail to detect shifts that are large enough to mimic unit roots. The response surface summarizing a conventional Monte Carlo highlights the effects on Dickey-Fuller (DF) and Augmented Dickey-Fuller (ADF) tests of the magnitudes of the autoregressive parameter, the break, the cumulative break, the estimation sample, and the percentage of the sample contaminated by the break. Diagnostic tests on the response surface support its specification. A recursive Monte Carlo computes sequences of rejection frequencies of DF and Chow tests and shows that these are low. Thus, care is required in interpreting unit-root tests since failure to reject does not entail that the null is true.

## 8.1 Introduction

There has been considerable interest in *unit-root tests* as part of a research strategy to pre-test for the degree of *integration* of economic time series. If a time series has all the latent roots of its autoregressive representation inside the unit circle, then it is denoted by $I(0)$; a series that requires differencing $d$ times to make it $I(0)$ is denoted by $I(d)$. [If $\alpha(L)y_t = \varepsilon_t$ is a $p$-th order autoregressive process where $\varepsilon_t$ is white noise, then $|\lambda^p - \sum \alpha^i \lambda^{p-i}| = 0$ yields the latent roots $\lambda_i$ of the (scalar) matrix polynomial. Thus, an $I(0)$ series has $|\lambda_i| < 1$ $\forall i$.] The case of central interest is whether a series is $I(0)$ or $I(1)$, since the underlying statistical distributions are different and empirical evidence suggests that many economic time series behave like $I(1)$ processes [see, e.g., Nelson and Plosser (1982)]. Several tests for unit roots have been proposed, including suggestions by Dickey and Fuller (1979), (1981), Sargan and Bhargava (1983), and Phillips and Perron (1988): see Dolado and Jenkinson (1987) for a survey. Here we focus on the Dickey-Fuller (DF) and Augmented Dickey-Fuller (ADF) tests, implemented by testing the null hypothesis $H_0$: $\beta = 0$ in the equation

$$\Delta y_t = \mu + \beta y_{t-1} + \varepsilon_t \quad \text{for } t = 1, \ldots, T, \tag{8.1}$$

where $\varepsilon_t \sim IN(0, \sigma_\varepsilon^2)$ and $y_0 = 0$. [The regression in (8.1) is extended by the inclusion of $\kappa$ lags of $\Delta y_t$ for an ADF($\kappa$) test.] Under the null, $\Delta y_t = \mu + \varepsilon_t$ and the alternative hypothesis is that $\beta < 0$. The $I(1)$ versus $I(0)$ hypothesis is tested using the conventional $t$-test on $\beta$. However, the usual critical values are inappropriate for $\mu = 0$, since the test statistic does not converge on a normal distribution asymptotically, but to a functional of a Wiener process. Let $W(r)$ denote a Wiener process for $0 \leq r \leq 1$, so that $W(r)$ is a continuous random walk on $(0, 1)$, then when $\mu = 0$:

$$t(\beta = 0) \xrightarrow{d} \frac{1}{2} \frac{W(1)^2 - 1}{[\int_0^1 W(t)^2 dt]^{\frac{1}{2}}}. \tag{8.2}$$

For *fixed* $r$, $W(r) \sim N(0, r)$, so that $W(1)^2$ in (8.2) is distributed as a $\chi^2(1)$. Critical values for the $t$-test that $\beta = 0$ in (8.1) are reported in, e.g., Fuller (1976) and Dickey and Fuller (1979). If $\mu$ is constant and nonzero when $\beta = 0$, then $t(\beta = 0)$ reverts to a standardized normal distribution [see Dickey and Fuller (1979, p. 429), West (1988), and Hylleberg and Mizon (1989)].

The problem we investigate here is when $\beta < 0$ but $\mu$ is nonzero and *non-constant* due to a structural break in the economic subsystem of which $\{y_t\}$ forms part. Our concern arises from the fact that even small step changes in a time series have similar autoregressive characteristics to those of an $I(1)$ process. Indeed, since Working (1934) it has been known that random walks can mimic many shapes, so it is unsurprising that autoregressive series with step changes can be mistaken for random walks. For example, Miller (1988) reports different test outcomes of pre- and post-floating exchange rate regimes for cointegration tests between savings and investment. In cases of structural shifts, therefore, the discriminatory power of DF and ADF tests between $I(1)$ and $I(0)$+shift may be low. Such a conjecture is consistent with empirical evidence reported by *inter alias* Perron and Phillips (1987) and Rappoport and Reichlin (1989). Section 8.2 analyzes the impact of a single shift in the intercept in a stationary model.

In Section 8.3, we investigate the rejection frequencies of DF and ADF tests using Monte Carlo and discover that even for small shifts in the intercept, the tests may be *biased* and reject the null *less often when it is false than when it is true*. A logistic response surface is fitted to discover the main determinants of the power loss relative to an equivalent autoregressive process without a regime shift. Given the important role of structural breaks and predictive failure in econometric modeling, in Section 8.4 we investigate the rejection frequencies of a variant of the commonly used Chow (1960) test. In practice, investigators often only report *full-sample* DF and ADF tests for scalar autoregressions, conditioning the test outcome on the implicit assumption that the parameters of (8.1) are constant over the sample period. In the present context, and we suspect in much empirical research, that implicit assumption is invalid.

Even though pre-tests for unit roots may only be a step in a multistage modeling strategy, there seem to be good grounds for applying rigorous testing procedures to (8.1), especially recursive methods. Hendry and Mizon (1990) report an empirical example which reveals substantial changes in inferences from unit-root tests on subsamples of quarterly data on money, prices, incomes, and interest rates in the UK. In the absence of constancy tests, it is difficult to ascertain which state of nature [$I(1)$ or $I(0)$+shift] led to any given non-rejection outcome. Thus, we undertake a *recursive* Monte Carlo study [see Hendry and Neale (1987)] to investigate the rejection frequencies of DF and Chow tests on data subsamples. We find uniformly low rejection frequencies, even when the break-point is known, for intercept shifts that are large enough to greatly lower the powers of the unit-root tests.

This issue is important not only because the statistical distributions alter between $I(0)$ and $I(1)$, but also because if $I(1)$ is *incorrectly* inferred when the process is $I(0)$ then two-step procedures for estimating cointegrating vectors, such as those proposed by Engle and Granger (1987), need not be consistent. To illustrate that the problem is nontrivial, *Figures 8.1(a)* and *8.1(b)* show the time series of two artificially generated variables. Which is $I(1)$ with constant parameters and which $I(0)$ but with a mean that is shifted by $\sigma_\varepsilon/2$ for a subperiod of the sample ?

## 8.2 A First Analysis of the Effects of Shifts on $I(1)$ Tests

A closely related issue to the effects of structural breaks on time series is whether economics variables are "trend stationary" or "difference stationary" [see, e.g., Nelson and Plosser (1982) for analysis and bibliography]. In the former

$$y_t = \mu_0 + \mu_1 t + \nu_t , \tag{8.3}$$

where $\nu_t$ is $I(0)$. If (8.3) holds

$$\Delta y_t = \mu_1 + \Delta \nu_t ,$$

and hence $\Delta y_t$ is at most $I(0)$. The latter corresponds to (8.1) when $\beta = 0$, so that $\Delta y_t$ is also $I(0)$. Thus, discriminating between (8.1) and (8.3) empirically has proved difficult.

Suppose a split time trend affected (8.3) so that $\mu_0$ and $\mu_1$ switched at time $T_0$ to $\mu_0^*$ and $\mu_1^*$. Although $y_t$ is still "trend stationary" for a *correct* specification of the trend, fitting (8.3) may not reveal that, and hence in a direct test against (8.1), the latter may

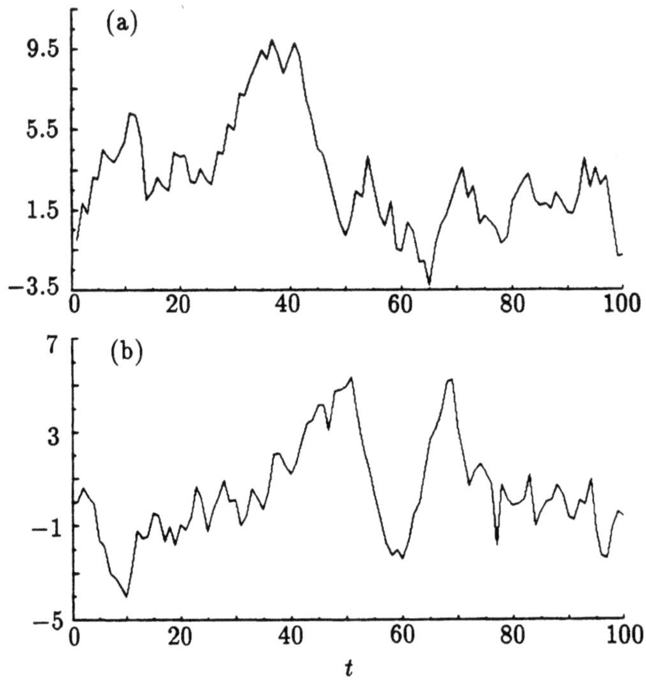


*Figure 8.1:* Artificial time series: Random walk or stationary + shift ?

be (incorrectly) selected [see Rappoport and Reichlin (1989) for an analysis of this case]. When $\beta = 0$, (8.1) can be interpreted as a limiting situation where the "trend" (i.e., $\Delta y_t$) changes every period. An entire spectrum of models exists between (8.1) and (8.3)—and these intermediate cases may be relevant for studying some economic time series.

Similar conclusions hold if only the intercept $\mu$ shifts in (8.1), even when $\beta < 0$ as we now show. Let a major institutional change or regime shift (such as the formation of an Oil Cartel) impinge on (8.1). We model this as a change in the intercept from $\mu$ to $\mu + \delta$:

$$\mu_t = \mu + \delta_t \quad \text{where } \delta_t = 0 \quad \text{for } t = 1, \ldots, T_0, \text{ and } \delta_t = \delta > 0 \quad \forall t > T_0 . \tag{8.4}$$

Thus, the time series undergoes a regime shift, which can be viewed as a surprise that persists after $T_0$. If $\beta = 0$, then $\delta$ involves a change in the *trend* of $y_t$, which is easy to detect, and $\{y_t\}$ does not behave like a trend-stationary process: We will not consider such a case any further. The case of interest here is $0 > \beta \geq -1$, so that the process is stationary when $\delta = 0$, but undergoes a change in its long-run mean when $\delta \neq 0$. By allowing the regime shift process itself to be stochastic (with a Poisson distribution, say), then $\{y_t\}$ can remain stationary despite the structural break. To control the Monte Carlo, however, we did not adopt this approach, but fixed both the timing and the magnitudes of the breaks deterministically.

When (8.1) and (8.4) comprise the data generation process, for $t \leq T_0$, then $E[y_t] = -\mu/\beta$; however, when $t \gg T_0$, $E[y_t] = -(\mu + \delta)/\beta$. The mean lag of the adjustment to the shift depends on the value of $\beta$. *Figure 8.2* illustrates one possibility for a computer-generated first-order autoregression like (8.1) with $\beta = -0.2$, $\mu = 1.0$, $\sigma_\epsilon^2 = 1$, $T = 60$, and $\delta = \sigma_\epsilon$. The break occurs at $T_0 = 30$, and the graph shows the time series of $y_t$ together with the fitted values up to $T_0$, and the one-step-ahead forecasts thereafter, based on the pre-break estimated parameters.

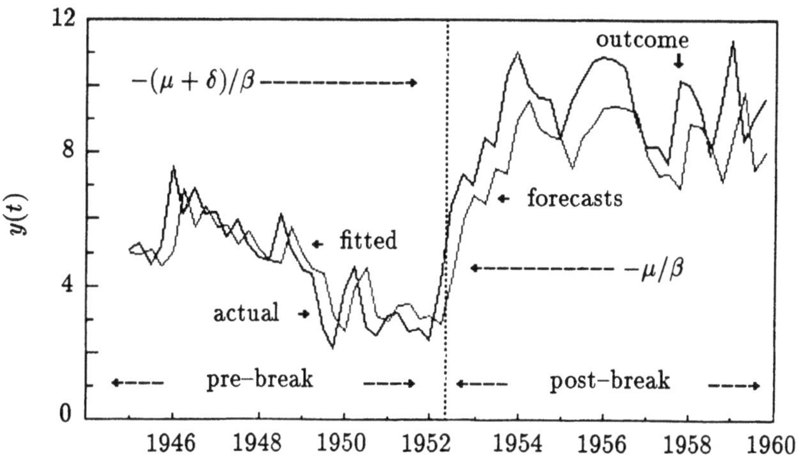


*Figure 8.2:* Outcomes and fitted or one-step forecast values for an autoregressive process with a shift in the intercept.

Two problems result from such a structural break:

1. Forecast errors from a model estimated on data up to $T_0$ will be larger than anticipated from $T_0$ onward—this is a *predictive failure* problem.

2. To minimize the mis-prediction in retrospect, coefficient estimates will alter if estimated from the whole sample period—we show below that the estimated $\beta$ is driven toward zero, creating a *unit-root* problem.

A model in *levels* of $y_t$ on $y_{t-1}$ will not capture the changed mean and will persistently mis-predict as *Figure 8.2* illustrates. If re-estimated over a sample period that includes some observations later than $T_0$, the estimate of $\beta$ will tend to zero so that the model can track the changed level of the data. A model in the *differences* of $y_t$, although it is misspecified when $\beta \neq 0$, will produce a few large forecast errors, then will continue to behave as expected. Thus, comparison of the fit may favor the latter even when $\beta < 0$ and (8.1) subject to (8.4) generated the data.

Consider testing (8.1) for a unit root using a variant of the DF statistic where an intercept is included, implemented (for analytical convenience) as a $t$-test for a unit root after an auxiliary regression of $y_t$ on a constant:

$$\Delta \hat{\varepsilon}_t = \varphi \hat{\varepsilon}_{t-1} + \omega_t \quad \text{where } \hat{\varepsilon}_t = y_t - \hat{c} \quad \text{when } \hat{c} = T^{-1} \sum_{t=1}^{T} y_t \,. \tag{8.5}$$

The null hypothesis of no unit root is rejected if $t_{\hat{\varphi}}$ is significantly negative (given appropriate critical values). Consider the extreme case where $\beta = -1$, so that

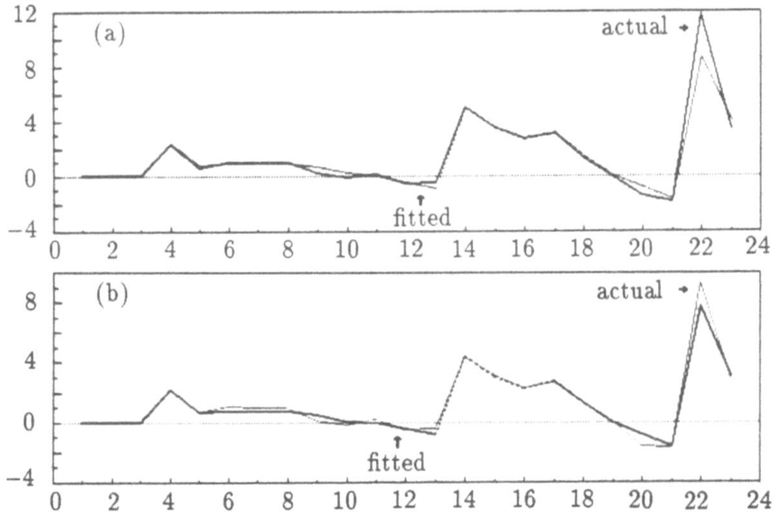


*Figure 8.3:* Logistic of the "power" of (a) the Dickey-Fuller test and (b) the ADF(1) test with fitted values.

$$y_t = \mu + \delta_t + \varepsilon_t \,.$$

Let $L = (T - T_0)/T$ denote the proportion of the sample that follows the break. The estimated intercept has an expectation at any point $T$ given by

$$E[\hat{c}] = \begin{cases} \mu & \text{for } T \leq T_0 \\ \mu + L\delta & \text{for } T > T_0 \,. \end{cases} \tag{8.6}$$

[We are indebted to Neil Ericsson for his help in clarifying the following derivation.] As $T \to \infty$ for a fixed $T_0$ such that $L \to 1$, then $E[\hat{c}] \to \mu + \delta$. Thus for $T > T_0$

$$\hat{\varepsilon}_t = y_t - \hat{c} \simeq \varepsilon_t + \delta_t - L\delta = \begin{cases} \varepsilon_t - L\delta & \text{for } t \leq T_0 \\ \varepsilon_t - (1 - L)\delta & \text{for } t > T_0 \,. \end{cases} \tag{8.7}$$

Both before $T_0$, and long afterward as $L \to 1$, $\{y_t - \hat{c}\}$ will be mean-zero white noise and the estimate of $\varphi$ in (8.5) will tend to $-1$. However, for $T$ greater than $T_0$:

$$\hat{\varphi} = (\sum_{t=2}^{T} \hat{\varepsilon}_{t-1}^2)^{-1}(\sum_{t=2}^{T} \hat{\varepsilon}_{t-1}\Delta\hat{\varepsilon}_t) = (\sum_{t=2}^{T} \hat{\varepsilon}_{t-1}^2)^{-1}(\sum_{t=2}^{T} \hat{\varepsilon}_{t-1}\hat{\varepsilon}_t) - 1\,. \tag{8.8}$$

The expectation of the numerator is given by (ignoring the $T^{-1}$ term which occurs at $T_0$)

$$\begin{aligned} E\left[\sum_{t=2}^{T} \hat{\varepsilon}_{t-1}\hat{\varepsilon}_t\right] & \qquad\qquad\qquad\qquad\qquad\qquad\qquad\qquad\qquad\qquad\qquad (8.9)\\ \simeq\; & E\left\{\sum_{t=2}^{T_0}(\varepsilon_{t-1} - L\delta)(\varepsilon_t - L\delta) + \sum_{t=T_0+1}^{T} [\varepsilon_{t-1} + (1-L)\delta][\varepsilon_t + (1-L)\delta]\right\} \\ =\; & (T_0 - 1)[L\delta]^2 + (T - T_0)[(1-L)\delta]^2 \\ \simeq\; & (T-1)[L^2 + L(1-2L)]\delta^2 = (T-1)L(1-L)\delta^2\,. \end{aligned}$$

Similarly, approximating the expectation of the denominator yields

$$\begin{aligned} E\left[\sum_{t=2}^{T} \hat{\varepsilon}_{t-1}^2\right] \;\simeq\; & E\left\{\sum_{t=2}^{T_0}(\varepsilon_{t-1} - L\delta)^2 + \sum_{t=T_0+1}^{T} [\varepsilon_{t-1} + (1-L)\delta]^2\right\} \\ =\; & (T-1)\sigma_\varepsilon^2 + (T_0 - 1)[L\delta]^2 + (T - T_0)[(1-L)\delta]^2 \\ \simeq\; & (T-1)[\sigma_\varepsilon^2 + L(1-L)\delta^2]\,. \end{aligned} \tag{8.10}$$

To a first approximation, therefore, for large values of $T$ and $T_0$

$$E[\hat{\varphi}] \simeq -1 + \frac{(T-1)L(1-L)\delta^2}{(T-1)[\sigma_\varepsilon^2 + L(1-L)\delta^2]} = -\frac{1}{1 + L(1-L)(\delta/\sigma_\varepsilon)^2}\,. \tag{8.11}$$

Five features of (8.11) merit note. Firstly, the denominator is unity if $L = 0$ or $L = 1$, replicating the earlier point that the effect of the break eventually wears off. Secondly, the denominator is symmetric in the proportion of the sample over which the break occurs, as it has the same value at $L = L^*$ and $(1 - L) = L^*$; thus, (8.11) is minimized at $L = 0.5$. Thirdly, the natural units for measuring the size of the break are $(\delta/\sigma_\varepsilon)$. Fourthly, the approximation in (8.11) is independent of $T$. Finally, the expected value of the least squares estimator of $\varphi$ is biased toward zero, i.e., toward a unit-root value. For example, at $L = 0.5$ and $(\delta/\sigma_\varepsilon) = 2$, so that the mean shifts in mid-sample by two error standard deviations, then

$$E[\hat{\varphi}] = -0.5.$$

At the same value of $L$ but with $\delta = \sigma_\varepsilon/2$, then $E[\hat{\varphi}] \simeq -0.94$, so this second size of shock has a very small effect. These implications will prove useful in interpreting the Monte Carlo evidence in Section 8.3, especially when formulating a response surface for the outcomes in dynamic models.

The implications of (8.11) for the power of unit-root tests are indirect, since we have not analyzed the properties of the estimated coefficient standard errors, which are needed to deduce the outcomes of a $t$-test. In a pilot simulation of this case, when $T = 30$ and a shift of size $\sigma/2$ occurs at $T_0 = 15$ and persists to the end of the sample, then the power of the DF test is greatly reduced from 99.9%, when there is no break, to 60.7% with the break [based on 10,000 Monte Carlo replications]; the ADF(1) test loses even more power. Thus, an econometrician using *only full sample* ADF or DF statistics to investigate nonstationarity in a white noise process with a single break may be led to the erroneous conclusion that the hypothesis of a unit root is not rejected.

More generally, for $\beta \neq 0$ the process generating $y_t$ in (8.1) can be written as

$$y_t = \sum_{j=0}^{t} \gamma^j \mu_{t-j} + \sum_{i=0}^{t} \gamma^i \varepsilon_{t-i} \tag{8.12}$$

where $\gamma = (1 + \beta)$, so that $y_t = a_t + b_t$, say. Thus, $E[y_t] = a_t$ and when $|\gamma| < 1$ (so $H_0$ is false), but $\mu$ is constant, then $a_t \rightarrow -\mu/\beta$ as $t$ increases. Following a shift in $\mu$ to $\mu + \delta$ at $T_0$, $a_t \rightarrow -(\mu + \delta)/\beta$. Finally, if $\delta = 0$ again at $T_1$, then $a_t$ converges back to its original value. If the sample period of the observed data is $(1, \ldots, T)$ and $T_1 - T_0$ is the proportion $L$ of $T$, then the *sample mean* of $y$ over $t = 1, \ldots, T$ will be approximately equal to

$$-[L(\mu + \delta)/\beta + (1 - L)(\mu/\beta)] = -(\mu + L\delta)/\beta.$$

This approximation matched the sample means in the Monte Carlo experiments examined (*Figure 8.8* below illustrates). The behavior of $b_t$ for *estimated* $\beta$ is less clear analytically, since the residual then contains part of the unmodeled shock.

This example is simplistic, but reveals potential problems in interpreting unit-root tests. [As shown below, the DF and ADF tests are often biased if there is a structural break and a root less than unity, rejecting *less often* than under the null of a unit root and no structural break.] Certainly, if $y_t$ *had* a unit root then the structural shift need not preclude finding that. In more complex cases, such as structural shifts that involve changes to *slope* coefficients, reflecting learning and adjustment to financial or technological innovation, different conclusions could result. For instance, in processes that have a single unit root but undergo a slope shift in a coefficient, tests for nonstationarity may imply that these are $I(2)$. The problem is sufficiently intractable analytically that Monte Carlo seems a sensible approach. This conclusion is reinforced by our desire to study recursive estimation and testing methods to see if these can detect changes in parameter values.

## 8.3 A Distribution Sampling Study of Tests for Unit Roots

To investigate the impact of structural shifts on $I(1)$ versus $I(0)$ tests, we have conducted the following *distribution sampling* study of (8.1), in four stages:

1. We simulated the null distributions of the DF and ADF(1) tests of $H_0$: $\beta = 0$ in (8.1) when $\mu = 0$, for a range of values of the sample size $T \in [10, 350]$ to calculate critical values of the test statistics at the 5% level.

2. We computed the powers of the same tests when $\beta < 0$ and $\mu = 0$, but there were no structural breaks (so $\delta = 0$) to establish baselines for "power losses" due to potential structural shifts.

3. We calculated the rejection frequencies of the same tests for $\beta < 0$ and $\mu = 0$ but $\delta \neq 0$, so that the intercept is *non-constant*. The values of $\delta$ in case 3. were usually zero up to $T_0$, then $\sigma/2$ till $T_1$, and zero again thereafter, where $T_1$ and $T$ could coincide.

4. We estimated response surfaces for the powers of these unit-root tests both to summarize efficiently the experimental findings and to ascertain the main determinants of their performance.

We investigated the DF and ADF(1) tests of the unit-root hypothesis as representative of the tests most commonly applied in econometrics. Sample sizes up to 100 were considered as a reasonable order of magnitude to illustrate the potential difficulties of testing for $I(1)$, unaugmented by constancy and other diagnostic tests. Sample sizes of 350 were used as a check on the behavior of our Monte Carlo response surface in a very much larger sample. The simulations were undertaken on PC-NAIVE [see Hendry and Neale (1987)], and the graphs and response surface estimates were produced using PC-GIVE [see Hendry (1989)].

The experiment was an incomplete design, intended to highlight the salient determinants of the tests' rejection frequencies. Throughout, $\sigma_\varepsilon = 1$ and $\mu = 0$, and we considered three values of $\beta$; three sample sizes; five break periods; and three structural break magnitudes corresponding to step changes of 0, $\delta = \sigma_\varepsilon/2$, and $\delta = \sigma_\varepsilon$ over the relevant period $[T_0, T_1]$. It is easy to verify that $\sigma_\varepsilon = 1.0$ involves no loss of generality, in that $\delta$ can span the relevant set of experiments: As with (8.11) above, $\delta/\sigma_\varepsilon$ rather than $\delta$ alone determines the deviation, if any, in rejection frequency in 3. relative to 2. in the tests. Thus the design parameters were $(\beta, \delta, T, T_0, T_1)$ with the following values:

$\beta = -0.2, -0.1, 0.0$;
$T = 30, 100, 350$;
$\delta = 0.0$ outside of $[T_0, T_1]$ and $\sigma_\varepsilon/2$ or $\sigma_\varepsilon$ otherwise;
$[T_0/T \rightarrow T_1/T] =$
(i) $0.3 \rightarrow 0.5$; (ii) $0.5 \rightarrow 0.7$; (iii) $0.7 \rightarrow 0.9$; (iv) $0.3 \rightarrow 1$; (v) $0.5 \rightarrow 1$.

Such a design would have generated 99 experiments if fully implemented, noting that there are only 11 different combinations of $\delta$ and $[T_0, T_1]$. However, several invariances suggested by (8.11) were established as the experiment proceeded, and so only 23 "representative" experiments were actually conducted. These are shown in *Table 8.1*. In each sample, $M + T$ data points were generated with $M = 0$ when $\beta = 0$, but $M = 30$ observations were discarded to remove the impact of $y_0 = 0$ under $H_1$ (i.e., when $|\beta| < 1$).

Throughout, $N = 10,000$ replications were calculated to ensure well-determined test power estimates. Standard errors of rejection frequencies are given by

$$\mathrm{SE}(\hat{P}) \approx \sqrt{\hat{P}(1 - \hat{P})/100}, \tag{8.13}$$

where $\hat{P}$ denotes an estimated power. Such standard errors are always less than 0.5% and are under 0.25% for $P < 5\%$.

(a) Under the null that (8.1) is the data generation process with $\mu = \beta = 0$, the 5% critical values of the DF and ADF statistics were calculated for the residuals from a regression of $y_t$ on a constant. Thus, the intercept is estimated even though the population value happens to be zero when there is no break. Appendix A records the critical values of these DF and ADF tests used in the experiments: Any required intermediate values were calculated by linear interpolation. This set comprised experiments 1 to 3. Thus, 5% rejections are ensured under the null by the calculation of the critical values empirically: See rows 1, 2, and 3 in *Table 8.1*.

(b) The tests' powers were then computed using the critical values under the null hypothesis established in (a) when $\delta = 0$ but $\beta < 0$. This set comprised experiments 4, 5, 14, 18, and 22. As shown in *Table 8.1*, the tests' powers increase as $T$ increases, and as $|\beta|$ increases; in no case is the rejection frequency lower under the alternative than under the null in (a).

(c) As with (b), the tests' powers were calculated assuming the critical values in (a), but in this set (the remaining 15 experiments) $\delta \neq 0$. Cases (i)–(iii) were conducted once (at $\beta = -.1$) to establish that the exact period of the break-point was essentially irrelevant, whereas the proportion $L$ of the sample affected by the break was crucial. Thereafter, only cases (i), (iv), (v), and no break [denoted (o) below] were computed for $\delta = 1/2$. Three experiments were undertaken for $\delta = 1$ (all at $T = 30$ and $\beta = -0.2$). Finally, there were three experiments at $T = 350$ to test the constancy of the response surface fitted in (d) below. The experimental findings are again recorded in *Table 8.1*.

Since the rejection frequency is 5% under the null that $\mu = \beta = 0$, whereas in some cases under the alternative that $\mu + \delta \neq 0$, *fewer* than 5% occurred, the tests are *biased*: there is a lower probability of rejecting $H_0$ when it is false than when it is true. This is an example of the *implicit null* problem discussed by Mizon and Richard (1986).

Three main conclusions can be drawn from *Table 8.1*. Firstly, cases (i), (ii), and (iii) (experiments 6, 7, and 8) do indeed yield similar rejection frequencies. Next, the rejection frequency [denoted by $P$(DF), etc.] falls as $\delta$ rises (for $\beta \neq 0$), and rises as $|\beta|$ or $T$ increases. Finally, the DF and ADF tests perform about equally well or badly in almost every experiment. It is difficult to determine any finer details of the tests' performance from the table alone.

*Table 8.1:* The experimental design and test rejection frequencies.

| Experiment | Case | $\beta$ | $\delta$ | $T$ | $L$ | $P$(DF) % | $P$(ADF) % |
|---|---|---|---|---|---|---|---|
| 1 | (o) | 0.0 | 0.0 | 350 | 0.0 | 5.0 | 5.0 |
| 2 | (o) | 0.0 | 0.0 | 100 | 0.0 | 5.0 | 5.0 |
| 3 | (o) | 0.0 | 0.0 | 30 | 0.0 | 5.0 | 5.0 |
| 4 | (o) | −.1 | 0.0 | 100 | 0.0 | 33.6 | 31.4 |
| 5 | (o) | −.1 | 0.0 | 30 | 0.0 | 8.2 | 8.9 |
| 6 | (i) | −.1 | 0.5 | 100 | 0.2 | 12.7 | 12.6 |
| 7 | (ii) | −.1 | 0.5 | 100 | 0.2 | 12.3 | 12.1 |
| 8 | (iii) | −.1 | 0.5 | 100 | 0.2 | 12.4 | 11.9 |
| 9 | (iv) | −.1 | 0.5 | 100 | 0.7 | 6.3 | 5.5 |
| 10 | (v) | −.1 | 0.5 | 100 | 0.5 | 4.3 | 4.0 |
| 11 | (i) | −.1 | 0.5 | 30 | 0.17 | 5.7 | 6.0 |
| 12 | (iv) | −.1 | 0.5 | 30 | 0.67 | 2.9 | 2.9 |
| 13 | (v) | −.1 | 0.5 | 30 | 0.5 | 3.1 | 3.1 |
| 14 | (o) | −.2 | 0.0 | 100 | 0.0 | 88.9 | 80.1 |
| 15 | (i) | −.2 | 0.5 | 100 | 0.2 | 64.9 | 53.7 |
| 16 | (v) | −.2 | 0.5 | 100 | 0.5 | 44.0 | 33.1 |
| 17 | (iv) | −.2 | 0.5 | 100 | 0.7 | 52.8 | 41.4 |
| 18 | (o) | −.2 | 0.0 | 30 | 0.0 | 15.8 | 15.5 |
| 19 | (i) | −.2 | 1.0 | 30 | 0.17 | 4.6 | 5.1 |
| 20 | (iv) | −.2 | 1.0 | 30 | 0.67 | 1.2 | 1.0 |
| 21 | (v) | −.2 | 1.0 | 30 | 0.5 | 0.8 | 0.9 |
| 22 | (o) | −.1 | 0.0 | 350 | 0.0 | 99.9 | 99.8 |
| 23 | (v) | −.1 | 0.5 | 350 | 0.5 | 58.9 | 48.7 |

(d) To summarize this set of experiments, a response surface was fitted separately for each test, following the approach in Hendry (1984). The fitted equations were designed to have the following properties:

$$P[t(\beta = 0) < C|H_1] = P = \psi[\alpha, \beta, g(T), L, \delta], \tag{8.14}$$

where

$$P[t(\beta = 0) < C|H_0] = \psi(\alpha, 0, T, 0, 0) = \alpha, \quad \forall T \tag{8.15}$$

noting that $\alpha = 0.05$ in the present experiments.

$$\hat{P} \in (0, 1), \tag{8.16}$$

which is ensured by the logistic transformation $\mathcal{L}^*(P) = \log[P/(1 - P)]$. Since $\mathcal{L}^*(\alpha) = -2.944$ for $\alpha = 0.05$, setting $\mathcal{L}(P) = \mathcal{L}^*(P) + 2.944$ implies that $\psi(.)$ does not depend on $\alpha$. We enforced the condition that $\psi(.)$ only depended on $T$ through

$\beta$ when $\delta = 0$, so that $P \to 1$ as $T \to \infty$ when $\beta \neq 0$. Finally, we anticipated that $\psi(.)$ would decrease with increasing values of $\delta$ and $L$ (for $L < 0.5$), and decrease as $\beta \to 0$, noting that $\beta < 0$.

The forms of $\psi(.)$ and $\mathcal{L}(.)$ must be *jointly* selected to ensure that these properties are all satisfied. Let $\tau = TL = T_1 - T_0$ denote the total length of the break period, then following the analysis in Section 8.2, we constructed a measure $\theta$ of the total impact of the break:

$$\delta\theta = \sum_{i=0}^{\tau-1} (1+\beta)^i \delta .$$

Thus, $\theta = \tau$ when $\beta = 0$, and $\theta = (1-\gamma^\tau)/\beta$ otherwise. Since $L$ enters symmetrically in (8.11), and since the effect of $L$ on $P$ seemed independent of the period over which the break occurred, we constructed $K = \min\{L, 1-L\}$ as the relevant index of the break period. A consequence of using $K$ is that type (iv) becomes equivalent to $[T_0 \to T_1] = [70 \to 100]$. From (8.16), $\mathcal{L}^*(.)$ *guarantees* admissible predictions irrespective of the specification of $\psi(.)$, and so can be taken as linear in the arguments of $\psi(.)$. From (8.15), $\mathcal{L}(P)$ need not include an intercept. We used $g(T) = \lambda T$ and restricted $T$ to enter only as a product with $\beta$ and $\delta\theta$. Thus, the postulated response surface was

$$\mathcal{L}(P) = \rho_1 \beta T + \rho_2 \delta K + \rho_3 \delta\theta T , \tag{8.17}$$

and we anticipated $\rho_i < 0$ for $i = 1,2,3$. Following Cox (1970), since the Jacobian of the transformation from $P$ to $\mathcal{L}$ is $1/[P(1-P)]$, and the estimated probabilities have standard errors as in (8.13), every variable was adjusted for heteroscedasticity, with correction factors shown as $H_d$ and $H_a$ for DF and ADF, respectively, where $H = \sqrt{NP(1-P)}$, which can be estimated using $\hat{P}$ from the Monte Carlo. If (8.17) is correctly specified, then, after the $H$ transformations, the residual standard deviation should be unity.

The estimated equations obtained from the 23 experiments were

$$\begin{array}{llll} H_a\mathcal{L}(\text{ADF}) = & -\ 0.208H_a(\beta T) & -\ 6.15H_a(\delta K) & -\ 0.0016H_a(\delta\theta T) \\ & \quad (0.007) & \quad (0.58) & \quad (0.00022) \end{array} \tag{8.18}$$

$$\begin{array}{lll} R^2 = .995 \quad \sigma = 5.69 & F[3,20] = 1229.1 & \text{DW} = 1.63 \\ \text{Chow } F[6,14] = 0.62 & \text{normality } \chi^2(2) = 0.44 & \\ \text{AR 1-2 } F[2,18] = 0.01 & \text{RESET } F[1,19] = 0.43 & \text{W } F[6,12] = 0.54 \end{array}$$

$$\begin{array}{llll} H_d\mathcal{L}(\text{DF}) = & -\ 0.233H_d(\beta T) & -\ 6.23H_d(\delta K) & -\ 0.0019H_d(\delta\theta T) \\ & \quad (0.009) & \quad (0.66) & \quad (0.00014) \end{array} \tag{8.19}$$

$$\begin{array}{lll} R^2 = .996 \quad \sigma = 5.24 & F[3,20] = 1690.5 & \text{DW} = 1.35 \\ \text{Chow } F[6,14] = 0.60 & \text{normality } \chi^2(2) = 0.37 & \\ \text{AR 1-2 } F[2,18] = 1.72 & \text{RESET } F[2,18] = 1.08 & \text{W } F[6,13] = 0.94 . \end{array}$$

The legend is as follows: $R^2$ is the squared multiple correlation coefficient; $\sigma$ denotes the residual standard deviation; $F[3,20]$ tests for no relationship in the response surface; Chow $F[.,.]$ is the Chow test for constancy over the final six experiments; normality $\chi^2(2)$ is the Jarque and Bera (1980) test; AR 1-2 $F[.,.]$ tests for autocorrelated residuals across the experiments given the order in *Table 8.1* (a test for functional form specification relative to $\beta$); RESET $F[.,.]$ is the test in Ramsey (1969); W $F[.,.]$ denotes White's (1980) heteroscedasticity test; and (..) below coefficient estimates denote heteroscedastic-consistent standard errors.

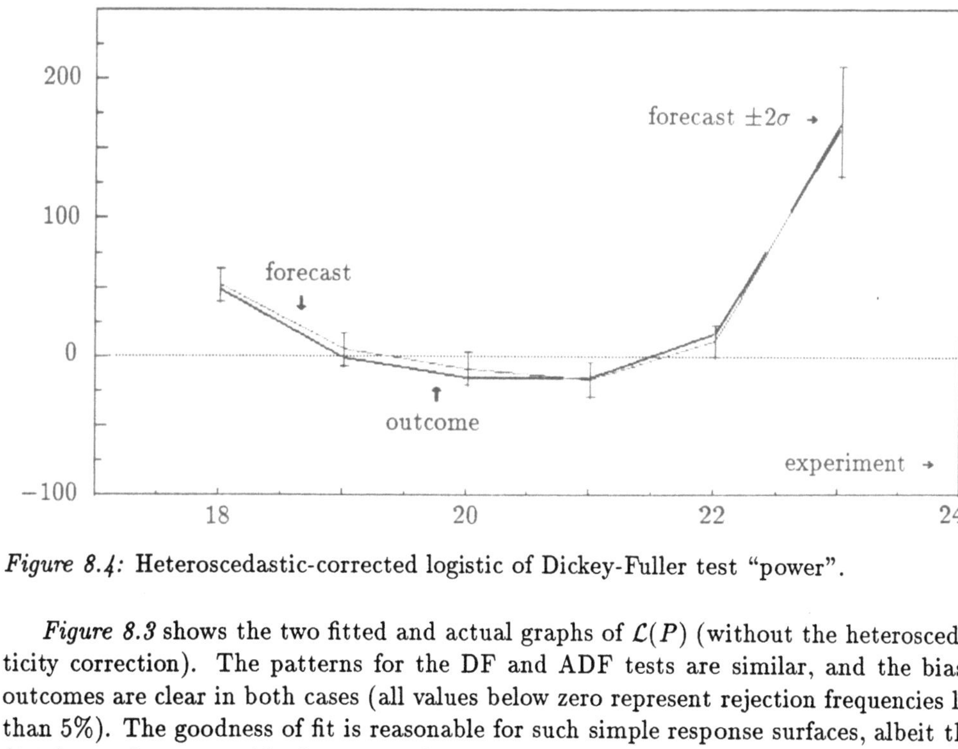


*Figure 8.4:* Heteroscedastic-corrected logistic of Dickey-Fuller test "power".

*Figure 8.3* shows the two fitted and actual graphs of $\mathcal{L}(P)$ (without the heteroscedasticity correction). The patterns for the DF and ADF tests are similar, and the biased outcomes are clear in both cases (all values below zero represent rejection frequencies less than 5%). The goodness of fit is reasonable for such simple response surfaces, albeit that (8.19) reveals some residual autocorrelation and both values of $\sigma$ are far from unity. The forecasts from (8.18) over the last six experiments, which include the only two cases where $\beta \neq 0$ and $T = 350$, are shown in *Figure 8.4*, and, despite the challenging nature of this test, the equation predicts very accurately. Two illustrative time series from processes with structural breaks are reported in *Figures 8.5* and *8.6*: Such data are easily mistaken for $I(1)$ time series. The empirical frequency distribution of the DF test (scaled to have a unit standard deviation) is shown in *Figure 8.7* for $\beta = -0.1$ and a break of type (ii) when $T = 100$. Although the null hypothesis is false, and the rejection frequency at the nominal

5% critical value is just over 4% (with a standard error of 0.2), the shape is nevertheless close to that anticipated under the null distribution.

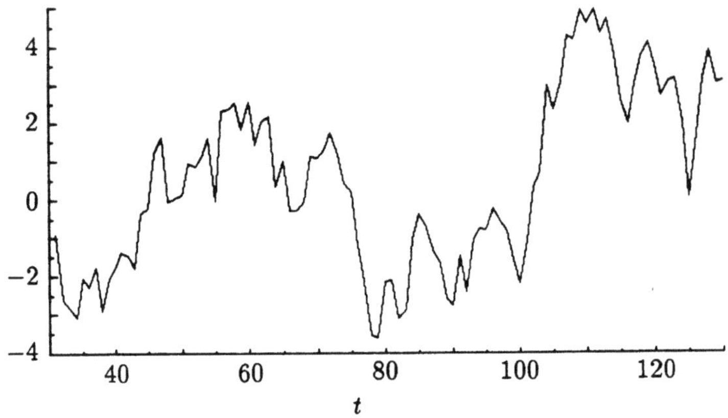


*Figure 8.5:* Time series of $y_t = 0.9y_{t-1} + \varepsilon_t$ with a step change of 0.5 at $t = 100$ to 120 (with 30 initial values); $\varepsilon_t \sim IN(0, 1)$.

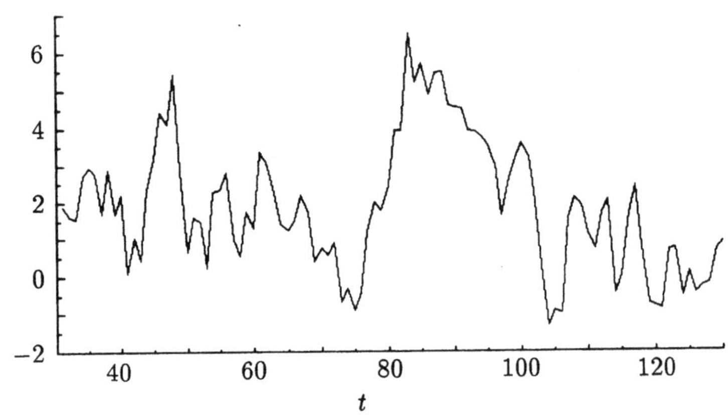


*Figure 8.6:* Time series of $y_t = 0.8y_{t-1} + \varepsilon_t$ with a step change of 0.5 at $t = 80$ to 100 (with 30 initial values); $\varepsilon_t \sim IN(0, 1)$.

*Figure 8.8* records the behavior across sample sizes in the same experiment of the mean estimate of $c$ from the regression of $y_t$ on a constant, together with ±2 ESE (the conventionally calculated OLS coefficient standard error, which is not an appropriate measure of uncertainty for that equation) and ±2 MCSD (the Monte Carlo sampling standard deviation based on the empirical distribution of the 10,000 values of the intercept). Relative to the *correct* MCSD, the coefficient change is small, matching the small magnitude of the shock, but being less consonant with the accompanying substantial power loss. The estimated intercept converges on $-L\delta/\beta = 0.2 \times 0.5/0.1 = 1.0$.

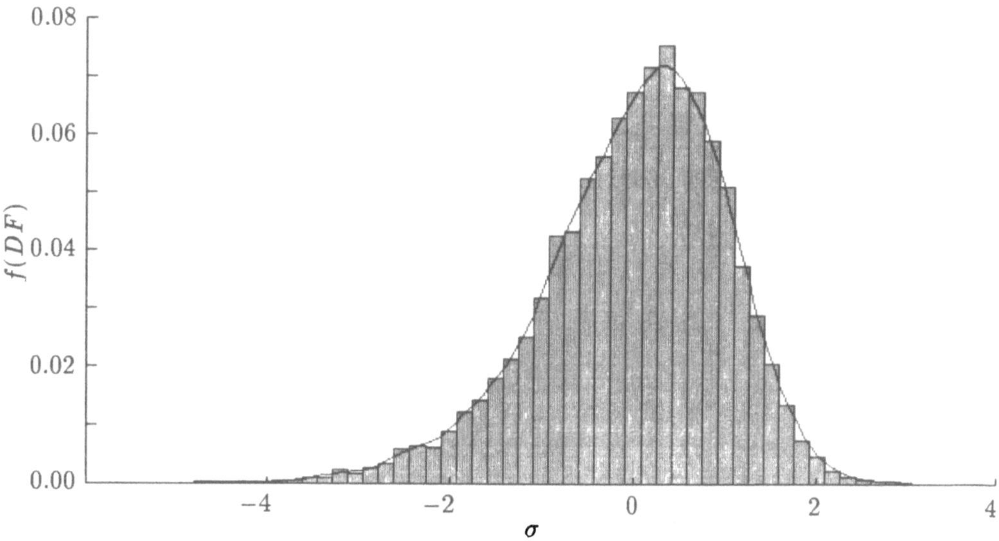


*Figure 8.7:* Standardized frequency distribution of the Dickey-Fuller test for $N = 10,000$; break of 0.5 from $T = 50$ to 70.

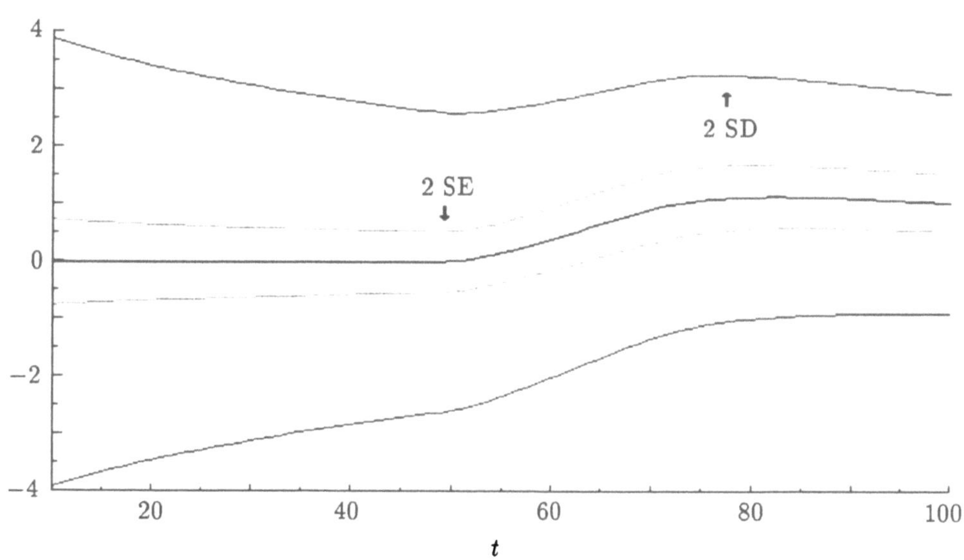


*Figure 8.8:* Recursive estimation of the intercept with a break of 0.5 at $T = 50$ to 70.

From (8.18) and (8.19), the following formula explains the power loss due to intercept shifts:

$$P(\mathrm{DF}) = \mathcal{A}/(1+\mathcal{A}), \tag{8.20}$$

where $\mathcal{A} = \exp[-0.23\beta T - 6.2\delta K - 0.002\delta\theta T - 2.944]$. Equation (8.20) (and a similar formula for ADF) satisfies most of the response surface requirements above, including anticipated signs. It highlights the respective roles of sample size $T$, the autoregression in the data $\beta$, the size of the shift $\delta$, its total impact $\theta$, and its relative duration $K$. The one apparent failure of (8.18) and (8.19) is the value of $\sigma$: Instead of unity, it exceeds 5. From the outcome of the White test, the residual variance is not a function of higher powers of the included regressors, and no other explanatory variables suggested themselves although 23 is a relatively small number of experiments and higher-order interactions of parameters with $T$ may matter. The precise specification of the response surface can be rejected on the basis that $\sigma \gg 1$, but that outcome is unsurprising when $N = 10,000$. As $N \to \infty$, *any discrepancy* between the true and the conjectured response surface would ensure that $\sigma \to \infty$ also, so the observed discrepancies are consistent with a reasonable specification. To illustrate this claim, *Figure 8.9* records the outcomes in terms of $P$ and the predictions from (8.20) (or the equivalent formula for the ADF test).

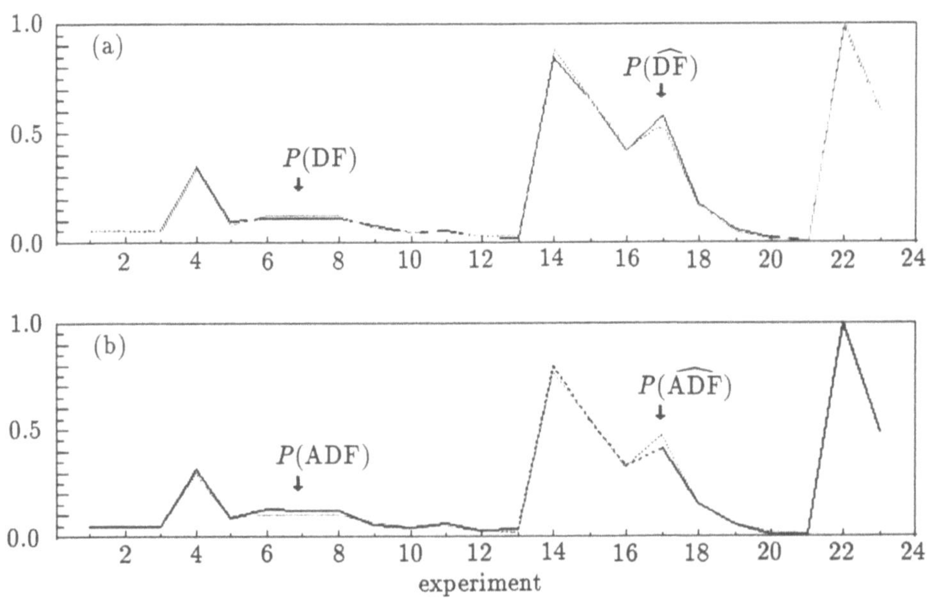


*Figure 8.9:* Fitted and actual values for the "powers" of the DF and ADF tests using equation (8.20).

Thus, the major remaining issue is whether an auxiliary test could detect the coefficient shift and hence enable discrimination between cases where a unit root finding is due to a structural break and where it is not. Chow tests were computed to investigate the size and power of a widely used test that might be employed to detect the possible existence of structural breaks despite the nonstandard setting of (believed) unit-roots. However, since the timing of structural breaks are usually not known, we decided to undertake all the simulations using the recursive techniques advocated in Hendry and Neale (1987).

## 8.4 Recursive Monte Carlo

[This section is drawn from the doctoral dissertation in preparation by the second author.] In this section, we focus on the DF and Chow statistics only, to investigate their behavior at all intermediate sample sizes from $T = 10$ to 100. The Chow test is of a "break-point" form, so the model is fitted up to each time period $T_a$, then over the whole period $T$, and the goodness of fit compared. The test outcomes are reported in graphs where the horizontal axis is $T_a$, so that for small values of $T_a$, long forecast horizons are entailed, and as $T_a$ tends to $T$, the horizon tends to zero. The three stages 1.–3. discussed in Section 8.3 were followed, but instead of 4., the rejection frequencies of the relevant tests as $T$ changes are summarized graphically. These graphs can be interpreted as nonparametric descriptions of a projection of a response surface where only $T$ is allowed to vary.

*Figures 8.10* and *8.11* record the null rejection frequencies of the DF and Chow tests when both $\beta$ and $\delta$ are zero. The DF test has an actual rejection frequency (called "size" as a shorthand below) close to its nominal size of 5%. Deviations outside the range 5% $\pm$ 0.44% are significantly discrepant, but it must be remembered that the original critical values are only recorded to three decimal digits and that intermediate values are interpolated (also see Appendix A). The Chow test rejection frequencies generally exceed 5% for this unit-root process, converging on the nominal size from about 8% as $T$ increases and the horizon falls. Even so, compared with the rejection frequencies we report below, the "excess size" is relatively negligible *and in the wrong direction* since the parameter constancy tests will transpire to have *low power* when $H_0$ is false.

*Figure 8.12* shows the bias in estimating $\beta$ when $\beta = 0$, together with $\pm$ 2 MCSD and $\pm$ 2 ESE. The bias decreases at about $1/T$, and the comparison between ESE and MCSD reveals that the ESE is usably accurate for MCSD despite the unit root in the data generation process.

*Figures 8.13* and *8.14* report the DF test rejection frequencies for $\beta = -0.1$ and $-0.2$, respectively, for four values of the break period. When $\delta = 0$, the power of the DF test to reject $H_0$: $\beta = 0$ increases steadily with $T$ in both figures, reaching over 30% for the former and almost 90% for the latter. However, when $\delta = \sigma_\epsilon/2$, the power loss is dramatic: for $\beta = -0.1$ and cases (iv) and (v), the test is biased by $T = 100$ and even for $\beta = -0.2$, the power is less than 50% in those cases. [Remember that $(0.3T \rightarrow T)$ and $(0.7T \rightarrow T)$ are equivalent in break length, and correspond to the same value of $K$. The rejection frequencies here match those in Section 8.3 at $T = 100$ with perhaps slightly more bias.] A degree of "power recovery" is visible for breaks of type (i), but overall the power loss remains large. These recursive results are also consistent with the response surface in

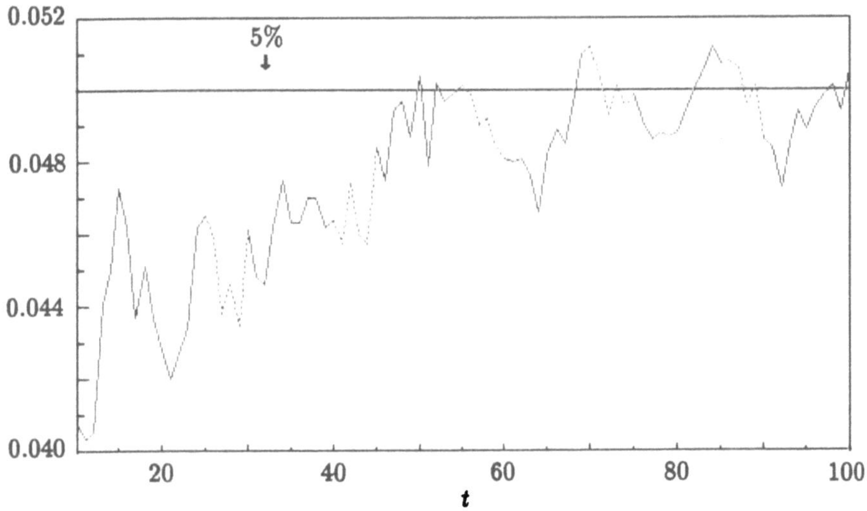


*Figure 8.10:* Unit-root recursive "$t$-test" rejection frequencies on $H_0$: $\beta = 0$.

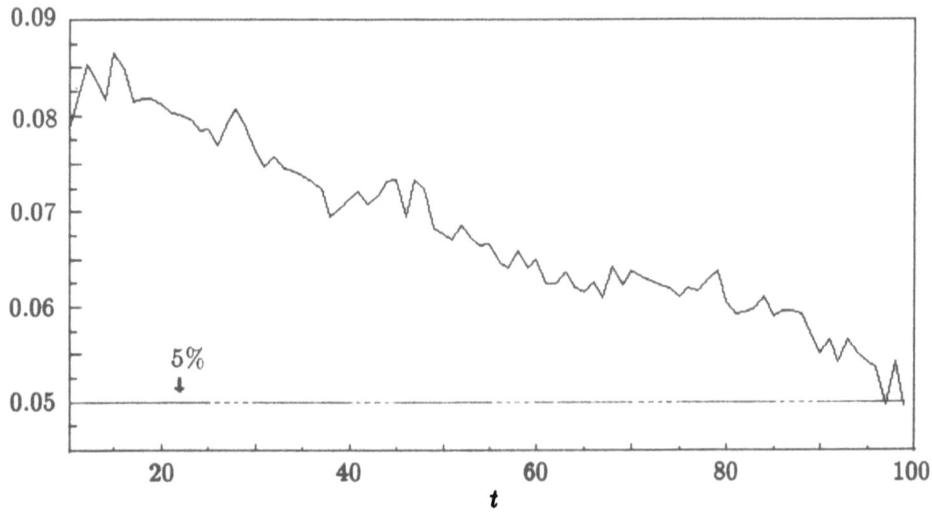


*Figure 8.11:* "Break-point" Chow test rejection frequencies on $H_0$: $\beta = 0$.

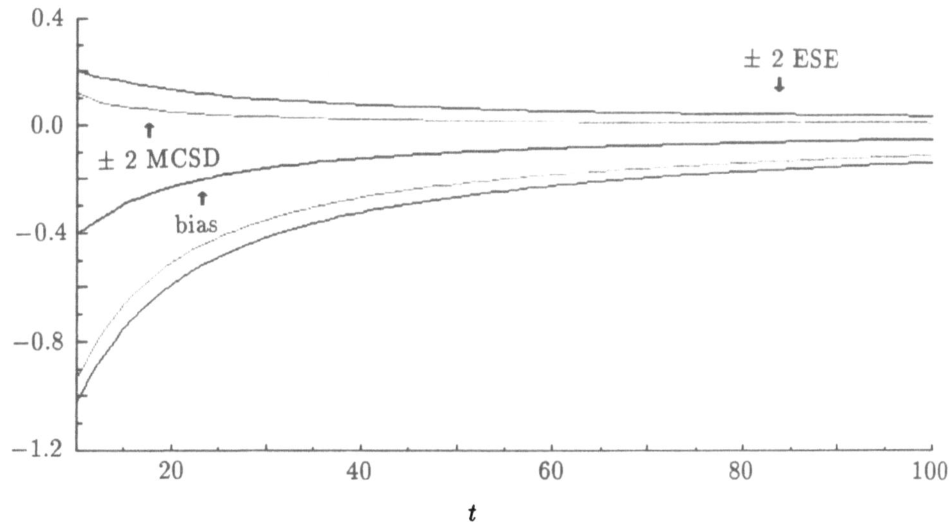


*Figure 8.12:* Recursive coefficient bias with ± 2 ESE and ± 2 MCSD: $\beta = 0$.

(8.19), which correctly characterizes the final period outcomes.

Naturally, the detectability of the break is a vital issue both for empirical research and for interpreting how big a shock $\delta = \sigma_\varepsilon/2$ really is. Firstly, the actual null rejection frequencies of the $F$-tests still exceed 5% for most values of $T$ but are nevertheless close to the correct nominal level by around $T = 80$. The excess size could be offset by increasing the critical value, but this would *lower* the test power when the null of constancy is false. As can be seen in *Figures 8.15* and *8.16*, the highest rejection frequency is about 12% for a break-point, which coincides with the structural shift, and otherwise is very low. These outcomes are when each test outcome is interpreted as the only test conducted. If the complete sequence of tests is to be judged using an overall 5% critical value, the test powers would again be much lower. Despite being designed to detect parameter change, the test is often *biased* toward the end of the sample, due to ignoring the heteroscedasticity created by the break. This highlights the dangers of computing only one Chow test at an arbitrary point, and suggests also computing the variance ratio test.

To investigate the conjecture that the intercept shift of $\sigma/2$ is simply too small to be detectable, we reran many of the experiments with $\delta = \sigma$. The Chow test rejection frequencies now peaked at around 25%, but otherwise were similar in profile to those in *Figures 8.15* and *8.16*, converging back toward 5% near the end of the sample. The DF test outcomes did not alter much either, and retained profiles similar to those in *Figure 8.13.* Extending this form of analysis to a range of values of $\delta$, *Figure 8.17* plots the power functions of the $F$-test for a break over $[70 \rightarrow 100]$ against the corresponding values of $\delta$, calculated at the break-point $T_0 = 70$ and at $T - 10$, for a type (iv) experiment with $\beta = -0.1$. The power rises rapidly at the break-point as $\delta$ increases, but shows little

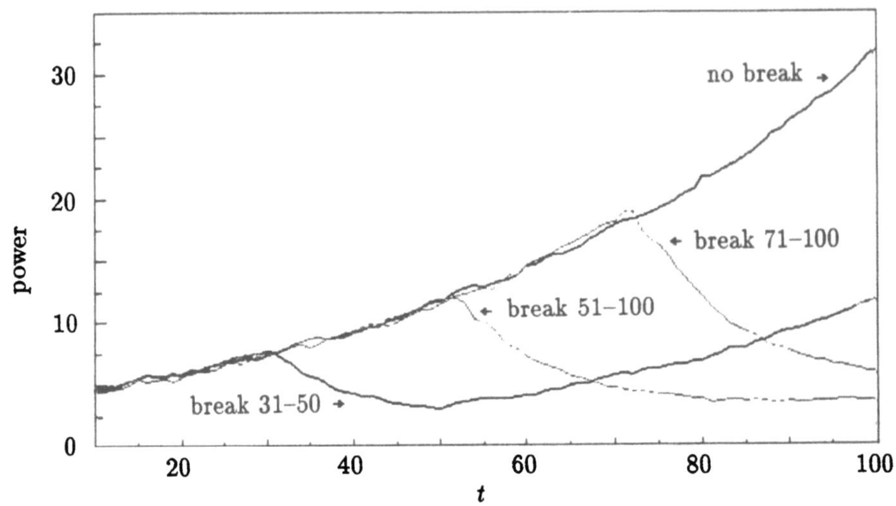


*Figure 8.13:* Recursive Dickey-Fuller "$t$-test" rejection frequencies of $H_0$: $\beta = 0$ when $\beta = -0.1$ with step breaks at the sample periods indicated.

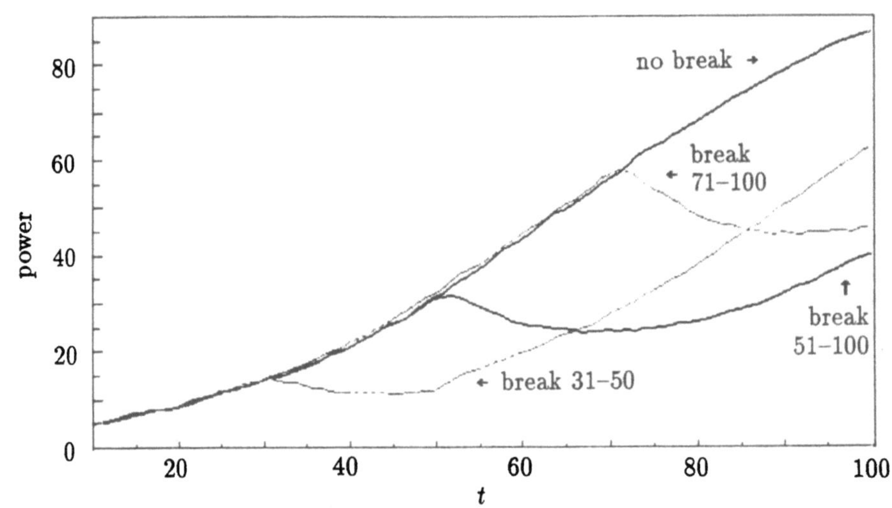


*Figure 8.14:* Recursive Dickey-Fuller "$t$-test" rejection frequencies of $H_0$: $\beta = 0$ when $\beta = -0.2$ with step breaks at the sample periods indicated.

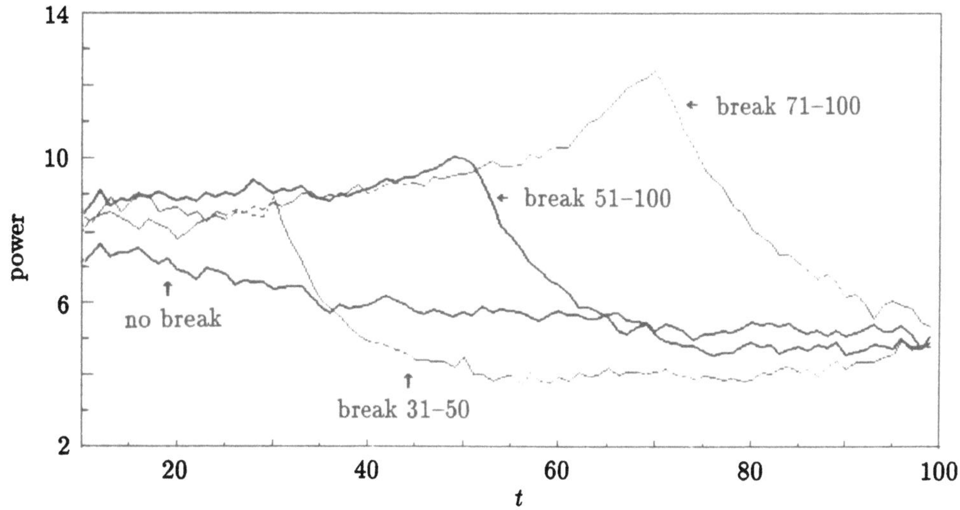


*Figure 8.15:* Recursive "break-point" Chow test rejection frequencies when $\beta = -0.1$.

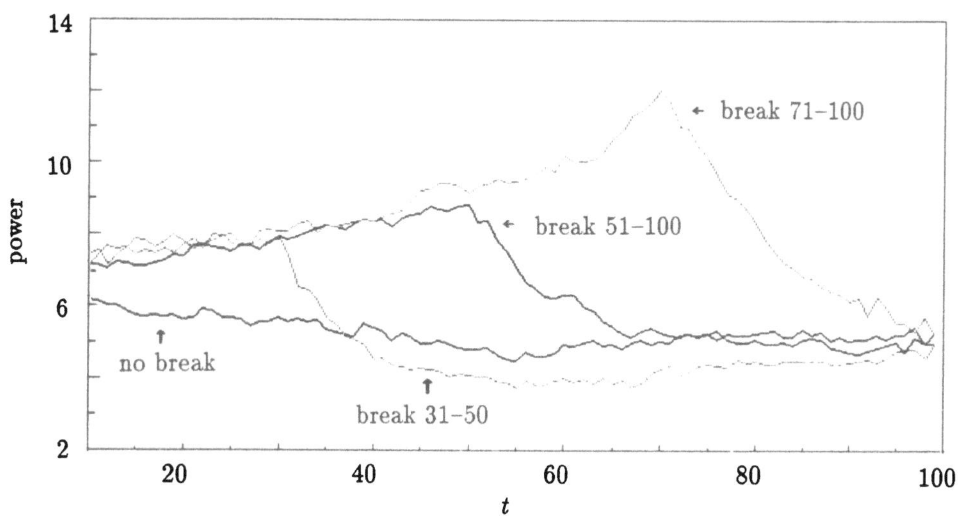


*Figure 8.16:* Recursive "break-point" Chow test rejection frequencies when $\beta = -0.2$.

movement for the test based on the last ten sample observations. These results contrast with Perron (1989), since small step shifts in dynamic processes, which are sufficient to induce apparent unit roots, can be very difficult to detect, so one would not know when to add dummy variables, even though breaks larger than $\sigma$ seem far easier to detect.

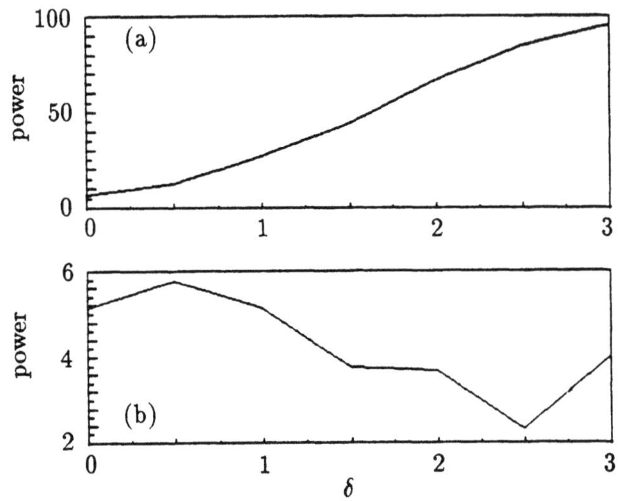


*Figure 8.17:* Chow test power function at (a) $T_0 = 70$ and (b) $T - 10$ as $\delta$ increases.

## 8.5 Conclusion

Firstly, *regime shifts can mimic unit roots* in stationary autoregressive time series. Secondly, such shifts may be very hard to detect using conventional parameter-constancy tests. Consequently, uncritical application of unit-root pretests without associated diagnostic tests for constancy of the supposed unit root may yield misleading conclusions empirically, and, even with conventional constancy tests, the existence of unit roots may be incorrectly assessed. Thirdly, the power function of unit-root tests is open to Monte Carlo response surface investigation despite their having nonstandard distributions. A number of salient determinants of rejection frequencies were established. The same approach could be applied to evaluate tests of parameter constancy when there are unit roots, in order to evaluate the factors influencing the distributions in that case. Finally, recursive Monte Carlo methods seem useful for investigating processes with structural breaks.

Did you guess which graph was which of *Figures 8.1(a)* and *8.1(b)*? In fact, *Figure 8.1(b)* is the one with the break.

## Acknowledgment

We are pleased to acknowledge that this research was financed in part by Grants B00220012 and R231184 from the Economic and Social Research Council. Helpful comments from Neil Ericsson and John Muellbauer are gratefully acknowledged.

## References

Chow, G.C. (1960), Tests of equality between sets of coefficients in two linear regressions. *Econometrica*, **28**, 591-605.

Cox, D.R. (1970), *The Analysis of Binary Data.* London: Chapman and Hall.

Dickey, D.A. and Fuller, W.A. (1979), Distribution of the estimators for autoregressive time series with a unit root. *Journal of the American Statistical Association*, **74**, 427-431.

Dickey, D.A. and Fuller, W.A. (1981), Likelihood ratio statistics for autoregressive time series with a unit root. *Econometrica*, **49**, 1057-1072.

Dolado, J.J. and Jenkinson, T. (1987), Cointegration: A survey of recent developments. University of Oxford Applied Economics D.P. 39.

Engle, R.F. and Granger, C.W.J. (1987), Cointegration and error correction: Representation, estimation and testing. *Econometrica*, **55**, 251-276.

Fuller, W.A. (1976), *Introduction to Statistical Time Series.* New York: John Wiley.

Hendry, D.F. (1984), Monte Carlo experimentation in econometrics, chapter 16 in Z. Griliches and M.D. Intriligator (eds.), *Handbook of Econometrics.* New York: North Holland.

Hendry, D.F. (1989), PC-GIVE: An interactive econometric modelling system. Oxford Institute of Economics and Statistics.

Hendry, D.F. and Mizon G.E. (1990), Evaluating dynamic econometric models by encompassing the VAR, forthcoming in P.C.B. Phillips and V.B. Hall (eds.), *Models, Methods and Applications of Econometrics.* Cambridge: Cambridge University Press.

Hendry, D.F. and Neale, A.J. (1987), Monte Carlo experimentation using PC-NAIVE. *Advances in Econometrics*, **6**, 91-125.

Hylleberg, S. and Mizon, G.E. (1989), A note on the distribution of the least squares estimator of a random walk with drift. *Economics Letters*, **29**, 225-230.

Jarque, C.M. and Bera, A.K. (1980), Efficient tests for normality, homoscedasticity and serial independence of regression residuals. *Economics Letters*, **6**, 255-259.

MacKinnon, J.G. (1990), Critical values for cointegration tests. Unpublished paper, San Diego, University of California.

Miller, S.M. (1988), Are saving and investment cointegrated? *Economics Letters*, **27**, 31-34.

Mizon, G.E. and Richard, J.-F. (1986), The encompassing principle and its application to non-nested hypothesis tests. *Econometrica*, **54**, 657-678.

Nelson, C.R. and Plosser, C.I. (1982), Trends and random walks in macroeconomic time series: Some evidence and implications. *Journal of Monetary Economics*, **10**, 139-162.

Perron, P. (1989), The great crash, the oil price shock and the unit root hypothesis. *Econometrica*, **57**, 1361-1402.

Perron, P. and Phillips, P.C.B. (1987), Does GNP have a unit root? A re-evaluation. *Economics Letters*, **23**, 139-145.

Phillips, P.C.B. and Perron, P. (1988), Testing for a unit root in time series regression. *Biometrika*, **75**, 335-346.
Ramsey, J.B. (1969), Tests for specification errors in classical linear regression analysis. *Journal of the Royal Statistical Society*, **B-31**, 350-371.
Rappoport, P. and Reichlin, L. (1989), Segmented trends and non-stationary time series. *Economic Journal*, **99**, 168-177.
Sargan, J.D. and Bhargava, A. (1983), Testing residuals from least squares regression for being generated by the Gaussian random walk. *Econometrica*, **51**, 153-174.
West, K.D. (1988), Asymptotic normality when regressors have a unit root. *Econometrica*, **56**, 1397-1417.
White, H. (1980), A heteroscedastic consistent covariance matrix and a direct test for heteroscedasticity. *Econometrica*, **48**, 817-838.
Working, H. (1934), A random difference series for use in the analysis of time series. *Journal of the American Statistical Association*, **29**, 11-24.

# Appendix A:
# 5% Critical Values of the DF and ADF(1) Tests

| $T$ | DF* | ADF(1)* | DF+ |
|---|---|---|---|
| 10 | −3.32 | −3.25 | −3.22 |
| 20 | −3.07 | −3.02 | −3.02 |
| 30 | −2.99 | −2.95 | −2.96 |
| 40 | −2.96 | −2.93 | −2.94 |
| 50 | −2.93 | −2.91 | −2.92 |
| 60 | −2.92 | −2.90 | −2.91 |
| 70 | −2.91 | −2.89 | −2.90 |
| 80 | −2.91 | −2.89 | −2.90 |
| 90 | −2.90 | −2.89 | −2.89 |
| 100 | −2.89 | −2.88 | −2.89 |
| 350 | −2.88 | −2.88 | −2.88 |

Critical values from $T = 150$ up to $T = 300$ are equal to the final table entry to two digits.

The entries given in the columns 2 and 3 (marked by *) are based on a minimum of 10,000 Monte Carlo replications, using the critical value corresponding to observation 9,501. The entries given in column 4 (marked by +) are calculated from the relevant response surface reported by MacKinnon (1990), namely,

$$C_{\mathrm{DF}} = -2.86 - 2.74/T - 8.4/T^2$$

Ordinary least squares estimation of a response surface of the same form as that used by MacKinnon but relating the critical values of the DF test to the sample size across our 11

Monte Carlo estimates yielded:

$$\hat{C}_{\mathrm{DF}} = \underset{(0.004)}{-\ 2.86} \ \underset{(0.25)}{-\ 3.58/T} \ \underset{(2.35)}{-\ 10.3/T^2}$$

with $R^2 = 0.9985$ and $\sigma = 0.0056$.

This equation and the tabulated numbers in column 2 both suggest that the Monte Carlo-based critical values are too large in absolute value at the smallest sample sizes. Such a finding could account for the underestimation of the rejection frequency in *Figure 8.10* when $T < 50$, and would induce some power loss at small sample sizes relative to the values based on MacKinnon's equation. The discrepancies relative to the results in column 4 are due in part to the estimator used for the critical values, which MacKinnon bases on averages of the neighboring observations where 5% are above and 95% are below.

# Part II

# Model Building in the Presence of Structural Change

## CHAPTER 9

# Forecast Encompassing and Model Evaluation

*Maozu Lu and Grayham E. Mizon*


## Summary

Tests for parameter constancy and the ability to encompass rival models are important parts of model evaluation. It is shown that the recently proposed forecast-encompassing test statistics have implicit null hypotheses that combine hypotheses of parameter constancy and complete parametric encompassing. An additional attraction of these tests is that they are easily calculated even for large-scale econometric models. However, just as minimum MSE is a necessary, but not sufficient, condition for model congruence, so the requirement that one model forecast variance-encompass another is a necessary condition for the first model to have constant parameters and encompass the other model.


## 9.1 Introduction

Evaluation of the forecast performance of rival models for the same phenomena has long been a topic of concern to both theoretical and applied econometricians. Indeed, in the area of macroeconometric modeling there have been a number of recent contributions, e.g., Fair (1985), Chong and Hendry (1986), and the survey by Wallis (1989). Since one of the main purposes of macroeconometric model building is to provide sound forecasts of the economic indices of interest, it is not surprising that the relative forecast performance of these models has been a prominent feature of model evaluation exercises. Both the *ex ante* and *ex post* forecast performances of models provide evidence on their coherence with observation, and as such are important in "proving" models. Hence forecast performance is a natural criterion for evaluating and comparing models.

However, there are many different phenomena which a model can be used to forecast the numerical value of, and there are numerous ways of assessing the quality of the generated forecasts. To narrow the scope of this chapter we will concentrate on the use of forecasts of modeled (endogenous) variables in the evaluation of model performance. Furthermore, although we will argue that forecast performance is not the sole (nor the best) indicator of model quality, in the context of evaluating large-scale macroeconometric models, it is both natural and potentially valuable because it is feasible. Model congruence as discussed in Hendry (1987) and Hendry and Mizon (1990)—which requires a model to be coherent with information in sample data, the properties of the measurement system, and an *a priori* theory, and to be able to encompass (i.e., perform at least as well as) rival models of the same phenomena—is not only difficult to achieve, but difficult to test in large macroeconometric models. The large number of instrumental variables relative to sample size and the presence of nonlinearities in these models are but two of the more obvious impediments to rigorous evaluation. The fact that forecasts are usually readily available, and that they can yield information about model quality, even though it is incomplete information, makes the role of forecasts in model evaluation worthy of analysis.

The main purpose of this chapter is to present a class of forecast-based encompassing test statistics, and to assess their potential role in model evaluation by comparing their implicit null hypotheses with those of existing test statistics for parametric encompassing and parameter constancy. Whilst the ideal would be to assess thoroughly each aspect of model congruence, this is rarely feasible for large models. The argument in favor of forecast-encompassing tests is that by using available information they are feasible tests and that they compare models on the basis of a *combination* of parametric-encompassing and parameter constancy hypotheses. Section 9.2 comments briefly on some of the commonly used methods for evaluating large macro models, and argues that there is a need for new and valid means to evaluate alternative models. The population-encompassing hypotheses associated with parametric encompassing, parameter constancy, and forecast encompassing are presented and analyzed in Section 9.3. Conclusions are drawn in Section 9.4.

## 9.2 Evaluation of Large-scale Macroeconometric Models

*Ex ante* forecasts from large macroeconometric models are rarely, if ever, purely dependent on the models, but usually reflect the judgments of the model proprietors. Thus, although such modifications to the pure model forecasts may produce more accurate forecasts, thus pleasing the modeling teams' clients, this means that only limited evidence is provided about model validity by the forecast track record [see Chong and Hendry (1986) and, e.g., Wallis *et al.* (1986, 1987)]. Indeed the fact that "add factors" and "type I and type II fixes" have assumed a permanent place in the production of macro forecasts is itself an indication of model non-congruence. In addition to requiring "pure" forecasts based on the model alone, accurate measures of forecast uncertainty, particularly in dynamic models, are needed to test model validity appropriately using *ex ante* forecasts. Much work remains to be done on this topic, though some progress has been made recently [see

Ericsson and Marquez (1989) for an analysis and assessment]. These points remain as caveats to much of the sequel.

One commonly used method for evaluating large dynamic models involves dynamic simulation, both within-sample and post-sample. This is indicated, for example, by the prominent role that simulation plays in the influential text by Pindyck and Rubinfeld (1981), in which it is argued that minimum root mean squared error of simulation residuals is an important tool in model evaluation. The potential problems of such a strategy were discussed by Hendry and Richard (1982), and it was shown to be an invalid model selection procedure by Chong and Hendry (1986). One problem arises from the fact that simulation errors are the cumulation of reduced form errors, and so simulation residuals are highly autocorrelated, which severely biases the conventional measures of their variability. More importantly though, the accuracy of a simulation is determined by which, and how many, variables are treated as strongly exogenous, and does not depend on whether or not they are valid exogenous variables. In addition, Pagan (1989) has shown that in situations such as within-sample prediction, and one-step-ahead post-sample forecasting, simulation residuals are a one-to-one mapping of the estimation residuals. Hence, even in the context of within-model evaluation (rather than cross-model validation), analysis of simulation residuals *per se* can reveal no information beyond that obtainable from the corresponding analysis of estimation residuals.

Tempting though it is to look for a single criterion adequate for model selection, this is not a fruitful search on which to embark. However, one criterion that has been used in this way is mean squared error (MSE), with models being ranked on the basis of their MSE's and the model with minimum MSE being the preferred model. Similarly, there are information criteria, such as the AIC of Akaike (1973), the Schwarz (1978) criterion, Mallows' (1973) $C_p$, the Hannan-Quinn (1979) criterion, as well as the final prediction error criterion, which have been used widely in selecting a preferred model from a set of models. The principal difficulty in using such criteria to select models is that they are at best necessary, but not sufficient, conditions for model congruence. Hence it is possible to select, using one of these criteria, a model that has serially correlated and heteroscedastic errors, invalid exogeneity, and nonconstant parameters. Indeed, for one model to have minimum MSE, it is neither necessary nor sufficient that it have valid exogeneity, constant parameters, or provide accurate forecasts. Thus very little is learned about a model's validity from its having minimum MSE, either within-sample or in the forecast period.

On the other hand, Hendry and Richard (1982) have shown that the ability of one model to encompass another will ensure MSE dominance of the former, *when both models are otherwise congruent*. The converse is not true though. Hence, for a model to have minimum MSE is not in isolation a good indicator of model validity. Furthermore, Ericsson (1988) in analyzing the relationship between parameter constancy and minimum mean squared forecast error (MSFE) has shown that (i) parameter constancy is neither necessary nor sufficient for minimum MSFE and (ii) parameter constancy and minimum MSFE jointly are not sufficient for the best forecasting model. Therefore, as suggested by Hendry and Richard (1982), Hendry and Mizon (1990), and Pesaran and Smith (1985) amongst others, model builders are living dangerously if they try to infer the statistical properties of a model solely from its MSFE. A more robust strategy is to assess a model's congruence with *all* available information. The role that forecast-encompassing tests can play in this, and, in particular their role relative to those of tests for parametric encompassing and parameter constancy, is now analyzed.

## 9.3 Parameter Constancy and Encompassing Hypotheses

Since a major purpose of this chapter is expository, the analysis will be undertaken in the context of normal linear models. This has the advantage of using a class of models for which the principal results on statistical inference are widely known, and for which it is relatively easy to present new results. However, most of the results presented below can be readily extended to more general models when appropriate assumptions and regularity conditions are adopted. Andrews and Fair (1988) provide a framework for doing precisely this for nonlinear dynamic simultaneous equations models, thus enabling analysis when observations are heterogeneous and temporally dependent.

Consider the following non-nested linear regression models:

$$M_1\colon\ y_t|\mathbf{w}_t \sim NI(\mathbf{x}_t'\boldsymbol{\beta}, \sigma^2) \tag{9.1}$$

$$M_2\colon\ y_t|\mathbf{w}_t \sim NI(\mathbf{z}_t'\boldsymbol{\gamma}, \tau^2), \tag{9.2}$$

when $(\mathbf{x}_t, \mathbf{z}_t) \subseteq \mathbf{w}_t$ and $t = 1, 2, \ldots, T$. For notational convenience below, let $\boldsymbol{\alpha}' = (\boldsymbol{\beta}', \sigma^2)$ and $\boldsymbol{\delta}' = (\boldsymbol{\gamma}', \tau^2)$. Note that the two models have common conditioning, and that the parameters are assumed constant so that there is only one regime. In this context the Wald-encompassing test statistics developed by Mizon (1984) and Mizon and Richard (1986), take the form

$$\eta(\tilde{\mathbf{b}}) = T\tilde{\varphi}'V(\tilde{\varphi})^+\tilde{\varphi} \xrightarrow{d}_{M_1} \chi^2(r), \tag{9.3}$$

when $\tilde{\varphi} = (\tilde{\mathbf{b}} - \mathbf{b}_{\hat{\alpha}})$, $\tilde{\mathbf{b}}$ is any statistic which is a function of $y$ and $\mathbf{w}$ that is of interest in the context of $M_2$, $\mathbf{b}_\alpha$ is the pseudo true value of $\tilde{\mathbf{b}}$ under $M_1$, $\hat{\boldsymbol{\alpha}}$ is a $\sqrt{T}$-consistent estimator of $\boldsymbol{\alpha}$ (e.g., the MLE of $\boldsymbol{\alpha}$), and $\tilde{\varphi}$ is such that $\sqrt{T}\tilde{\varphi} \xrightarrow{d}_{M_1} N[0, V(\tilde{\varphi})]$ with $V(\tilde{\varphi})$ having rank $r$ and a generalized inverse $V(\tilde{\varphi})^+$. The population-encompassing hypothesis underlying the test statistic $\eta(\tilde{\mathbf{b}})$ is $\varphi = (\mathbf{b} - \mathbf{b}_\alpha) = \mathbf{0}$, when $\mathbf{b} = E_\delta(\tilde{\mathbf{b}})$ and $\mathbf{b}_\alpha = E_\alpha(\tilde{\mathbf{b}})$ with $E_\alpha$ and $E_\delta$ denoting the expectation operator (or if appropriate plim) under $M_1$ and $M_2$, respectively. If $M_1\mathcal{E}M_2$ (i.e., $M_1$ encompasses $M_2$) then $\varphi = \mathbf{0}$ for any $\mathbf{b}$, and $M_1$ parsimoniously encompasses the completing model $M_c$ (i.e., $M_1\mathcal{E}_pM_c$ ), when $M_c$ is such that $M_i \subset M_c$ for $i = 1, 2$.

In the case that $\mathbf{w}_t' = (\mathbf{x}_t', \mathbf{z}_t')$, which will now be analyzed in more detail, an obvious choice for $M_c$ is

$$M_c\colon\ y_t|\mathbf{w}_t \sim NI(\mathbf{x}_t'\boldsymbol{\lambda} + \mathbf{z}_t'\boldsymbol{\mu}, \omega^2), \tag{9.4}$$

for which it follows that $M_1 \mathcal{E} M_2 \equiv M_1 \mathcal{E}_p M_c \equiv H_p : \boldsymbol{\mu} = \mathbf{0}$ [see Hendry and Richard (1987)]. For this example $\boldsymbol{\mu} = \mathbf{0}$ is equivalent to $(\boldsymbol{\gamma} - \boldsymbol{\gamma}_\alpha) = \mathbf{0}$ so that the hypothesis that $M_1 \mathcal{E} M_2$ can be tested in practice using the complete parametric-encompassing test statistic $\eta(\hat{\boldsymbol{\gamma}})$ [when $\hat{\boldsymbol{\gamma}} = (Z'Z)^{-1}Z'\mathbf{y}$ is the OLS estimator of $\boldsymbol{\gamma}$], which is equivalent to the traditional $F$-test statistic for the hypothesis $\boldsymbol{\mu} = \mathbf{0}$. The parameters of $M_1$, $M_2$, and $M_c$ can also be expressed as functions of the parameters of the joint distribution of $y$, $\mathbf{x}$, and $\mathbf{z}$ (each of which is assumed to have a zero mean without loss of generality) and this proves useful later. Hence, $\boldsymbol{\gamma} = \sum_{zz}^{-1} \sum_{zy}$ and $\boldsymbol{\gamma}_\alpha = \Pi\boldsymbol{\beta}$ when $\Pi = \sum_{zz}^{-1} \sum_{zx}$, $\boldsymbol{\beta} = \sum_{xx}^{-1} \sum_{xy}$, and $\sum_{ij}$ for $i, j = y, x, z$ are the population second moment matrices from the joint distribution of $y$, $\mathbf{x}$, and $\mathbf{z}$. Also, noting that $\boldsymbol{\mu} = [I - \Pi P]^{-1}[\boldsymbol{\gamma} - \Pi\boldsymbol{\beta}]$ when $P = \sum_{xx}^{-1} \sum_{xz}$, the equivalence between $\boldsymbol{\mu} = \mathbf{0}$ and $(\boldsymbol{\gamma} - \boldsymbol{\gamma}_\alpha) = \mathbf{0}$ is easily seen. Finally, note that the relationship between $\mathbf{x}$ and $\mathbf{z}$ can be written as $\mathbf{x}_t = \Pi'\mathbf{z}_t + \mathbf{v}_t$ with $\mathbf{v}_t \sim NI(\mathbf{0}, \Omega)$ when $\Omega = \sum_{xx} - \sum_{xz} \sum_{zz}^{-1} \sum_{zx}$.

It is now possible to exploit the generality and flexibility of this encompassing framework to consider hypotheses relevant for cross-model, cross-regime, and cross-model *and* cross-regime comparisons.

### 9.3.1 Single regime

Within a single regime (i.e., constant parameters across all observations) the following hypotheses are relevant in the comparison of $M_1$ and $M_2$:

$$H_p : (\boldsymbol{\gamma} - \Pi\boldsymbol{\beta}) = (\boldsymbol{\gamma} - \boldsymbol{\gamma}_\alpha) = \mathbf{0} \quad \text{(parameter-encompassing)} \tag{9.5}$$

$$H_v : (\tau^2 - \sigma^2 - \boldsymbol{\beta}'\Omega\boldsymbol{\beta}) = 0 \quad \text{(variance-encompassing)} \tag{9.6}$$

$$H_d : \sigma^2 < \tau^2 \quad \text{(variance-dominance)}. \tag{9.7}$$

$H_p$ is the hypothesis that $M_1$ provides a complete parametric encompassing of $M_2$ as discussed above and in Mizon (1984) and Mizon and Richard (1986), and is a sufficient condition for variance-encompassing $H_v$. The variance-encompassing statistic for testing $H_v$ is asymptotically equivalent to the Cox (1961) generalized likelihood ratio test statistic, and the Davidson-MacKinnon (1981) $J$-test statistic. Noting that $\Omega$ is a covariance matrix, and so positive semi-definite, means that $H_v$ implies variance dominance $H_d$. So variance dominance, like minimum MSE, is a necessary, but not sufficient, condition for encompassing [see Hendry and Richard (1982)]. Other choices of $\mathbf{b}$ (and hence $\tilde{\mathbf{b}}$) lead to further encompassing hypotheses that can be tested in comparing $M_1$ and $M_2$, but $H_p$ and $H_v$ are the most commonly tested hypotheses in the context of a single regime.

An important example of there being more than one regime occurs when the available data are partitioned into the sample period $t = 1, 2, \ldots, T$ and the forecast period $t = T+1, T+2, \ldots, T+n$. Clearly, if the length of the forecast period $n$ is at least equal to the number of regressors in $\mathbf{x}$ and $\mathbf{z}$ ($k$ and $\ell$, respectively), and observations on the "forecast" period values of $y$, $\mathbf{x}$, and $\mathbf{z}$ are available, then the hypotheses $H_p$ and $H_v$ can be tested again, this time using the forecast data. It will often not be possible to perform these tests though, either because $n < k + \ell$ or because the data for the exogenous variables in the forecast period are not made available. However, when more than one regime is allowed it becomes both possible and important to test for parameter constancy.

### 9.3.2 Two regimes

The importance of parameter constancy tests in econometric modeling has been stressed by numerous authors, e.g., Chow (1960), Hendry (1979), and more recently by Anderson and Mizon (1989). For economic and econometric modeling to be fruitful it is essential that the phenomena being analyzed are capable of characterization by some invariants. Such invariants will be the focus of attention in modeling, whether the statistical methods used are parametric, semi-parameteric, or nonparametric, though in this chapter only parametric modeling is considered. Parameter estimates which within sample are sensitive to minor changes in the sample period are of less value than those which are invariant to such changes, and are an indication of model misspecification. Hence parameter constancy tests, whether implemented via recursive estimation [see, e.g., Harvey (1981)], or regime-shift tests such as those described in Chow (1960), are potentially powerful for *proving* models. In addition, parameter constancy tests have recently been shown by Hendry (1988) to play a critical part in discriminating between feedback (error correction) and feedforward (rational expectations) models.

Economic structural change may induce impulse changes, gradual changes, or the presence of one or more regimes in an econometric model. In addition to there being potential uncertainty about the form of the impact of structural change on a model, there can be uncertainty *a priori* about the timing of its impact. Anderson and Mizon (1989) discuss of alternative ways to represent structural change in econometric models, as well as methods for detecting the timing of its impact. In this chapter attention is confined to the possibility of there being a regime shift between sample and forecast periods—this being the most relevant consideration when the use of model forecasts in model evaluation is being analyzed. Hence the potential break-point—the end of the sample period—is known. This does not imply that within-sample parameter constancy checks are less important, but simply reflects the fact that the focus of this chapter is on the use of forecasts in model evaluation.

When one model (e.g., $M_1$) is to be tested for a regime shift, at a known break-point, the following hypotheses are of interest:

$$H_{cp} : \Delta\boldsymbol{\beta} = \boldsymbol{\beta}_2 - \boldsymbol{\beta}_1 = \mathbf{0} \tag{9.8}$$

$$H_{cv} : \sigma_2^2 - \sigma_1^2 = 0 \tag{9.9}$$

$$H_c : H_{cp} \cap H_{cv} \tag{9.10}$$

$$H_{chow} : X_2\Delta\boldsymbol{\beta} = \mathbf{0} \tag{9.11}$$

$$H_{pred}: \text{MSFE} - \sigma_1^2 = 0 \quad \text{or} \quad \text{MSFE}/\sigma_1^2 = 1\,. \tag{9.12}$$

Since the primary interest is in forecast tests, the two regimes are the sample period $t = 1, 2, \ldots, T$ (denoted regime 1) and the forecast period $t = T+1, T+2, \ldots, T+n$ (denoted regime 2). The combination of the two regimes, when parameter constancy is assumed, is denoted regime 0. Subscripts on parameters denote the regime for which that parameter is constant. Hence $H_{cp}$ is the hypothesis that the regression coefficients are constant across the sample and forecast periods, for which an appropriate test statistic, provided $H_{cv}$ holds, is the analysis of variance (ANOVA) $F$-statistic—see Chow (1960). $H_{cv}$ is the hypothesis of common error variances across the sample and forecast periods, for which an appropriate test statistic is the variance ratio $F$-test, which is distributed independently of the ANOVA $F$-statistic—see Phillips and McCabe (1983). These two hypotheses together constitute $H_c$, the hypothesis of parametric constancy. For further discussion of the relationship between these hypotheses, the corresponding test statistics, and possible changes in the parameters of the joint distribution of $y$ and $\mathbf{x}$, see Anderson and Mizon (1983).

It is often the case that the forecast period does not have enough observations to allow $\beta_2$ and $\sigma_2^2$ to be estimated (i.e., $n < k$ when $k$ is the number of regressors in $\mathbf{x}$), and so tests have been proposed that rely on parameter estimates using the $T$ observations in regime 1 and the $T+n$ observations in both regimes when parameter constancy is assumed. The Chow (1960) prediction test is one such statistic, and its implicit null hypothesis is given by $H_{chow}$. Note that, $H_{chow}$ hypothesizes that there is no change in the conditional mean of $y$ across regimes 1 and 2, i.e., $E_\alpha(y_t - \mathbf{x}_t'\hat{\beta}_1) = 0$ for $t \in \{T+1, T+2, \ldots T+n\}$ when it is maintained that $H_{cv}$ holds (i.e., $\sigma_1^2 = \sigma_2^2$). This hypothesis would need modification if the model being analyzed was dynamic (e.g., $\mathbf{x}_t$ includes lagged values of $y_t$), to deal appropriately with the conditional expectation. Also note that $H_{chow}$ can be expressed as $E_\alpha(\hat{f}_t) = E_\alpha(f_t) = 0$ for $t \in \{T+1, T+2, \ldots, T+n\}$ when $f_t$ and $\hat{f}_t$ are the forecast error $(y_t - \mathbf{x}_t'\beta_1)$ and forecast residual $(y_t - \mathbf{x}_t'\hat{\beta}_1)$, respectively, so that $\hat{f}_t = f_t + \mathbf{x}_t'(\beta_1 - \hat{\beta}_1)$. Hence using the fact that $\hat{\mathbf{f}}_t = (\mathbf{y}_2 - X_2\hat{\beta}_1)$ has covariance matrix $V_\alpha(\hat{\mathbf{f}}) = [\sigma_2^2 I_n + \sigma_1^2 X_2(X_1'X_1)^{-1}X_2']$—the second term being $V_\alpha(X_2\hat{\beta}_1)$—enables the Chow test statistic to be written as $\hat{\mathbf{f}}_t'\hat{V}(\hat{\mathbf{f}})^{-1}\hat{\mathbf{f}}_t/n$ when $\hat{V}(\hat{\mathbf{f}}) = \hat{\sigma}_1^2[I + X_2(X_1'X_1)^{-1}X_2']$ and $H_{cv}$ is maintained. The Chow test statistic is distributed as a central $F(n, T-k)$ variate under $H_{chow} \cap H_{cv}$.

If the sampling variability of $\hat{\beta}_1$ is ignored (it is asymptotically negligible) then $\hat{\sigma}_1^2 I_n$ can be used instead of $\hat{V}(\hat{\mathbf{f}})$ to yield another test statistic frequently used in this situation, namely, $(\hat{\mathbf{f}}'\hat{\mathbf{f}})/(n\hat{\sigma}_1^2)$, which is known as the $\chi^2$ prediction test [see Hendry (1974)] and has a limiting $\chi^2(n)/n$ distribution under $H_{pred}$. The implicit null hypothesis of this test statistic, $H_{pred}$, involves $\text{MSFE} = E_\alpha(y_t - \mathbf{x}_t'\beta_1)^2$ for $t \in \{T+1, T+2, \ldots, T+n\}$ which is the mean squared prediction error *with* $\beta_1$ *treated as known*. Hence, noting that $\text{MSFE} = (\mathbf{x}_t'\Delta\beta)^2 + \sigma_2^2$, when the expectation is with respect to the distribution of $y$ conditional on the realized value of $\mathbf{x}$ without parameter constancy assumed, it is clear that the prediction $\chi^2$ test statistic will be sensitive to changes in $\beta$ *and* $\sigma^2$. Despite the fact that this test statistic is closely related to the MSFE of a model, and hence might be thought of as a potential model selection criterion, this is not a modeling strategy to be recommended—see Kiviet (1986). The fact that the prediction test statistic ignores the sampling variability associated with $\hat{\beta}_1$ is one of its weaknesses. Schmidt (1974) and Baillie (1979) though have provided an easily computable second-order approximation to the exact MSFE for dynamic linear models, and Ericsson and Marquez (1989) have shown, on the basis of Monte Carlo simulation results, that the approximation works well for closed autoregressive models. However, further research is needed before such approximations can be incorporated into a modified $\chi^2$ prediction test statistic that allows for the uncertainty associated with parameter estimation and is applicable for dynamic linear models with exogenous variables.

The above analysis of parameter constancy within $M_1$ obviously can be repeated for $M_2$, but since all that is involved is the replacement of $M_2$ and its parameters for $M_1$ and its parameters, the details will not be presented. Instead, attention is now turned to simultaneous inter-model and inter-regime comparisons, which are the essence of forecast-encompassing tests.

Two obvious hypotheses that combine the inter-model and inter-regime hypotheses considered above are:

$$H_{fp}\colon (\gamma_2 - \Pi_1\beta_1) = \Delta\gamma + (\gamma_1 - \gamma_{\alpha 1}) = (\gamma_2 - \gamma_{\alpha 2}) + \Delta\gamma_\alpha = \mathbf{0} \tag{9.13}$$

$$H_{fv}\colon \tau_2^2 - \sigma_1^2 - \beta_1'\Omega_1\beta_1 = (\tau_2^2 - \tau_1^2) + (\tau_1^2 - \sigma_1^2 - \beta_1'\Omega_1\beta_1) = 0\,. \tag{9.14}$$

$H_{fp}$ is a forecast parameter-encompassing hypothesis, and is seen to consist of the sum of $H_p$: $(\gamma - \gamma_\alpha) = \mathbf{0}$ and $H_{cp}$: $\Delta\gamma = \mathbf{0}$. Similarly, $H_{fv}$ is a forecast variance-encompassing hypothesis, which is the sum of $H_v$ and $H_{cv}$. In that these hypotheses are combining inter-model and inter-regime comparisons they are of potential interest, but to compute test statistics for these hypotheses it will be necessary to estimate the parameters in $\boldsymbol{\alpha}$ and $\boldsymbol{\delta}$ for *both* regimes. However, this will require $n > k+\ell$, which will often not be the case, but, even if it were, it would then be preferable to test the constituent hypotheses separately rather than as a linear combination. Hence $H_{fp}$ and $H_{fv}$ have no more significance than that of illustrating the fact that forecast-encompassing hypotheses are a composite of parameter constancy and parametric encompassing.

To avoid the difficulty associated with $n < k + \ell$ it is possible to consider the following forecast-encompassing hypothesis:

$$H_f : Z_2(\gamma_1 - \Pi_1\beta_1) = \mathbf{0}\,, \tag{9.15}$$

the motivation for which comes from asking whether $M_1$ can explain the forecast generated by $M_2$. In other words the hypothesis is that there is no difference between the means of the two forecasts $Z_2\hat{\gamma}_1$ and $Z_2\gamma_{\hat{\alpha}1}$, when $\gamma_{\hat{\alpha}1} = (Z_1'Z_1)^{-1}Z_1X_1\hat{\beta}_1$. The advantage of this hypothesis is that the encompassing statistic for testing it is easily calculated using sample period data and model forecasts. Indeed, if $n > \ell$ the resulting test statistic will be identical to that for $H_p$ within-sample. However, the advantage is not a real one, for if the sample data are available it would again be preferable to test $H_p$ directly. Hence it seems appropriate to consider alternative forecast-encompassing hypotheses.

The hypotheses above have been developed separately, and have been somewhat *ad hoc* in nature. More is revealed about each hypothesis, and their relationship to each other and the hypotheses for inter-model and inter-regime comparisons, by casting them as restrictions on a congruent general model. Within this framework each hypothesis is seen as a parsimonious encompassing hypothesis relative to the general model, and so provided that the general model is congruent the resulting forecast-encompassing tests will be valid (i.e., unaffected by misspecification errors, apart of course from the type I errors associated with the misspecification testing of the general model). The relevant general, or completing, model for both inter-model and inter-regime comparisons is $M_c^*$:

$$\begin{bmatrix} \mathbf{y}_1 \\ \mathbf{y}_2 \end{bmatrix} = \begin{bmatrix} X_1 & Z_1 & 0 & 0 \\ X_2 & Z_2 & X_2 & Z_2 \end{bmatrix} \begin{bmatrix} \boldsymbol{\lambda}_1 \\ \boldsymbol{\mu}_1 \\ \Delta\boldsymbol{\lambda} \\ \Delta\boldsymbol{\mu} \end{bmatrix} + \begin{bmatrix} \boldsymbol{\epsilon}_1 \\ \boldsymbol{\epsilon}_2 \end{bmatrix} \tag{9.16}$$

with $\Delta\boldsymbol{\lambda} = \boldsymbol{\lambda}_2 - \boldsymbol{\lambda}_1$, $\Delta\boldsymbol{\mu} = \boldsymbol{\mu}_2 - \boldsymbol{\mu}_1$ and $\boldsymbol{\lambda}_i$ and $\boldsymbol{\mu}_i$, $i = 1,2$ defined as $\boldsymbol{\lambda}$ and $\boldsymbol{\mu}$ above but for each regime separately. It is also assumed that $\boldsymbol{\epsilon}_i \sim NI(\mathbf{0}, \omega_i^2 I)$ so that the errors are homoscedastic and serially independent within regimes, and independent across regimes. Indeed, the congruence of $M_c^*$ here involves (i) valid conditioning, so that $\mathbf{x}$ and $\mathbf{z}$ are weakly exogenous variables for the parameters of interest; (ii) no serial correlation in the errors $\boldsymbol{\epsilon}$; (iii) no heteroscedasticity in the errors within regimes; (iv) no ARCH [Engle (1982)] effects; (v) no functional form misspecification; and (vi) the errors being innovations relative to the appropriate information set. This implies that $M_c^*$ is appropriately specified, i.e., contains the relevant parameters and includes a correct representation of the regime shift. Within this framework it is relevant to consider the conditional independence hypotheses $\sum_{zy} = \sum_{zx} \sum_{xx}^{-1} \sum_{xy}$ and $\sum_{xy} = \sum_{xz} \sum_{zz}^{-1} \sum_{zy}$, which lead to $M_1$ and $M_2$, respectively, and the hypotheses asserting parameter constancy. For $M_1$ to encompass $M_2$ (i.e., $M_1 \mathcal{E} M_2$), $M_1$ should be a congruent model with constant parameters, which requires $\Delta\boldsymbol{\lambda} = \mathbf{0}$, $\Delta\boldsymbol{\mu} = \mathbf{0}$, $\boldsymbol{\mu}_1 = \mathbf{0}$, (and hence $\boldsymbol{\mu}_2 = 0$) and $\omega_1^2 = \omega_2^2$. This case implies that $M_1$ is an acceptable simplification of $M_c^*$, i.e., $M_1 \mathcal{E}_p M_c^*$. Whilst less demanding forms of encompassing, in which $M_1 \mathcal{E} M_2$ within either, or both, regimes *separately*, without there being parameter constancy (i.e., $\boldsymbol{\mu}_i = \mathbf{0}$, $i = 1,2$, $\Delta\boldsymbol{\lambda} \neq \mathbf{0}$ and $\omega_1^2 \neq \omega_2^2$) might be considered, they are less interesting because a model with such properties will suffer from predictive failure.

Note that $n > k + \ell$ will be required for all the parameters of $M_c^*$ to be estimable, otherwise only $\boldsymbol{\lambda}_1$, $\boldsymbol{\mu}_1$, $\omega_1^2$, $X_2\Delta\boldsymbol{\lambda}$ and $Z_2\Delta\boldsymbol{\mu}$ will be estimable. Hence only if $n > k + \ell$ will it be possible to test fully the hypothesis that $M_1$ with constant parameters parsimoniously encompasses $M_c^*$. When this case is being considered the important issue concerns the choice of an appropriate procedure for testing $M_1 \mathcal{E}_p M_c^*$. Although parameter nonconstancy is allowed for in $M_c^*$, the rejection of parameter constancy in $M_c$ will usually be *prima facie* evidence that all constant parameter models nested within $M_c^*$, such as $M_1$ and $M_2$, are noncongruent, and so would probably lead to the consideration of alternative general models $M_c^*$ which do appear to have constant parameters. Even if the investigator chooses to model the parameter nonconstancy via time-varying coefficient specifications, parameter constancy is still required, but for a different set of parameters. It therefore seems natural to test for parameter constancy first, and, only if this is not rejected, to proceed to parameter- and variance-encompassing hypotheses using all the data. Within the context of $M_c^*$ this provides a "natural" ordering of the constituent hypotheses of $M_1 \mathcal{E} M_2$ , which suggests a sequence of tests that will yield a separate induced test of the composite encompassing hypothesis—namely, (i) test $\omega_1^2 = \omega_2^2$, and if this is not rejected test (ii) $\Delta\boldsymbol{\lambda} = \mathbf{0}$ and $\Delta\boldsymbol{\mu} = \mathbf{0}$ conditional on $\omega_1^2 = \omega_2^2$, and if neither of these is rejected test (iii) $H_p$: $\boldsymbol{\mu} = \mathbf{0}$ treating the sample and forecast periods as a single regime. Even if there is a good reason to expect a regime shift at time $T + 1$, and there is an appropriate interpretation of it, it still would be sensible to test the hypothesis of parameter constancy before proceeding to $H_p$- and $H_v$-type hypotheses.

An alternative testing procedure to the three step one just described is implicit in the discussion of the hypothesis of forecast model encompassing in Ericsson (1988), in which it is hypothesized that the forecasts from $M_2$ have no power to explain the forecast error of $M_1$. If $M_1$ is of primary interest to an investigator and hence it has been thoroughly tested within-sample (so that, for example, $\boldsymbol{\mu}_1 = \mathbf{0}$ has not been rejected), and its own forecasting performance has been evaluated with $\Delta\boldsymbol{\lambda} = \mathbf{0}$ not having been rejected, the next hypothesis to consider in the context of $M_c^*$ is $\boldsymbol{\mu}_2 = \mathbf{0}$ (which of course is equivalent to $\Delta\boldsymbol{\mu} = \mathbf{0}$ when $\boldsymbol{\mu}_1 = \mathbf{0}$). Ericsson (1988) shows that a test of this hypothesis can be implemented by regressing $\mathbf{y}_0$ on $X_0$ and $Z^*$, when $Z^{*\prime} = (I_T, Z_2')$, and testing the hypothesis that the coefficients of $Z^*$ are zero via a statistic whose null distribution is $F(\ell, T + n - k - \ell)$. These two procedures are alternative ways for the investigator who is primarily interested in $M_1$ to evaluate its performance in the context of $M_c^*$—i.e., when rival models and parameter nonconstancy are being considered. Whilst the procedure described by Ericsson (1988) may well mimic accurately what many investigators do in practice, the former "ideal" procedure has the advantage of testing whether the underlying general model is congruent with constant parameters. An important feature of the "ideal" procedure is that the first hypothesis to be considered is tested using *all* the data (sample *and* forecast data on $y$, $\mathbf{x}$, *and* $\mathbf{z}$), and only when a hypothesis has been tested but not rejected is it imposed for the subsequent testing of the remaining hypotheses. However, such "ideal" procedures may not be feasible, and in particular it will often be the case that $n < k + \ell$, and/or observations on the exogenous variables are not available to the investigator, despite having been used in the generation of the forecasts from each model.

Concentrating attention now on situations where the forecast period contains insufficient observations to estimate all parameters of $M_c^*$, the following modification provides an appropriate general model:

$$M_c^{**}\colon \begin{bmatrix} \mathbf{y}_1 \\ \mathbf{y}_2 \end{bmatrix} = \begin{bmatrix} X_1 & Z_1 & 0 \\ X_2 & Z_2 & I_n \end{bmatrix} \begin{bmatrix} \boldsymbol{\lambda}_1 \\ \boldsymbol{\mu}_1 \\ \boldsymbol{\nu} \end{bmatrix} + \begin{bmatrix} \boldsymbol{\epsilon}_1 \\ \boldsymbol{\epsilon}_2 \end{bmatrix}, \tag{9.17}$$

when $\boldsymbol{\nu} = X_2\Delta\boldsymbol{\lambda} + Z_2\Delta\boldsymbol{\mu}$. Again, to ensure that subsequent inference is valid, it is necessary that $M_c^{**}$ be a congruent model. The model to be assessed within this framework is $M_1$ with constant parameters, which implies the restrictions $\boldsymbol{\nu} = \mathbf{0}$ and $\boldsymbol{\mu}_1 = \mathbf{0}$ on $M_c^{**}$. Clearly this joint hypothesis can be directly tested using an $F$-statistic, but as usual if a joint hypothesis is rejected it is not known whether the rejection was "caused" largely by a subset of the constituent hypotheses being strongly contradicted by the empirical evidence. In particular, the rejection of the joint hypothesis implying $M_1$, does not yield information as to whether parameter constancy, or the hypothesis that $M_1$ parametrically encompasses $M_2$, would be separately rejected. In fact, in this context it seems "natural" to test the hypothesis of parameter constancy ($\boldsymbol{\nu} = \mathbf{0}$) first, and if this is not rejected to test for parameteric encompassing. The Chow prediction test statistic applied to the regression of $y$ on $\mathbf{x}$ and $\mathbf{z}$ would be the appropriate statistic for testing for parameter constancy.

Whilst such a test statistic would be easy to compute in theory, it does require information on $y$, $\mathbf{x}$, and $\mathbf{z}$ for the two regimes, and particularly for large-scale macroeconometric models this may not be the case. This leads again to the consideration of forecast-encompassing test statistics, since in using the available forecast information they are likely to be feasible. It is important though to know how they fit into the above testing framework.

Two such hypotheses have been proposed, and their properties will now be discussed. Firstly, an obvious check on the validity of $M_1$ is to test whether its forecast errors are orthogonal to $Z_2$. This is the essence of the hypothesis of forecast model encompassing:

$$H_{fme}\colon\ E_\theta[Z_2'(\mathbf{y}_2 - X_2\boldsymbol{\beta}_1)] = Z_2'(Z_2 - X_2P_2)\boldsymbol{\mu}_2 + Z_2'X_2\Delta\boldsymbol{\beta} = \mathbf{0}\,, \tag{9.18}$$

when $\boldsymbol{\theta}$ is a vector consisting of all the parameters of $M_c^*$ so that $E_\theta$ denotes expectation with respect to $M_c^*$. Note that, if use is made of the relationship between the parameters of the distribution of $y$ conditional on $\mathbf{x}$ and $\mathbf{z}$ as defined in the joint distribution of these variables, then it can be shown that $H_{fme}$ is equivalent to $Z_2'Z_2[(\boldsymbol{\gamma}_2 - \boldsymbol{\gamma}_{\alpha 2}) + \Pi_2\Delta\boldsymbol{\beta}] = \mathbf{0}$. Hence this forecast-encompassing hypothesis also combines elements of parameter constancy ($\Delta\boldsymbol{\beta} = \mathbf{0}$) and parametric encompassing [$\boldsymbol{\mu}_2 = \mathbf{0}$ or equivalently $(\boldsymbol{\gamma}_2 - \boldsymbol{\gamma}_{\alpha 2}) = \mathbf{0}$], which is a desirable characteristic if it is not possible to test separately these hypotheses. If $\Delta\boldsymbol{\beta} = \mathbf{0}$ is known, or assumed, then $H_{fme}$ is the hypothesis that $M_1$ completely parametrically encompasses $M_2$ in the forecast period, i.e., $(\boldsymbol{\gamma}_2 - \boldsymbol{\gamma}_{\alpha 2}) = \mathbf{0}$ provided that $n > k + \ell$, or that $Z_2(\boldsymbol{\gamma}_2 - \boldsymbol{\gamma}_{\alpha 2}) = \mathbf{0}$ if $n < k + \ell$. As mentioned above, Ericsson (1988), in first proposing a form of $H_{fme}$, considered the case in which $\boldsymbol{\mu}_1 = \mathbf{0}$, $\Delta\boldsymbol{\lambda} = \mathbf{0}$, and $n > k + \ell$ so that a test of the hypothesis can be implemented by regressing $\mathbf{y}_0$ on $X_0$ and $Z^*$, using an $F$-test statistic for the hypothesis that the coefficients of $Z^*$ are zero. However, whenever $n < k + \ell$ the Ericsson form of this hypothesis can be tested by regressing $\mathbf{y}_0$ on $X_0$ and $D' = (0, I_n)$, and testing the hypothesis that the coefficients of $D$ (namely, $Z_2\boldsymbol{\mu}_2$ which is $\boldsymbol{\nu}$ when $\boldsymbol{\mu}_1 = \mathbf{0}$ and $\Delta\boldsymbol{\lambda} = \mathbf{0}$) are zero. Note though, that in the context of $M_c^{**}$, if $\boldsymbol{\mu}_1 = \mathbf{0}$ is imposed (either *a priori* or after not being rejected within sample), the hypothesis that the coefficients of $D$ are zero may be rejected because $X_2\Delta\boldsymbol{\lambda} \neq \mathbf{0}$ or $Z_2\boldsymbol{\mu}_2 \neq \mathbf{0}$, and that it is possible for $X_2\Delta\boldsymbol{\lambda} \neq \mathbf{0}$ even though $X_2\Delta\boldsymbol{\beta} = \mathbf{0}$ has not been rejected, since $X_2\Delta\boldsymbol{\lambda} = \mathbf{0}$ and $X_2\Delta\boldsymbol{\beta} = \mathbf{0}$ are only equivalent if $\boldsymbol{\mu}_1 = \boldsymbol{\mu}_2 = \mathbf{0}$. It is therefore preferable to test the hypothesis $M_1\mathcal{E}_p M_c^{**}$ via the procedure that tests (i) $\boldsymbol{\nu} = \mathbf{0}$, and if this is not rejected (ii) $\boldsymbol{\mu} = \mathbf{0}$. The alternative procedure of testing sequentially the hypotheses (i) $X_2\Delta\boldsymbol{\beta} = \mathbf{0}$, (ii) $\boldsymbol{\mu}_1 = \mathbf{0}$, and (iii) $Z_2\boldsymbol{\mu}_2 = \mathbf{0}$ has the drawback of not using *all* the available information in testing the first hypothesis in the sequence, thus opening up the possibility of invalid conditional inferences beyond those arising from type I error. Furthermore, it must be emphasized that although it is relatively easy to implement tests for forms of $H_{fme}$ that involve combinations of $H_p$ and $H_{cp}$, and as such (subject to adopting an appropriate testing procedure) they can have desirable properties, it has to be remembered that $H_{fme}$ is necessary, *not sufficient*, for $H_p \cap H_{cp}$.

Another feature of $H_{fme}$ is that tests of it require data on $X_2$ and $Z_2$, which may not always be available, particularly for large-scale macroeconometric models. When this is the case a second form of forecast-encompassing hypothesis is relevant:

$$H_{fve}\text{: } E_\theta[\boldsymbol{\gamma}_1' Z_2'(\mathbf{y}_2 - X_2\boldsymbol{\beta}_1)] = \boldsymbol{\gamma}_1' Z_2'(Z_2 - X_2P_2)\boldsymbol{\mu}_2 + \boldsymbol{\gamma}_1' Z_2' X_2 \Delta\boldsymbol{\beta} = \mathbf{0}\,, \tag{9.19}$$

which is equivalent to $\boldsymbol{\gamma}_1' Z_2' Z_2[\Pi_2\Delta\boldsymbol{\beta} + (\boldsymbol{\gamma}_2 - \boldsymbol{\gamma}_{\alpha 2})] = \mathbf{0}$. This hypothesis asserts that the forecast error from $M_1$ [namely, $\mathbf{f}_1 = (\mathbf{y}_2 - X_2\boldsymbol{\beta}_1)$] is orthogonal to the forecast of $M_2$, namely, $Z_2\hat{\boldsymbol{\gamma}}_1$ (which is equivalent in mean to $Z_2\boldsymbol{\gamma}_1$, though the former has a nondegenerate sampling distribution under $M_c^{**}$). This is analogous to the hypothesis that the best combined forecast of $\mathbf{y}_2$ is one that gives a weight of unity to that from $M_1$ ($X_2\hat{\boldsymbol{\beta}}_1$) and a weight of zero to that from $M_2$ ($Z_2\hat{\boldsymbol{\gamma}}_1$). This is, in fact, the way that Chong and Hendry (1986) motivated their forecast-encompassing test statistic, the behavior of which they analyzed in the constant parameter case, i.e., $\Delta\boldsymbol{\beta} = \mathbf{0}$, $\Delta\boldsymbol{\gamma} = \mathbf{0}$, and $\omega_1^2 = \omega_2^2$. However, the form of their test statistic, a "$t$"-test of the hypothesis that the coefficient of $Z_2\hat{\boldsymbol{\gamma}}_1$ is zero in the regression of $\hat{\mathbf{f}}_1 = (\mathbf{y}_2 - X_2\hat{\boldsymbol{\beta}}_1)$ on $Z_2\hat{\boldsymbol{\gamma}}_1$, is still applicable for the more general hypothesis given in (9.19). Noting that $\boldsymbol{\gamma}_1' Z_2' Z_2[\Pi_2\Delta\boldsymbol{\beta} + (\boldsymbol{\gamma}_2 - \boldsymbol{\gamma}_{\alpha 2})]$ is asymptotically equivalent to $\boldsymbol{\gamma}_1' Z_2' X_2\Delta\boldsymbol{\beta} + \boldsymbol{\gamma}_2' Z_2' Z_2(\boldsymbol{\gamma}_2 - \boldsymbol{\gamma}_{\alpha 2}) - \Delta\boldsymbol{\gamma}' Z_2' Z_2(\boldsymbol{\gamma}_2 - \boldsymbol{\gamma}_{\alpha 2})$, it then follows that $H_{fve}$ is asymptotically equivalent to

$$\boldsymbol{\gamma}_1' Z_2' X_2\Delta\boldsymbol{\beta} - \Delta\boldsymbol{\gamma}' Z_2' Z_2(\boldsymbol{\gamma}_2 - \boldsymbol{\gamma}_{\alpha 2}) - \frac{1}{2}(\tau_2^2 - \sigma_2^2 - \boldsymbol{\beta}_2'\Omega_2\boldsymbol{\beta}_2) \tag{9.20}$$

using *Lemma 2* of Mizon and Richard (1986). Hence, $H_{fve}$ combines $H_{chow}$ type parameter constancy (namely, $X_2\Delta\boldsymbol{\beta} = \mathbf{0}$ and $Z_2\Delta\boldsymbol{\gamma} = \mathbf{0}$) and variance-encompassing $H_v$ within the forecast period (i.e., $\tau_2^2 - \sigma_2^2 - \boldsymbol{\beta}_2'\Omega_2\boldsymbol{\beta}_2 = 0$). Therefore, the forecast-encompassing test statistic proposed by Chong and Hendry (1986) is (i) numerically easy to calculate; (ii) feasible to compute, even for large-scale macroeconometric models, since it only requires model forecasts (albeit one-step-ahead forecasts in dynamic models); and (iii) valuable in model evaluation because it has an implicit null hypothesis that combines parameter constancy and variance encompassing. Despite these considerable advantages though, it remains a hypothesis that is necessary, *but not sufficient*, for $M_1\mathcal{E}_p M_c^*$.

Finally, note that:

$$\begin{aligned} \text{MSFE}_1 &= E_\theta(y_t - \mathbf{x}_t'\boldsymbol{\beta}_1)^2 \\ &= \{\mathbf{z}_t'[(\boldsymbol{\gamma}_2 - \boldsymbol{\gamma}_{\alpha 2}) + \Pi_2\Delta\boldsymbol{\beta}]\}^2 + (\Delta\boldsymbol{\beta} - P_2\boldsymbol{\mu}_2)'\Omega_2(\Delta\boldsymbol{\beta} - P_2\boldsymbol{\mu}_2) + \omega_2^2 \end{aligned} \tag{9.21}$$

and

$$\text{MSFE}_2 \quad = \quad E_\theta(y_t - \mathbf{z}_t'\boldsymbol{\gamma}_1)^2 = (\mathbf{z}_t'\Delta\boldsymbol{\gamma})^2 + \boldsymbol{\lambda}_2'\Omega_2\boldsymbol{\lambda}_2 + \omega_2^2 \tag{9.22}$$

for $t \in \{T+1, \ldots, T+n\}$ so that

$$\text{MSFE}_1 - \text{MSFE}_2 = \tag{9.23}$$
$$[\Pi_2\Delta\boldsymbol{\beta} + (\boldsymbol{\gamma}_2 - \boldsymbol{\gamma}_{\alpha 2}) - \Delta\boldsymbol{\gamma}]'\mathbf{z}_t\mathbf{z}_t'[\Pi_2\Delta\boldsymbol{\beta} + (\boldsymbol{\gamma}_2 - \boldsymbol{\gamma}_{\alpha 2}) + \Delta\boldsymbol{\gamma}] + \boldsymbol{\beta}_1'\Omega_2(\boldsymbol{\beta}_1 - 2\boldsymbol{\lambda}_2).$$

Analysis of (9.23) reveals that it is not possible, in general, to sign $\text{MSFE}_1 - \text{MSFE}_2$ uniquely. In fact, only if there are constant parameters *and* one model encompasses the other, is it possible to determine the sign of this difference. For example, if $\Delta\boldsymbol{\beta} = \mathbf{0}$, $\Delta\boldsymbol{\gamma} = \mathbf{0}$, and $(\boldsymbol{\gamma} - \boldsymbol{\gamma}_\alpha) = \mathbf{0}$, then $\text{MSFE}_1 - \text{MSFE}_2 \leq \mathbf{0}$, and also $E(\widehat{\text{MSFE}}_1 - \widehat{\text{MSFE}}_2) \leq \mathbf{0}$ when $\widehat{\text{MSFE}}_1 = (\mathbf{y}_2 - X_2\boldsymbol{\beta}_1)'(\mathbf{y}_2 - X_2\boldsymbol{\beta}_1)/n$ and $\widehat{\text{MSFE}}_2 = (\mathbf{y}_2 - Z_2\boldsymbol{\gamma}_1)'(\mathbf{y}_2 - Z_2\boldsymbol{\gamma}_1)/n$. Therefore, even within the simple framework of $M_c^*$, choosing a model because it has minimum $\widehat{\text{MSFE}}$ does not ensure that it is a congruent model—i.e., does not ensure that it would parsimoniously encompass $M_c^*$. In particular, it does not guarantee that it has constant parameters, or that it encompasses rival models that are nested within $M_c^*$. However, if $M_1\mathcal{E}_pM_c^*$ when $M_1$ has constant parameters then it will dominate $M_2$ on the criterion of MSFE. Another aspect of this last result is revealed by considering

$$H_{msfe}\text{: } E(\text{MSFE}_1 - \text{MSFE}_2) = -\boldsymbol{\beta}'\Omega\boldsymbol{\beta}, \tag{9.24}$$

which corresponds to the forecast-encompassing hypothesis $H_b^*$ in Ericsson (1988). A comparison of (9.20) and (9.24) shows that if $\Delta\boldsymbol{\beta} = \mathbf{0}$ and $\Delta\boldsymbol{\gamma} = \mathbf{0}$ then $H_{fve}$ and $H_{msfe}$ are equivalent.

## 9.4 Conclusion

In this chapter the potential value of the recently proposed forecast-encompassing test statistics in model evaluation has been assessed. It has been shown that, just as minimum MSE (or MSFE) is necessary (but not sufficient) for model congruence, so the requirement that $M_1$ forecast variance-encompasses $M_2$ is a necessary condition for $M_1$ to have constant parameters and $M_1\mathcal{E}M_2$. Although forecast variance-encompassing tests are easily calculated, and have the advantage of having an implicit null hypothesis, which is a combination of parameter constancy and parametric encompassing, it is preferable to test these last two hypotheses separately when this is feasible.

We have concentrated on model evaluation when there are two regimes, the sample and forecast periods, respectively. In practice it will be advisable to check for the existence of more than one structural break—by, for example, using recursive estimation techniques, such as those implemented in PC-GIVE [see Hendry (1989)]. In addition, tests should also be performed to assess a model's congruence with other sources of information, e.g., tests for serial correlation, heteroscedasticity, non-normal skewness and kurtosis in residuals, and wrong functional form. Whilst this chapter has sought to promote practical and feasible forecast-encompassing tests, particularly in the context of the evaluation of large-scale macroeconometric models, it has not sought to undermine the use of "ideal" standards in model evaluation.

Finally, it should be pointed out that further research is needed to provide a detailed extension of the results in this contribution to cover nonlinear and dynamic econometric models; Mariano and Brown (1983) provide useful analysis of this problem. In addition, the recent results of Andrews and Fair (1988) provide a framework within which to develop these results.

## Acknowledgments

The financial support for this research from the ESRC Macroeconomic Research Consortium (ESRC grant B01250024) is gratefully acknowledged.

## References

Akaike, H. (1973), Information theory and an extension of the maximum likelihood principle, pp. 267-281 in B.N. Petrov and F. Csaki (eds.), *2nd International Symposium on Information Theory.* Budapest: Akademiai Kiado.

Anderson, G.J. and Mizon, G.E. (1983), Parameter constancy tests: Old and new. Discussion Paper in Economics and Econometrics, No. 8325, Southampton University, forthcoming in *Oxford Bulletin of Economics and Statistics.*

Anderson, G.J. and Mizon, G.E. (1989), What can statistics contribute to the analysis of economic structural change? pp. 3-21 in P. Hackl (ed.), *Statistical Analysis and Forecasting of Economic Structural Change.* Berlin: Springer-Verlag.

Andrews, D.F. and Fair, R.C. (1988), Inference in nonlinear econometric models with structural change. *Review of Economic Studies*, **55**, 615-640.

Baillie, R.T. (1979), The asymptotic mean squared error of multistep prediction from the regression model with autoregressive errors. *Journal of the American Statistical Association*, **47**, 175-184.

Chong, Y.Y. and Hendry, D.F. (1986), Econometric evaluation of linear macro-economic models. *Review of Economic Studies*, **53**, 671-690.

Chow, G.C. (1960), Tests of equality between sets of coefficients in two linear regressions. *Econometrica*, **28**, 591-605.

Cox, D.R. (1961), Tests of separate families of hypotheses, pp. 105-123 in *Proceedings of the Fourth Berkeley Symposium on Mathematical Statistics and Probability, Volume 1.* Berkeley: University of California Press.

Davidson, R. and MacKinnon, J.G. (1981), Several tests for model specification in the presence of alternative hypotheses. *Econometrica*, **49**, 781-793.

Engle, R.F. (1982), Autoregressive conditional heteroscedasticity, with estimates of the variance of the United Kingdom inflation. *Econometrica*, **50**, 987-1008.

Ericsson, N.R. (1988), Parameter constancy, mean squared forecast errors, and measuring forecast performance: An exposition. Mimeo.

Ericsson, N.R. and Marquez, J.R. (1989), Exact and approximate multi-period mean square forecast errors for dynamic econometric models. International Finance Discussion Papers No. 348, Board of Governors of the Federal Reserve System.

Fair, R.C. (1985), Evaluating the predictive accuracy of models, chapter 33 in Z. Griliches and M.D. Intriligator (eds.), *The Handbook of Econometrics.* Amsterdam: North-Holland.

Hannan, E.J. and Quinn, B.G. (1979), The determination of the order of an autoregression. *Journal of the Royal Statistical Society*, **B-41**, 190-195.

Harvey, A.C. (1981), *The Econometric Analysis of Time Series.* Oxford: Philip Allan.

Hendry, D.F. (1974), Stochastic specification in an aggregate demand model of the United Kingdom. *Econometrica*, **42**, 559-578.

Hendry, D.F. (1979), Predictive failure and econometric modelling in macroeconomics: The transactions demand for money, pp. 217-242 in P.A. Ormerod (ed.), *Economic Modelling.* London: Heinemann Education Books.

Hendry, D.F. (1987), Econometric methodology: A personal perspective, chapter 10 in T.F. Bewley (ed.), *Advances in Econometrics.* Cambridge: Cambridge University Press.

Hendry, D.F. (1988), The encompassing implications of feedback versus feedforward mechanisms in econometrics. *Oxford Economic Papers*, **40**, 132-149.

Hendry, D.F. (1989), PC-GIVE: An interactive econometric modelling system (version 6.0). Institute of Economics and Statistics, University of Oxford.

Hendry, D.F. and Mizon, G.E. (1990), Procrustean econometrics: Or stretching and squeezing the data, in C.W.J. Granger (ed.), *Modeling Economic Series. Readings in Econometric Methodology.* Oxford: Oxford University Press.

Hendry, D.F. and Richard, J.-F. (1982), On the formulation of empirical models in dynamic econometrics. *Journal of Econometrics*, **20**, 3-33.

Hendry, D.F. and Richard, J.-F. (1987), Recent developments in the theory of encompassing, in B. Cornet and H. Tulkens (eds.), *Contributions to Operations Research and Econometrics. The XXth Anniversary of CORE.* Cambridge: MIT Press.

Kiviet, J.F. (1986), On the rigour of some misspecification tests for modelling dynamic relationships. *Review of Economic Studies*, **53**, 241-261.

Mallows, C.L. (1973), Some comments on $C_p$. *Technometrics*, **15**, 661-676.

Mariano, R.S. and Brown, B.W. (1983), Prediction-based tests for misspecification in non-linear simultaneous systems, in S. Karlin, T. Amemiya and L. Goodman (eds.), *Studies in Econometrics, Time Series and Multivariate Statistics.* Orlando: Academic Press.

Mizon, G.E. (1984), The encompassing approach in econometrics, pp. 135-172 in D.F. Hendry and K.F. Wallis (eds.), *Econometrics and Quantitative Economics.* Oxford: Basil Blackwell.

Mizon, G.E. and Richard, J.-F. (1986), The encompassing principle and its application to testing non-nested hypotheses. *Econometrica*, **54**, 657-678.

Pagan, A. (1989), On the role of simulation in the statistical evaluation of economic models. *Journal of Econometrics*, **40**, 125-139.

Pesaran, M.H. and Smith, R.P. (1985), Evaluation of macroeconometric models. *Economic Modelling*, **2**, 125-134.

Phillips, G.D.A. and McCabe, B.P. (1983), The independence of tests for structural change in regression models. *Economics Letters*, **12**, 283-287.

Pindyck, R.S. and Rubinfeld, D.L. (1981), *Econometric Models and Economic Forecasts*, 2nd edition. New York: McGraw-Hill.

Schmidt, P. (1974), The asymptotic distribution of forecasts in the dynamic simulation of an econometric model. *Econometrica*, **42**, 303-309.

Schwarz, G. (1978), Estimating the dimension of a model. *The Annals of Statistics*, **6**, 461-464.
Wallis, K.F. (1989), Macroeconomic forecasting: A survey. *Economic Journal*, **99**, 28-61.
Wallis, K.F. *et al.* (1984, 1985, 1986, 1987), Models of the UK economy, *Annual Reviews by the ESRC Macroeconomic Modelling Bureau*. Oxford: Oxford University Press.

CHAPTER 10

# An Introduction to Time-Varying Parameter Cointegration

*Clive W.J. Granger and Hahn S. Lee*

## Summary

The idea of cointegration is extended by allowing the coefficients in a dynamic model to vary over time. Each variable in the system is individually nonstationary, but there may exist a common factor or "permanent component" of the system. A richer concept of equilibrium in dynamic structure is introduced, and an application of the varying parameter regression (VPR) model to cointegrating regression is developed for testing the existence of such a relationship. Some properties of the time-varying parameter (TVP) cointegrated process and of the estimation procedure are suggested, and an empirical example is illustrated by using US data on prices and wages.

## 10.1 Introduction

Much of traditional time series econometrics have considered just stationary series, yet it is an empirical fact that many actual macroeconomic time series appear not to be stationary but to have the property that the differenced series can be classified as stationary. Thus the level of these series is the accumulation of stationary components, and may be denoted as $I(1)$. A series that needs to be differenced $d$ times to achieve stationarity is denoted $I(d)$, so that a stationary series can be denoted $I(0)$.

A pair of $I(1)$ series, $x_t$ and $y_t$, when plotted on the $(X, Y)$ plane may appear to lie generally around a line $y = Ax + m$. If $x_t$ and $y_t$ are a pair of macroeconomic series, the economy might appear to *prefer* to stay near to this line and this preference may

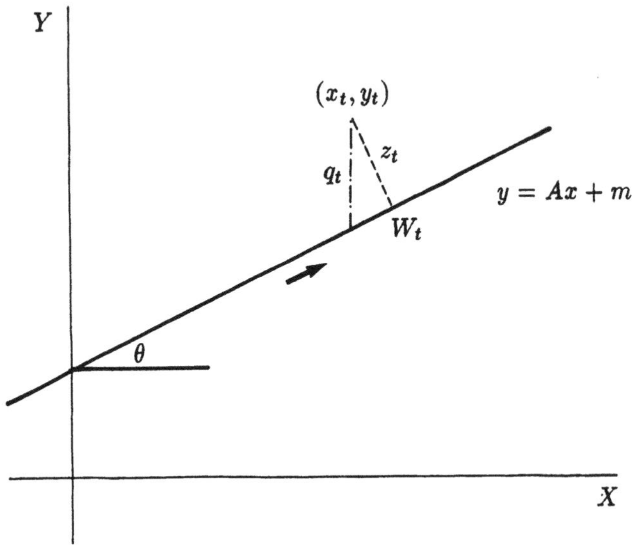


*Figure 10.1:* Illustration of $I(1)$ series having an attractor.

reflect the working of some markets or the use of effective government policies. This line can be thought of as an attractor and might be associated with some types of long-run equilibrium. A sufficient condition for the line to be an attractor is that the signed minimum distance, $z_t$, from a point $(x_t, y_t)$ to the line is $I(0)$.

A simple way to generate series having this property is first to generate a pair of independent series $W_t$ and $z_t$ such that $W_t$ is $I(1)$ and $z_t$ is $I(0)$, and then to form

$$\left.\begin{aligned} x_t &= W_t - z_t \sin\theta \\ y_t &= AW_t + m + z_t \cos\theta \end{aligned}\right\} , \tag{10.1}$$

where $\tan\theta = A$. For example, $W_t$ and $z_t$ could each be generated by

$$\left.\begin{aligned} W_t &= W_{t-1} + \varepsilon_t \\ z_t &= \alpha z_{t-1} + \xi_t \end{aligned}\right\} , \tag{10.2}$$

with $|\alpha| < 1$, and $\varepsilon_t$ and $\xi_t$ independent white noises. With this construction, it may be noted that

$$\begin{aligned} q_t &= y_t - Ax_t - m \\ &= z_t(\cos\theta + A\sin\theta) = z_t/\cos\theta \,. \end{aligned} \tag{10.3}$$

*Figure 10.1* shows the relationship of the variables in the generating mechanism. The generating mechanism (10.1) has two factors, $W_t$ and $z_t$. $W_t$ can be associated with the "permanent component" of the system, and $z_t$ with the "transitory component" provided that an extra identifying condition is imposed, such as $W_t$, $z_t$ being independent series.

If $x_t$ and $y_t$ are both $I(1)$, but there is a linear combination which is $I(0)$, the series are said to be cointegrated, as discussed in Granger (1986). A simple method of testing for cointegration is to estimate $A$ and $m$ from the regression

$$y_t = Ax_t + m + q_t\,, \tag{10.4}$$

where $q_t$ is the residual, and then to use a Dickey-Fuller-type procedure to see if $\hat{q}_t$ is $I(0)$. This method is discussed in Engle and Granger (1987), but more powerful methods are now available [see Engle and Yoo (1989) for a survey of the literature].

There are two important implications of cointegration. First, there must be a common factor representation, with a single $I(1)$ factor. The generating mechanism (10.1) provides an example, with $W_t$ being this common factor. Second, there will be an error-correction mechanism of the form

$$\begin{aligned} \Delta x_t &= \gamma_1 z_{t-1} + \text{lags}\,(\Delta x_t, \Delta y_t) + \text{residual} \\ \Delta y_t &= \gamma_2 z_{t-1} + \text{lags}\,(\Delta x_t, \Delta y_t) + \text{residual}\,, \end{aligned}$$

where the residuals are jointly white noises and a required condition is that at least one of $\gamma_1$, $\gamma_2$ should be nonzero. These implications are discussed and interpreted in Granger (1986).

These now standard cointegration models use constant parameters to consider a stable long-run relationship among economic time series data. In a dynamic economy, however, one may expect to encounter structural shifts as taste, technology, or policy changes over time. Thus the possibility of time-varying parameter cointegration needs to be considered. The natural generalization of constant parameter cointegration is that there exists a sequence $A_t$ and $m_t$ such that the signed distance $z_t$ from $(x_t, y_t)$ to the line $y_t = A_t x_t + m_t$ is $I(0)$—or its generalization TVP-$I(0)$ defined in Section 10.2, while $x_t$ and $y_t$ are individually $I(1)$—or the generalized TVP-$I(1)$ also defined below.

In the generating mechanisms of the form (10.1), the "permanent component" of the system, $W_t$, and the "transitory component", $z_t$, could be generated without change, but $A$ and $m$ might be replaced by $A_t$ and $m_t$, respectively, giving an attractor that varies through time. There may be other nonstationarities in the system, such that $\Delta W_t$ and $z_t$ are each generated by AR processes having time-varying coefficients. It is clearly difficult to cope with all possibilities with a limited quantity of data.

It can be noted from (10.3) that $z_t$ is $I(0)$ if $q_t = y_t - A_t x_t - m_t$ is $I(0)$, since $z_t = q_t \cos\theta_t$ where $\tan\theta_t = A_t$. It is generally true that the product of a zero mean $I(0)$ and an independent $I(1)$ variable is $I(0)$ in mean, or short memory as defined in Section 10.2 [see Granger and Hallman (1988)]. Even if $A_t$ is $I(0)$, $\cos\theta_t$ is bounded in variance and has other $I(0)$ properties. Thus, a test that $q_t$ is $I(0)$ is sufficient for $z_t$ being $I(0)$.

The question discussed in this chapter is how to test if a pair of series are cointegrated with time-varying parameters, and then how to estimate the attractor. In Section 10.2, the basic theory of time-varying processes and the properties of TVP cointegration processes are discussed, and Section 10.3 suggests various TVP cointegration models. There are two basically different cases, in which $A_t$ and $m_t$ change deterministically and in which at least one sequence is stochastic. The testing procedures of TVP cointegration, by applying the Kalman filter technique, are discussed in Section 10.4. Section 10.5 presents a brief conclusion and suggestion for further research.

## 10.2 Nonstationary Processes

It is shown in Cramér (1961) that any univariate process $x_t$ with unconditional mean zero and bounded variance for all $t$ will have generalized Wold representation of the form

$$x_t = \sum_{j=0}^{\infty} C_{jt}\varepsilon_{t-j}\,, \tag{10.5}$$

where $C_{jt}$ is a deterministic double sequence with the constraint that $\sum_{j=0}^{\infty} C_{jt}^2$ is finite and bounded above for all $t$, and $\varepsilon_t$ is white noise with mean zero.

It is possible to associate an evolutionary spectrum with such a process, as discussed by Priestley (1965, 1981), of the form

$$f_t(\omega) = \frac{1}{2\pi}\,|\sum_j C_{jt}e^{i\omega j}|^2\,. \tag{10.6}$$

The evolutionary spectrum has the same physical interpretation as the spectrum of a stationary process, i.e., that it describes a distribution of power over frequency. But, while the latter is determined by the behavior of the process over time, the former represents specifically the spectral content of the process in the neighborhood of the time instant $t$ [see Priestley (1981), p. 825].

**Definition 10.1** *If a series with no deterministic component has an evolutionary spectrum, $f_t(w)$, that is bounded above and positive for all t and $\omega$, the process is said to be time-varying parameter $I(0)$, denoted TVP-$I(0)$.*

An example of such a process is

$$x_t = \rho_t x_{t-1} + \varepsilon_t\,,$$

where $|\rho_t| < 1$ for all $t$, which corresponds to that $C_{0t} = 1$, $C_{jt} = \rho_t\rho_{t-1}\cdots\rho_{t-j+1}$. It will have the property that it is short memory in mean, in the sense that if $f_{t,h}$ is the optimum linear forecast of $x_{t+h}$ made at time $t$, then

$$E[f_{t,h}^2] \to 0 \quad \text{as } h \uparrow \infty\,, \quad \text{for all } t\,. \tag{10.7}$$

This follows directly from the above result, as $f_{t,h} = \sum_{j=0}^{\infty} C_{j+h,t}\,\varepsilon_{t-j}$ has mean zero and variance that decreases to zero as $h$ increases.

The natural generalization of $I(1)$ to TVP-$I(1)$ is a process $x_t$ such that an $N$-vector process $x_t$ with no deterministic components has a generalized form of Wold representation

$$x_t = C_t(B)\,W_t\,, \tag{10.8}$$

where $W_t = \frac{1}{1-B}\,\varepsilon_t = \sum_{s=1}^{t}\varepsilon_s$ and $C_t(B)$ is a time-varying matrix of linear functions of the lag operator $B$, and

1. $E(\varepsilon_t) = 0$ for all $t$.
2. $E(\varepsilon_t\varepsilon_s') = 0$ for all $t \neq s$.
3. $E(\varepsilon_t\varepsilon_t') = \Omega$ for all $t$.
4. $C_t(0) = I_N$ for all $t$.
5. If $C_t(B) = \sum_j C_{jt}B^j$, then $\sum_j C_{jt}\Omega C_{jt}' < \infty$ so that the variance of $\tilde{x}_t = C_t(B)\,\varepsilon_t$ is finite.

The TVP-$I(1)$ process $x_t$ will be long memory in mean, so that the condition (10.7) does not hold. The concepts of short and long memory in mean, as generalizations of $I(0)$ and $I(1)$, are useful for discussion of nonlinear cointegration [see Granger and Hallman (1988)].

The moving average polynomial $C_t(B)$ can always be expressed as

$$C_t(B) = C_t(1) + C_t^*(B)(1-B)$$

by rearranging the terms. Using this expression, we can rewrite (10.8) as

$$x_t = [C_t(1) + C_t^*(B)(1-B)]\,W_t = C_t(1)\,W_t + C_t^*(B)\,\varepsilon_t\,. \tag{10.9}$$

TVP cointegration, as defined in Section 10.1, will occur if there exists a vector $\alpha_t$ such that

$$\alpha_t'C_t(1) = 0 \quad \text{for } all\ t\,, \tag{10.10}$$

so that the rank of $C_t(1)$ is less than $N$ for all $t$. When this rank is the same for all $t$, the cointegrating rank can be defined as $r = \text{rank}\{C_t(1)\}$. The attractor process (TVP equilibrium error process) will then be derived as

$$z_t = \alpha_t' x_t\,, \tag{10.11}$$

provided that $\alpha_t'C_t^*(B)\varepsilon_t$ is TVP-$I(0)$. Using the equation (10.9), the following property of TVP cointegrated process can be obtained:

**Proposition 10.1** *Suppose there is TVP cointegration for a vector $x_t$ of TVP-$I(1)$ processes such that $C_t(B) \neq C_s(B)$ for some $t \neq s$ in equation (10.8). Then $x_t$ will not be cointegrated "on average".*

*Proof:* Using the equation $C_t(B) = C_t(1) + C_t^*(B)(1-B)$ and (10.10), it is straight forward to show that if $x_t$ is TVP cointegrated then the spectrum of $z_t$ is

$$f_{z_t}(\omega) = \alpha_t' C_t^*(z) \Omega \overline{C_t^*(z)} \alpha_t \quad \text{where } z = e^{-i\omega}, \tag{10.12}$$

where $f_{z_t}(\omega)$ may be considered as an evolutionary spectrum. When $x_t$ are TVP cointegrated, the equation (10.10) holds for all $t$ such that $\alpha_i \neq \alpha_j$ for some $i \neq j$, and we know that $z_t = \alpha_t' x_t$ is TVP-$I(0)$. Let $\tilde{z}_t = \tilde{\alpha}' x_t$ where $\tilde{\alpha}$ is a constant independent of $t$. Using the equation (10.9), the integrated part of the average spectrum of $\tilde{z}_t$ will then be

$$\frac{1}{N} \sum_{t=1}^{N} \frac{\tilde{\alpha}' C_t(1) \Omega C_t(1)' \tilde{\alpha}}{|1-z|^2}. \tag{10.13}$$

As $\Omega$ is a positive definite matrix, this is just a sum of nonnegative quadratic forms. Hence, for this integrated part to vanish at the zero frequency, we require that $\tilde{\alpha}' C_t(1) = 0$ for all $t$, which is clearly not possible when $C_t(B) \neq C_s(B)$ for some $t \neq s$ so that $C_t(1) \neq C_s(1)$ in general. □

If $C_t^*(B)$ is absolutely summable (i.e., $\sum_{j=0}^{\infty} |C_{jt}^*| < \infty$), then the TVP cointegrated process $x_t$ has a "common factor representation", which is just the equation (10.9). [If $C_t(B)$ is the moving average representation polynomial matrix implied by a finite ARMA model with stable roots, then $C_t(B)$ is 1-summable (i.e., $\sum_j |C_{jt}| < \infty$), hence it follows that $C_t^*(B)$ is absolutely summable; see Stock (1987).] Equation (10.9) shows that the TVP cointegrated process $x_t$ can be written in terms of a reduced number of common $I(1)$ factors $W_t$, plus an additional transitory component $C_t^*(B)\varepsilon_t$. It is now clear that the bivariate generating mechanism (10.1) is a special case of (10.9).

## 10.3 Various TVP Cointegration Models

Consider initially the case with $m$ constant, and equal to zero for convenience, so that a sequence $A_t$ exists such that $q_t = y_t - A_t x_t$ is $I(0)$. The TVP cointegration relationship with $m = 0$ will then be

$$y_t = A_t x_t + q_t, \tag{10.14}$$

where $q_t$ is $I(0)$. Two simple but interesting, relevant cases are

1. $A_t$ is a deterministic, known sequence such as a constant plus $\cos \lambda t$, $t^\delta$ or some parametric function of time, or a seasonal sequence, and

2. $A_t = a + bD_t$, where $D_t$ is some known, observed variable.

The equation (10.14) can be rewritten as

$$y_t = ax_t + b\tilde{x}_t + q_t\,, \tag{10.15}$$

where $\tilde{x}_t = f(t)x_t$ with $f(t)$ being a deterministic function of $t$ or an observed variable $D_t$. Considering the fact that the product of a pair of independent $I(1)$ series is long memory, each case is easily tested by just treating $\tilde{x}_t$ as another $I(1)$ variable, provided that $D_t$ is $I(1)$. The remaining parameters are just constants and so can be estimated super-efficiently as usual. Hence the standard testing procedures can be used in this case. It is also clear that $x_t$ and $y_t$ are not cointegrated with constant parameters unless $\tilde{x}_t$ is $I(0)$. However, if $D_t$ is $I(0)$ with mean zero, the situation is very different, as $\tilde{x}_t$ will then be $I(0)$ or at least short memory in mean. In this case, $\tilde{x}_t$ will not be involved in the cointegration and $x_t$ and $y_t$ will be cointegrated with constant parameters.

If there are structural changes at known times, this can be included in the regression model (10.4), using appropriate zero-one dummies. In practice, however, the possibility of changes at unknown times or, more generally, with $A_t$ changing in unknown ways—because a causal variable such as $D_t$ is unobservable—needs to be considered. In this case, a sensible, parsimonious specification has to be applied to approximate the time series behavior of $A_t$. An obvious way to proceed is to use the assumption that $A_t$ is given by a constant plus a low-order ARIMA process, for example, an AR(1) process $\tilde{A}_t$ given by

$$\tilde{A}_t = \rho\tilde{A}_{t-1} + \xi_t\,, \tag{10.16}$$

where $\xi_t$ is i.i.d. with $E(\xi_t) = 0$, or by a more general stochastic model. Such a scheme is especially suitable whenever little information is available on the exact cause of parameter instability and smooth rather than sharp variation in coefficients are desired. Moreover, this kind of specification has the advantage that it can be estimated by using the well-known Kalman filter algorithm. This kind of procedure should be considered as an attempt to model unobserved deterministic or stochastic causes of variation in $A_t$. If $|\rho| < 1$ in (10.16), then one is back to an $I(0)$ time-varying parameter and this was shown above not to affect any constant parameter cointegration that may exist between $x_t$ and $y_t$.

The case that may be expected to occur frequently is when $\rho = 1$, so that any parameter changes may persist with no inherent tendency to return to constant values. In this case, the Kalman filter estimate of $A_t$ will contain a unit-root and hence may cause a singularity that will make it difficult to distinguish between a real and a spurious cointegration. Suppose that $x_t$ and $y_t$ are independent $I(1)$ series both with positive values. Then the choice of $A_t$ such that $\hat{A}_t = R_t + \lambda_t$, where $R_t = y_t/x_t$ and $\lambda_t$ is $I(0)$ with mean zero, gives

$$\hat{q}_t = y_t - \hat{A}_t x_t = \lambda_t x_t\,,$$

which is short memory and indistinguishable from any $I(0)$ in mean series when $\lambda_t$ is unobserved. Hence this choice of $A_t$ will result in a spurious, apparent TVP cointegration. The difficulty associated with this situation occurs because an actual TVP cointegration model has similar properties to this spurious case. Suppose now that $x_t$ and $y_t$ are generated by

$$y_t = A_t x_t + q_t \tag{10.17}$$

with $x_t = x_{t-1} + \varepsilon_t$ and $A_t = A_{t-1} + \eta_t$, where $q_t$ is $I(0)$, independent of $x_t$ and $A_t$, and $\varepsilon_t$ and $\eta_t$ are independent white noises. Then it is clear that there is a TVP cointegration, with $x_t$ being the $I(1)$ common factor or permanent component. Dividing (10.17) by $y_t$, assumed to be positive, gives

$$R_t = A_t + q_t/x_t ,$$

which results in a similar equation for $\hat{A}_t$ as occurs in the spurious case, with $\lambda_t \equiv -q_t/x_t$.

To examine the potential problems in this situation, a small simulation experiment is designed. The data used are generated by

$$y_t = A_t x_t + q_t ,$$

where

$$\begin{aligned} x_t &= x_{t-1} + \xi_{xt} \\ A_t &= a + b\delta_t \\ \delta_t &= \delta_{t-1} + \xi_{\delta t} \\ q_t &= \rho q_{t-1} + \xi_{qt} \end{aligned}$$

with $\xi_{xt}, \xi_{\delta t}, \xi_{qt} \sim IN(0,1)$. The parameter $A_t$ is assumed to follow a random walk process except for a constant term. The Kalman filter estimate $\hat{A}_t$ of $A_t$ is derived to examine the relationships between $A_t$ & $\hat{A}_t$, and $R_t$ & $\hat{A}_t$. The cointegration test for each pair of series is carried out, and the $t$-statistics in the cointegrating regression are reported in *Table 10.1* for various values of $\rho$. Five hundred replications of this design are computed in each case with 200 observations.

When there is cointegration, one might expect the super-efficiency kind of property as in constant cointegration case. In TVP cointegration framework, the following proposition describes a property of the Kalman filter estimate $\hat{A}_t$ of $A_t$.

**Proposition 10.2** *Suppose that there is TVP cointegration. If the estimated residual process $\hat{q}_t$ is $I(0)$ or short memory in mean, then the estimate $\hat{A}_t$ and the true parameter $A_t$ are cointegrated where $A_t$ is generated by a random walk process.*

*Proof:* Using (10.17), the residual process $\hat{q}_t$ can be written as

$$\hat{q}_t = y_t - \hat{y}_t = A_t x_t + q_t - \hat{A}_t x_t = (A_t - \hat{A}_t)x_t + q_t . \tag{10.18}$$

Now, as $q_t$ is $I(0)$ or short memory, the RHS of (10.18) can be short memory only if $(A_t - \hat{A}_t)$ is $I(0)$ with mean zero. That is $A_t$ and $\hat{A}_t$ should be cointegrated. □

*Table 10.1:* Cointegration test for $A_t$ & $\hat{A}_t$ and $\hat{A}_t$ & $R_t$.

| Percentile | | 5 % | 10 % | 25 % | 50 % | 75 % | 90 % | 95 % |
|---|---|---|---|---|---|---|---|---|
| $\rho = 0.0$ | $A_t$ & $\hat{A}_t$ | 8.26 | 9.15 | 10.36 | 11.91 | 13.67 | 15.27 | 16.49 |
| | $\hat{A}_t$ & $R_t$ | 10.95 | 11.97 | 13.33 | 14.10 | 14.93 | 17.70 | 18.83 |
| $\rho = 0.5$ | $A_t$ & $\hat{A}_t$ | 6.72 | 7.31 | 8.07 | 8.93 | 10.14 | 11.21 | 12.59 |
| | $\hat{A}_t$ & $R_t$ | 10.85 | 11.66 | 13.17 | 14.10 | 15.13 | 17.24 | 18.95 |
| $\rho = 1.0$ | $A_t$ & $\hat{A}_t$ | 0.24[a] | 0.77[a] | 1.67[a] | 2.62[a] | 3.45 | 4.20 | 4.92 |
| | $\hat{A}_t$ & $R_t$ | 10.35 | 11.24 | 13.08 | 14.08 | 14.90 | 17.61 | 18.81 |

Dickey-Fuller statistics are shown with minus signs omitted for simplicity.
[a] Indicates no cointegration at 5% level.

This statement is supported by the simulation result. As shown in *Table 10.1*, if TVP cointegration occurs, i.e., $0 < \rho < 1$, $A_t$ and $\hat{A}_t$ are cointegrated so that $\hat{A}_t$ provides a good approximation to the true $A_t$. On the other hand, when $\rho = 1$ so that there is no TVP cointegration, $\hat{A}_t$ appears to be cointegrated with $R_t$ but not with $A_t$, resulting in the spurious TVP cointegration mentioned above. Hence the two-step procedure—first, estimating $A_t$ through the Kalman filter, and then testing for unit-root on the residual process—seems irrelevant to testing for TVP cointegration in this situation. The potential difficulty associated with this problem motivates an alternative approach, which will be discussed in the next section.

## 10.4 Testing Procedures for TVP Cointegration

As mentioned in Section 10.3, if there are structural changes of known break-points, this information can be incorporated in the cointegrating regression model by using appropriate dummy variables. This section begins with an example illustrating how the structural changes affect the usual cointegration test. The example also shows how the usual testing procedures for cointegration, e.g., the Engle-Granger two-step procedure, can be employed in this case to figure out such shifting relationships. The choice of the example is motivated by an earlier empirical study.

### 10.4.1 Shifting regimes model

Engle and Granger (1987) have presented an evidence against cointegrating relationship between prices and wages in the US. They, however, suggested that there might be a cointegrating relationship if another variable such as productivity were considered. Following the basic Sargan wage-bargaining model, a time trend is included as a proxy for the productivity factor, providing a regression model of the form

$$y_t = m_0 + m_1 t + A x_t + q_t \,. \tag{10.19}$$

*Table 10.2:* Cointegration test for $W_t$ and $P_t$.

| Sample period | $T$ | DF | ADF(1) | ADF(4) |
|---|---|---|---|---|
| 1947(i)–1986(xii) | 480 | −2.80 | −2.71 | −3.02 |
| 1947(i)–1972(xii) | 312 | −2.41 | −2.51 | −2.09 |
| 1973(i)–1979(xii) | 84 | −3.06 | −2.76 | −2.38 |
| 1980(i)–1986(xii) | 84 | −4.57[a] | −4.03[a] | −5.06[b] |
| with dummies | 480 | −4.87[a] | −4.54[a] | −4.12 |

[a] No cointegration at 5% level.
[b] No cointegration at 1% level.

The data are the consumer price index and the index of production worker hourly wage in manufacturing sector from 1947(i) to 1986(xii). The test statistics (DF, ADF) are shown in *Table 10.2*, which indicates that we do not reject the null of no cointegration even at 10 % level. That is, the prices and wages are not cointegrated in the model (10.19).

In fact, there have been many empirical studies that consider the changes in the level of economic time series data or the changes in the relationship between them due to the oil shocks [see, e.g., Perron (1987)]. As shown in *Figures 10.2* and *10.3*, we can reasonably suspect that regimes change would happen in the relationship between wages and prices. Accordingly, two dummies are included in (10.19), resulting in the model of the form

$$W_t = m_0 + m_1 D_{1t} + m_2 D_{2t} + m_3 t + A P_t + \text{residual} \tag{10.20}$$

$$\text{where} \quad D_{1t} = \begin{cases} 1 & \text{if } 1973(\text{i}) \le t \le 1979(\text{xii}) \\ 0 & \text{otherwise} \end{cases}$$

$$\text{and} \quad D_{2t} = \begin{cases} 1 & \text{if } t \ge 1980(\text{i}) \\ 0 & \text{otherwise}\,. \end{cases}$$

Following the two-step procedure, the unit-root test for the residual is carried out. As shown in *Table 10.2*, the DF and ADF statistics have increased significantly so that we can reject the null of no cointegration at 5 % significance level. [Since we have a regression model with two dummy variables and a trend term, it is required to find appropriate critical values for this case. Monte Carlo simulation suggests that the critical values in Engle and Yoo (1987) for $N = 5$ be appropriate. That is, the trend term and the dummies contribute to the left tail distribution as if they were additional $I(1)$ variables in the EG two-step test.] That is, the prices and wages are cointegrated when we consider the regime shifts. The unit-root tests for $W_t$ and $P_t$ are also carried out for the whole sample with dummies as well as for each subsample period. The results are summarized in *Table 10.3*, indicating that the unit-root hypothesis is not rejected even if the structural breaks are considered. This contradicts the result in Perron (1987), who found evidences against the unit-root hypothesis for most economic variables.

In this example, $A$ is assumed to be constant while $m_t \equiv m_0 + m_1 D_{1t} + m_2 D_{2t} + m_3 t$ is changing through time. As $m_t$ is a deterministic function of time, the result indicates

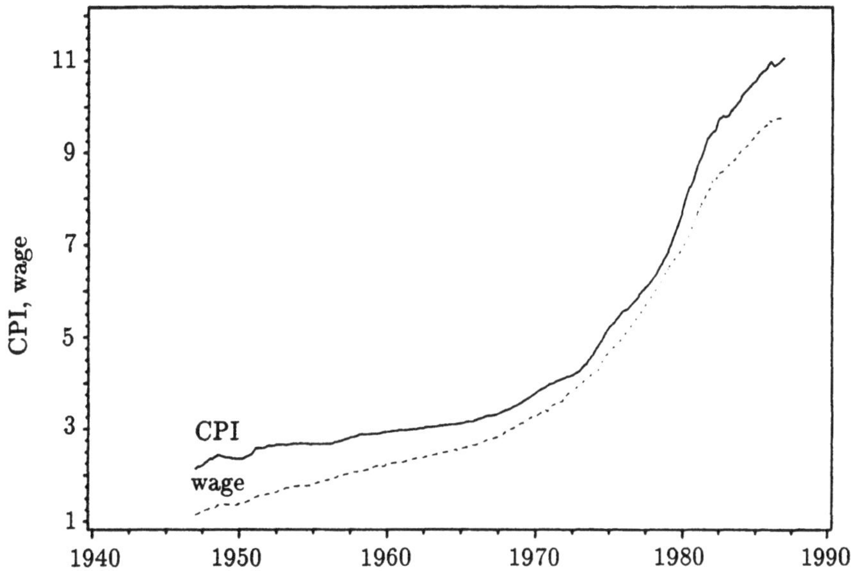


*Figure 10.2:* Prices and wages [1947(i)–1986(xii)].

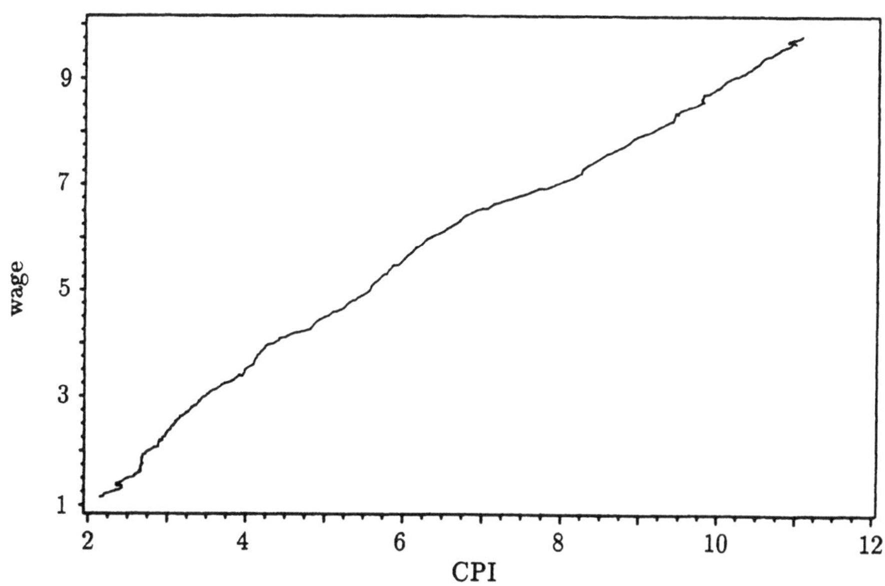


*Figure 10.3:* Relationship between prices and wages.

*Table 10.3:* Unit-root test for $W_t$ and $P_t$.

| Sample period | $T$ | $W_t$ DF | ADF(1) | ADF(4) | $P_t$ DF | ADF(1) | ADF(4) |
|---|---|---|---|---|---|---|---|
| 1947(i)–1986(xii) | 480 | 1.42 | 1.17 | −0.17 | 2.49 | 0.63 | 0.00 |
| 1947(i)–1972(xii) | 312 | 2.64 | 2.85 | 3.03 | 3.07 | 1.64 | 0.46 |
| 1973(i)–1979(xii) | 84 | 1.93 | 1.75 | 1.51 | 4.81 | 2.51 | 0.47 |
| 1980(i)–1986(xii) | 84 | 1.89 | 2.21 | 2.83 | 3.08 | 2.25 | 2.20 |
| With dummies | 480 | −2.86 | −2.80 | −2.10 | −2.03 | −1.14 | −0.86 |
| Sample period | $T$ | $\Delta W_t$ DF | ADF(1) | ADF(4) | $\Delta P_t$ DF | ADF(1) | ADF(4) |
| 1947(i)–1986(xii) | 479 | −14.87 | −8.07 | −3.14 | −7.56 | −5.61 | −3.48 |
| 1947(i)–1972(xii) | 311 | −16.55 | −10.91 | −6.37 | −11.44 | −7.90 | −4.64 |
| 1973(i)–1979(xii) | 83 | −9.23 | −6.18 | −3.16 | −4.00 | −1.74 | −0.44 |
| 1980(i)–1986(xii) | 83 | −8.76 | −4.65 | −1.43 | −4.33 | −4.42 | −3.05 |
| With dummies | 479 | −19.81 | −11.36 | −4.45 | −9.70 | −7.38 | −4.62 |

For the unit-root test, a trend term is included, so that the critical value of $\hat{\tau}_\tau$ from Fuller (1976, p. 373) should be applied.
For $\Delta W_t$ and $\Delta P_t$, a constant term is included, hence the appropriate critical values are given by $\hat{\tau}_\mu$.
The critical values for the model with dummy variables are found to be similar to the Engle-Granger critical values for $N = 4$ (for $P_t$, $W_t$), and $N = 3$ (for $\Delta P_t$, $\Delta W_t$).

that the prices and wages share the same stochastic trend or "permanent component", which is the basis for the cointegration concept. Hence, the standard implication and interpretation of cointegration can be considered.

### 10.4.2 Random walk parameter model

In the context of a dynamic stochastic structure of real world, a model of particular interest is the one in which little information is available on the exact cause of parameter variation. In this case, the strategy in using regression models with stochastically varying coefficients seems a reasonable attempt to model the unobserved deterministic or stochastic cause of parameter changes. As discussed in the previous section, when TVP cointegration occurs, such a scheme is capable of approximating the stochastic behavior of the true relationship through Kalman filter algorithm. However, this case contains the possibility of a spurious TVP cointegration, which makes it difficult to distinguish a real and a spurious relationship.

To motivate an alternative approach to testing for TVP cointegration and hence to detect the spurious relationship, the relative forecasting accuracy is examined for each case where (i) the DGP is given such that the regression model with constant parameters is correctly specified, (ii) the DGP is given such that TVP cointegration model is correctly specified, and (iii) the data $x_t$ and $y_t$ are generated by independent random walk processes

*Table 10.4:* Relative forecasting accuracy of TVP cointegration model.

| DGP | Alternative model | 5 % | 10 % | 25 % | 50 % | 75 % | 90 % | 95 % |
|---|---|---|---|---|---|---|---|---|
| (1) | Const. parameter | −19.30 | −11.45 | −4.42 | −1.76 | −0.50 | 0.27 | 0.64 |
| | Naive forecast | −6.25 | −2.92 | 1.76 | 5.09 | 7.12 | 8.24 | 9.00 |
| (2) | Const. parameter | −16.12 | −10.24 | −4.48 | −1.69 | −0.46 | 0.36 | 0.77 |
| | Naive forecast | −4.74 | −2.26 | 1.72 | 4.84 | 6.83 | 7.81 | 8.52 |
| (3) | Const. parameter | −0.72 | −0.41 | 0.24 | 1.08 | 2.52 | 3.80 | 4.96 |
| | Naive forecast | −0.52 | −0.20 | 0.66 | 1.69 | 3.38 | 6.43 | 8.28 |
| (4) | Const. parameter | −0.72 | −0.39 | 0.23 | 1.03 | 2.50 | 3.80 | 4.94 |
| | Naive forecast | −0.52 | −0.20 | 0.66 | 1.70 | 3.37 | 6.39 | 8.31 |
| (5) | Const. parameter | −22.56 | −18.08 | −7.77 | −2.89 | −1.19 | 0.11 | 0.31 |
| | Naive forecast | −22.85 | −18.23 | −7.94 | −2.82 | −1.18 | −0.13 | 0.19 |

In each case, the constant parameter-forecasting model is given by a least squares regression such that

$$\Delta y_t = \hat{A}\Delta x_t + \text{residual}$$

and the forecasting model estimated by TVP cointegration is given by

$$y_t = \hat{A}_t x_t + \text{residual}\,.$$

For cases (1) through (4), the data used for $x_t$ is generated by $\Delta x_t = 0.6\,\Delta x_{t-1} + \varepsilon_{xt}$, $\varepsilon_{xt} \sim N(0,1)$. The data for $y_t$ is given by the generating mechanism:

(1) $\Delta y_t = 2\,\Delta x_t + \varepsilon_t$
(2) $\Delta y_t = 0.4\,\Delta y_{t-1} - 0.1\Delta x_{t-1} + \varepsilon_t$
(3) $y_t = A_t x_t + u_t$ where $A_t = 2 + \delta_t$ for $\delta_t = \delta_{t-1} + \xi_t$ and $u_t = 0.4\,u_{t-1} + \varepsilon_t$
(4) $y_t = A_t x_t + u_t$ where $u_t = 0.7\,u_{t-1} + 0.1\,u_{t-2} + \varepsilon_t$ with $A_t$ and $\delta_t$ as in (3) .

For (5), the data $x_t$ and $y_t$ are generated by independent random walk processes.

[see *Table 10.4*].

One hundred replications of this design are computed in each case, with 100 observations used for fitting the initial model and 30 observations used recursively for one-step-ahead forecasting, to generate one-step-ahead forecast errors. The forecasting accuracy of the model estimated by TVP cointegration is compared with the constant parameter regression model and naive time series forecasting model, i.e., the random walk model, in terms of expected squared forecast errors. If expected squared error is the criterion for the evaluation of two competing forecasts, a valid test of one-step-ahead forecast errors can easily be derived under assumptions such that the individual forecasts are unbiased and the forecast errors are not autocorrelated. Consider the sample correlation coefficient between $(e^c + e^t)$ and $(e^c - e^t)$ such that

$$r = \frac{\sum(e_i^c + e_i^t)(e_i^c - e_i^t)}{\sqrt{\sum(e_i^c + e_i^t)^2 \sum(e_i^c - e_i^t)^2}}\,,$$

where $e_i^c$ and $e_i^t$ are one-step-ahead forecast errors from the constant parameter model and TVP cointegration model, respectively. Note that the two expected squared errors will be equal if and only if this pair of random variables are uncorrelated. Now, the usual test for zero correlation such that

$$t = \frac{r}{\sqrt{(1-r^2)/(n-2)}} \tag{10.21}$$

can be employed to test for equality of expected squared forecast errors. It can also be noticed that $t > 0$ implies that the TVP model performs better than the constant parameter model.

*Table 10.4* reports the distribution of the $t$-statistics. As is expected, it is clear that there are always gains to using the correctly specified model. That is, when the data are generated by DGP(1) so that the constant model is correctly specified, it performs better than TVP model as indicated by negativity of the $t$-statistics. On the other hand, if the data are given such that the TVP model is correct as in DGP(3), the $t$-statistics are mostly positive implying that it performs better than the constant model. One can notice that the constant parameter model performs better than TVP model even when it is not correctly specified as in DGP(2). One might not expect such a difference in the forecasting performances of two misspecified models, but this can be explained by the well-observed fact that increasing the in-sample fit of a model does not necessarily lead to a better post-sample forecasting. Also the distribution of the $t$-statistics for the DGP(3) and (4) in *Table 10.4* are similar to each other, indicating a robustness of the procedure to the specification of the residual process when TVP cointegration occurs. The naive forecast based on the random walk model is also compared with the TVP model, indicating some gains to using the information in $x_t$ to forecast $y_t$. The results support the strategy of using the relative forecasting accuracy for detecting the spurious TVP cointegration and hence suggest that it can also be applied in testing TVP cointegration.

In a bivariate case discussed so far, the testing procedure for TVP cointegration based on the forecasting performance can now be summarized as follows:

1. Test that both $x_t$ and $y_t$ are $I(1)$ or TVP-$I(1)$.
2. Run a TVP regression to estimate $m_t$ and $A_t$ in the model

$$y_t = m_t + A_t x_t + q_t \,. \tag{10.22}$$

3. Find the best forecasting model among alternative models with constant coefficients.
4. Compute the $t$-statistic from (10.21) to see whether the TVP cointegration model performs better.

In the first step, a test based on the estimation of evolutionary spectrum might be suggested, but it is virtually impossible to cope with all possibilities with a limited quantity of data available on the usual macroeconomic time series. In the case where the "permanent component", $W_t$, and the "transitory component", $z_t$, are generated without change as in (10.2), while only $A$ and/or $m$ in the generating mechanism (10.1) are allowed to vary over time, the usual tests for unit-root such as the DF and ADF testing procedure can be applied.

For the second step, the parameter $A_t$ is assumed to change very slowly over time following the random walk coefficient model. The residual process $q_t$ is assumed to follow AR(1) process, which can easily be extended to any ARMA($p$,$q$) model of finite order. This follows from the fact that any ARMA process can be expressed by a multivariate AR(1) model in a state space form [see Harvey (1981, p. 103)]. The AR(1) assumption seems too strong, but is based upon the conjecture that the super-efficiency kind of property can be expected when TVP cointegration occurs, as discussed in the previous section. This conjecture is supported by the simulation result above (see *Table 10.4*), where $q_t$ is generated by an AR(2) process, but is assumed to be an AR(1) in estimation. The distribution of the $t$-statistics are similar to each other, indicating that the specification of the residual process appears not to affect the result.

The restriction to constant coefficient models in the third step appears to be too strong. Suppose that a nonlinear relationship is present between $x_t$ and $y_t$. Then forecasting models with constant coefficients are misspecified, so that one might falsely accept the TVP cointegration, as varying parameters will usually better approximate the nonlinear relationship. However, such a relationship is worth finding as an approximation to the true relationship between $x_t$ and $y_t$, unless one can figure out the nonlinear form, hence should be distinguished from the pure spurious relationship discussed in the previous section.

In the fourth step, one is confronted with a trade-off between the size and power of the test. The choice of usual significance level, say, 5 % level, appears to result in a conclusion that is not very powerful. It is required to further investigate the distributional property of the $t$-statistics in (10.21).

This approach to testing for TVP cointegration is applied to examine the relationship between prices and wages discussed above. *Table 10.5* reports the $t$-statistics from (10.21) for each subsample period of interest. Even though there is no strong evidence found for cointegration in the first two subsample periods [1947(i)–1972(xii), 1973(i)–1979(xii)] using the usual Engle-Granger two-step procedure (see *Table 10.2*), an investigation of the relative forecasting accuracy within each regime indicates that the error-correction model performs better than any other constant parameter model considered, which in turn is outforecast by the TVP model. This result might be considered as an evidence for the existence of cointegrating relationship in each regime.

For the forecasting periods outside each regime, however, the error-correction model does not predict best among the alternative models considered, except for the periods just after the first oil shock [1947(i)–1976(xii), 1947(i)–1977(xii)]. In fact, TVP model turns out to perform generally better than any constant parameter model, as indicated by the positivity of the $t$-statistics, although the $t$-values are not significant using the usual significance level.

As discussed above, the choice of usual significance level, say, 5 % or 10% level, might

*Table 10.5:* Relative forecasting accuracy of TVP forecasting model.

| Sample period | Alternative model | $t$-statistics |
|---|---|---|
| 1947(i)–1972(xii) | ECM | −0.475 |
| 1973(i)–1979(xii) | ECM | −1.359 |
| 1980(i)–1986(xii) | ECM | −1.088 |
| 1947(i)–1976(xii) | ECM | −0.977 |
| 1947(i)–1977(xii) | ECM | −1.311 |
| 1947(i)–1978(xii) | ECM | 0.437 |
| 1947(i)–1981(xii) | VARD | −2.240 |
| 1947(i)–1982(xii) | VARD | 0.147 |
| 1947(i)–1983(xii) | VARD | 0.476 |
| 1973(i)–1981(xii) | VARD | −2.365 |
| 1973(i)–1982(xii) | VARD | 0.802 |
| 1973(i)–1983(xii) | VARD | 1.098 |

The constant parameter forecasting models considered are least squares regression model in difference (OLSD), VAR form for $\Delta W_t$ (VARD), and error-correction model (ECM) such that

$$\begin{aligned}\text{OLSD:}\quad \Delta W_t &= \text{const} + \delta\Delta P_t + \text{residual}\\ \text{VARD:}\quad \Delta W_t &= \text{const} + \hat{a}\Delta P_{t-1} + \hat{b}\Delta W_{t-1} + \text{lags}(\Delta W_{t-1}, \Delta P_{t-1}) + \text{residual}\\ \text{ECM:}\quad \Delta W_t &= \text{const} + \hat{\gamma} z_{t-1} + \text{lags}(\Delta W_t, \Delta P_t) + \text{ residual},\end{aligned}$$

where $z_t = W_t - \hat{\alpha} - \hat{\beta} P_t$ is the residual from the cointegrating regression.

result in a test that is not very powerful so that one is confronted with a trade-off between the size and power of the test. Hence we have to reserve any conclusion about the use of the relative forecasting accuracy in testing for TVP cointegration, unless a further investigation into the distributional property of the test statistics is reached. However, the positivity of the $t$-statistics, implying that root-mean-squared error of TVP forecasting model is smaller than that from any constant alternatives, suggests that the varying relationships between prices and wages might better be captured by TVP cointegration technique whenever there is any structural break.

### 10.4.3 TVP-error correction model

To motivate another approach to TVP cointegration testing, consider a pair of series $x_t$ and $y_t$ which are jointly generated as a function of possibly correlated white noises, $\varepsilon_{1t}$ and $\varepsilon_{2t}$, according to the following model:

$$y_t - \beta x_t = u_{1t}, \qquad u_{1t} = u_{1,t-1} + \varepsilon_{1t} \tag{10.23}$$

$$y_t - A_t x_t = u_{2t}, \qquad u_{2t} = \rho u_{2,t-1} + \varepsilon_{2t}\,. \tag{10.24}$$

The reduced form for this system becomes

$$x_t = \frac{1}{A_t - \beta}(u_{1t} - u_{2t}) = \frac{1}{A_t - \beta}u_{1t} - \frac{1}{A_t - \beta}u_{2t}$$

$$y_t = \frac{\beta}{A_t - \beta}(u_{1t} - u_{2t}) + u_{1t} = \frac{A_t}{A_t - \beta}u_{1t} - \frac{\beta}{A_t - \beta}u_{2t},$$

which indicates that $x_t$ and $y_t$ share two common factors, $u_{1t}$ and $u_{2t}$, with $u_{1t}$ being the permanent component of the system and $u_{2t}$ being the transitory component. As equation (10.24) describes a particular time-varying relationship of the random variables, which results in a stationary process, the variables $x_t$ and $y_t$ are TVP cointegrated, and the question is then whether it is possible to detect the relationship and to estimate $A_t$.

The model (10.23) and (10.24) can be expressed in the AR representation as

$$\begin{aligned} \Delta x_t &= \frac{1-\rho}{A_t - \beta}y_{t-1} - \frac{A_t - \rho A_{t-1}}{A_t - \beta}x_{t-1} + \eta_{1t} \\ \Delta y_t &= \frac{\beta(1-\rho)}{A_t - \beta}y_{t-1} - \frac{\beta(A_t - \rho A_{t-1})}{A_t - \beta}x_{t-1} + \eta_{2t} \end{aligned}$$

or equivalently

$$\Delta x_t = \phi_t[(1-\rho)y_{t-1} - \psi_t x_{t-1}] + \eta_{1t} \tag{10.25}$$

$$\Delta y_t = \beta\phi_t[(1-\rho)y_{t-1} - \psi_t x_{t-1}] + \eta_{2t}, \tag{10.26}$$

where $\phi_t = 1/(A_t - \beta)$ and $\psi_t = A_t - \rho A_{t-1}$, and the $\eta_t$'s are combinations of the $\varepsilon_t$'s and $A_t$'s. In general, TVP cointegrated process $x_t$ will have an AR model of the form

$$A_t^*(B)(1-B)x_t = -\gamma_t\alpha_t' x_{t-1} + d_t(B)\eta_t \tag{10.27}$$

so that the lagged values of $\Delta x_t$'s are included in the model with appropriate assumptions imposed on the behavior of $\gamma_t$ and $\alpha_t$. The AR representation (10.27) can also be interpreted as a TVP error-correction model with $z_t = \alpha_t' x_t$ being the attractor process of the system.

In this specification, however, if we assume $A_t$ in (10.24) to follow a random walk process, then $\phi_t$ in (10.25) and (10.26) can not be well approximated either by another random walk process or by any other low-order ARIMA process in general. Suppose now that the stochastic parameter $A_t$ is governed by a pattern of variation such that

$$A_t = A_{t-1} \times \xi_t, \tag{10.28}$$

where $\xi_t$ is assumed to be i.i.d. with $E(\xi_t) = 1$ and $\xi_t > 0$ for all $t$. Notice that the stochastic formulation of $A_t$ is set up such that the disturbance term enters multiplicatively. With initial condition $A_0 = 1$, the equation (10.28) can be rewritten as

$$A_t = \prod_{i=1}^{t} \xi_i .$$

Hence one can show that

$$E(A_t|\xi_1, \ldots, \xi_{t-1}) = A_{t-1} .$$

That is, the varying coefficient generated by equation (10.28) also follows a martingale process, hence possesses some inertia as in the random walk parameter model. We argue that unless there were strong *a priori* grounds for imposing an additive disturbance term as in the random walk parameter model, the specification (10.28) would be more appropriate in describing the evolution of parameters in the sense that $1/A_t$ can well be approximated by a similar pattern of parameter variation.

If the stochastic formulation of $A_t$ is given by (10.28), substituting $1/\delta_t$ for $A_t$ will then result in

$$\delta_t = \delta_{t-1} \times \frac{1}{\xi_t},$$

which can be rewritten as

$$\delta_t = \delta_{t-1} \times \varepsilon_t, \tag{10.29}$$

where $\varepsilon_t = 1/\xi_t$. From the initial condition $A_0 = 1$, it is easy to show that $\delta_0 = 1$ and that $\delta_t$ follows a sub-martingale process with $E(\varepsilon_t) > 1$.

This indicates that whenever the true parameter $A_t$ is assumed to follow a martingale process as in equation (10.28), we can approximate it by estimating $\delta_t$ through its consistent counterpart (10.29). For a pair of regression models with stochastically varying coefficients such that

$$x_t = \delta_t y_t + v_t \tag{10.30}$$

$$y_t = A_t x_t + u_t, \tag{10.31}$$

where the variables $x_t$ and $y_t$ represent the observables in the system, this specification of parameter variation might produce better approximation than any low-order ARIMA schemes, hence is suitable whenever little information is available on the dynamic structure of time series model. Under appropriate assumptions, the Kalman filter algorithm can be applied in this case to estimate the parameter of interest $A_t$, which is shown in Lee (1989) and is not attempted here.

## 10.5 Concluding Remarks

In a dynamic stochastic real world, we might reasonably expect that equilibrium relationship is changing over time as technology, taste, and society change. Consequently, the fact that a set of economic variables are not linearly cointegrated does not necessarily imply that there is no equilibrium relationship among them. In fact, the regression model may be subject to misspecification, since it ignores the effect of any changes in the unobserved components on the behavior of the observable variables in the model. Hence, rather than restricting to the linear, stable relationship, an introduction of the time-varying coefficients in the cointegrating regression is suggested, giving us a richer concept of cointegration.

Although many interesting results are obtained by introducing time-varying parameters in the cointegration concept, much work still remains to be done. Suppose a theory suggests some equilibrium relationships among a set of economic time series data, but that the cointegration technique does not provide any evidence of equilibrium relationship while the TVP cointegration procedure does. Then it might reflect the fact that the linear cointegrating regression model is misspecified and that we have omitted variables that should be included in the model. Hence, it is reasonable to try first to figure out the omitted variables available, rather than to be just satisfied with the changing equilibrium relationships.

Second, since the choice of the dependent variable for the TVP cointegrating regression is arbitrary, we may get conflicting results in practice for the same set of variables, depending upon the direction in which the regression is performed. As we usually do not know how the unobservable components affect the behavior of the observable data in the model, we may have to consider the model in both directions. Moreover, we need to consider the distributional properties of the attractor process and the power properties of the test. This also implies that we still have some practical problems to investigate.

## References

Cramér, H. (1961), On some classes of non-stationary stochastic processes, pp. 57-78 in J. Neyman (ed.), *Proceedings of the 4th Berkeley Symposium on Mathematical Statistics and Probability.* San Francisco: University of California Press.

Engle, R.F. and Granger, C.W.J. (1987), Cointegration and error correction: Representation, estimation and testing. *Econometrica*, **55**, 251-276.

Engle, R.F. and Yoo, B.S. (1987), Forecasting and testing in co-integrated systems. *Journal of Econometrics*, **35**, 143-159.

Engle, R.F. and Yoo, B.S. (1989), Cointegrated economic time series: A survey with new results. Discussion paper, San Diego: University of California.

Fuller, W.A. (1976), *Introduction to Statistical Time Series.* New York: John Wiley.

Granger, C.W.J. (1986), Developments in the study of cointegrated economic variables. *Oxford Bulletin of Economics and Statistics*, **48**, 213-228.

Granger, C.W.J. and Hallman, J. (1988), The algebra of $I(1)$. Finance and Economics Discussion Series 45, Board of Governors of the Federal Reserve System.

Harvey, A.C. (1981), *Time Series Models.* Oxford: Philip Allan.

Lee, H.S. (1989), A note on the log random walk parameter model. Manuscript, Department of Economics, San Diego: University of California.

Perron, P. (1987), The great crash, the oil shock and the unit root hypothesis. Department of Economics, University of Montreal, forthcoming in *Econometrica.*

Priestley, M.B. (1965), Evolutionary spectra and non-stationary processes. *Journal of the Royal Statistical Society*, **B-27**, 385-392.

Priestley, M.B. (1981), *Spectral Analysis and Time Series.* London: Academic Press.

Stock, J.H. (1987), Asymptotic properties of least squares estimators of cointegrating vectors. *Econometrica*, **55**, 1035-1056.

CHAPTER 11

# Disequilibrium Modeling, Switching Regressions, and Their Relationship to Structural Change

*G.S. Maddala*

## Summary

Disequilibrium and self-selection models do not directly deal with structural change. But both are switching regression models with endogenous switching and the techniques of analysis for switching regression models can be used to study structural change. The chapter discusses the different uses of these models in the modeling of structural change. One class of such models is the Markov switching model which has been used to analyze exchange rates, stock prices, and nonstationary time series. These are models with exogenous switching. The other class of models, with endogenous switching, can be fruitfully applied to analyze structural change which follow policy changes that eliminate opportunities of self-selection that economic agents have.

## 11.1 Introduction

This chapter reviews developments in econometric disequilibrium modeling and switching regressions, and outlines the relationship between this literature and the methods used for the analysis of structural change. The basic structure of disequilibrium econometric models can be described as one involving switching regressions with endogenous switching. A commonly used method for the analysis of structural change is also one of switching regressions where the switch point is usually unknown and needs to be estimated from the data. There is, thus, some similarity in the methodologies used. The purpose of this chapter is to outline some modifications in the switching regression methods of analyzing structural change that take into account gradual adjustments to, and expectations about, policy changes that are usually the main source of structural change.

## 11.2 Disequilibrium Models and Switching Regressions

The classical switching regression model consists of two regression equations

$$y_t = \beta_1' X_{1t} + u_{1t} \qquad \text{in regime 1} \tag{11.1}$$

$$y_t = \beta_2' X_{2t} + u_{2t} \qquad \text{in regime 2.} \tag{11.2}$$

The vectors $X_{1t}$ and $X_{2t}$ are exogenous variables. The dependent variable $y_t$ is generated either from regime 1 or from regime 2 but not from both. Suppose we define the indicator variable $I_t$ as

$$I_t = \begin{cases} 1 & \text{if } y_t \text{ comes from regime 1} \\ 0 & \text{if } y_t \text{ comes from regime 2.} \end{cases}$$

If $I_t$ is observed, we say that the sample separation is known. In this case equations (11.1) and (11.2) can be estimated by ordinary least squares using the observations of $y_t$ in the two regimes. The indicator variable $I_t$ may not be observed, but its determinants are. In this case we define another variable $Z_t$ as

$$Z_t = \beta_3' X_{3t} + u_{3t} \tag{11.3}$$

and define

$$I_t = \begin{cases} 1 & \text{if } Z_t > 0 \\ 0 & \text{otherwise.} \end{cases}$$

In this case we have probabilistic sample separation. Consider the model given by (11.1), (11.2), (11.3) and let $(u_1, u_2, u_3) \sim N(0, \Sigma)$, where

$$\Sigma = \begin{bmatrix} \sigma_{11} & \sigma_{12} & \sigma_{13} \\ \sigma_{12} & \sigma_{22} & \sigma_{23} \\ \sigma_{13} & \sigma_{23} & 1 \end{bmatrix}.$$

Since $Z_t$ can be observed only as a dichotomous variable, we assume $\text{Var}(Z_t) = 1$. If $\sigma_{13} = \sigma_{23} = 0$, we have a switching regression model with exogenous switching. Otherwise, we have endogenous switching [see Maddala (1983, p. 284)]. With endogenous switching, even if the sample separation is known, equations (11.1) and (11.2) cannot be estimated by ordinary least squares.

The disequilibrium model

$$D_t = \beta_1' X_{1t} + u_{1t} \qquad \text{demand function} \tag{11.4}$$

$$S_t = \beta_2' X_{2t} + u_{2t} \qquad \text{supply function} \tag{11.5}$$

$$Q_t = \min(D_t, S_t) \tag{11.6}$$

is a switching regression model with endogenous switching. Because, whether $Q_t$ is on the demand function or on the supply function depends on whether $D_t - S_t < 0$ or $> 0$, respectively, that is, whether $(\beta_1' X_{1t} + u_{1t}) - (\beta_2' X_{2t} + u_{2t}) < 0$ or $> 0$. In this case $u_{3t} = u_{1t} - u_{2t}$. Thus, $\sigma_{13} = \sigma_{11} - \sigma_{12}$ and $\sigma_{23} = \sigma_{12} - \sigma_{22}$.

Another modification of the switching regression model is that by Lee and Porter (1984) who consider the availability of a dichotomous indicator $w_t$ that provides sample separation, but this indicator is imperfect. They postulate a transition probability matrix

| | $w = 1$ | $w = 0$ |
|---|---|---|
| $I = 1$ | $p_{11}$ | $p_{10}$ |
| $I = 0$ | $p_{01}$ | $p_{00}$ |

If $p_{11} = p_{01}$, the indicator $w_t$ does not provide any information on sample separation. If $p_{11} = 1$ and $p_{01} = 0$, the indicator $w_t$ provides perfect sample separation. In the case $p_{11} \neq 1$ and $p_{01} \neq p_{11}$ $(> p_{01})$, we have imperfect sample separation. Lee and Porter examine the transportation prices charged by the Joint Executive Committee railroad cartel from 1880 to 1886, using the price equation

$$\log p_t = \beta_0 + \beta_1' x_t + \beta_2 I_t + u_t \,,$$

where $I_t = 0$ if there is a price war and $I_t = 1$ if there is cooperation. Thus, there is a regime switch; $w_t$ is an indicator obtained from the *Railroad Review*, which reported in each sample period whether or not a price war was occurring.

The disequilibrium model given by (11.4), (11.5), and (11.6) is a switching regression model with endogenous switching, and the estimated method to be used depends on whether sample separation is available or not. One source of sample separation often considered is that of $\Delta p_{t+1}$ or the change in price. Several models using this information have been proposed [see Maddala (1983) and Quandt (1988) for a more detailed discussion]. However, before any modeling of disequilibrium is done, it is important to consider the sources of disequilibrium. This will dictate both the proper formulation of the model and the estimation procedure to be used. Maddala (1983, p. 326) considers a classification of disequilibrium models based on

- imperfect adjustment of prices, and
- controls on the movement of prices.

There are some conceptual problems associated with the argument of imperfect adjustment of prices and condition (11.6). These have been discussed at length in Maddala (1983, pp. 319-326) and will not be reviewed here. In the case of controlled prices, the market is sometimes in equilibrium and sometimes in disequilibrium.

In these models of disequilibrium, there are regime shifts but the models do not fall in the category of analysis of structural change. It is a case of partial observability rather than changes in the economic structure. The methods of estimation, however, can be used to analyze structural change, as discussed in the following sections.

## 11.3 $D$-Methods for Analyzing Structural Change

Broadly speaking, structural change is defined as a change in the relative weights of significant components of the national product and expenditures. Some examples are a shift from agriculture to industry in the developing countries, the growth of the service sector relative to the manufacturing sector in the US during the last 20 years, and a rise in imports relative to exports (trade deficit). This chapter studies the structural change in a more limited sense, namely, changes in the parameters of a given model following changes in policies or occurrence of some major events. Though the model under consideration would usually be a simultaneous equations model, for ease of exposition, the discussion is confined to a simple regression model.

Consider the switching regression model given by equations (11.1) and (11.2) where $u_{1t} \sim IN(0, \sigma_1^2)$ and $u_{2t} \sim IN(0, \sigma_2^2)$. Goldfeld and Quandt (1973a) assume that there is an identifiable variable $Z_t$ such that if $Z_t \geq C$, then the observations are generated by (11.2). They then define the step function,

$$D(Z_t) = 1 \quad \text{if} \quad Z_t < C\,.$$

Later, instead of considering the step function, they consider $D(Z_t)$ to be a cumulative normal given by

$$D(Z_t) = \begin{cases} 1 & \text{if} \quad Z_t < C + \varepsilon_t \\ 0 & \text{if} \quad Z_t \geq C + \varepsilon_t\,, \end{cases}$$

where $\varepsilon_t \sim IN(0,1)$. The model considered by Goldfeld and Quandt is an exogenous switching model because $\varepsilon_t$ are assumed to be independent of $u_{1t}$ and $u_{2t}$. Goldfeld and Quandt give an extension of this $D$-method to simultaneous equations systems, but the extension is still in the framework of exogenous switching. Lee *et al.* (1979) extend the $D$-method to switching simultaneous systems with endogenous switching. They derive the valid instrumental variables to use, the asymptotic covariance matrix of the suggested estimator and test whether the coefficients in the earnings functions are significantly different between the two groups: those with college education and those without.

The $D$-method combines the equations in the two regimes into a single equation. This enables us to test for equality of specific coefficients in the two regimes. The alternative to this procedure would be to use a likelihood ratio test every time the significance of a particular coefficient needs to be tested.

## 11.4 Sources of Regime Shifts in Switching Regressions

The example considered in Lee *et al.* (1979) is one of structural change based on cross-section data. The example considered by Lee and Porter (1984) refers to time series data with two regimes, and each observation can be classified into one or the other regime. A problem often considered is the estimation of break-points where the regime shift took place. For instance, regime 1 could be for $t = 1, 2, \ldots, T_0$ and regime 2 for the remaining observations. Based on the time series data available, the problem is to estimate the break-point $T_0$ and the parameters $\beta_1$, $\beta_2$, $\sigma_1^2$, and $\sigma_2^2$.

A more pertinent question to ask is why the structural change took place? Very often this is a consequence of some policy change. In this case, the break-point $T_0$ is known. However, the adjustment to the new regime does not take place instantly. If there are considerable time lags in the adjustment process, one might not be able to detect any structural change in the regimes with the limited time series data. On the other hand, if the policy change is anticipated, the break-point detected may occur before $T_0$, the time when the policy change took place. A third possibility is that economic agents expect a policy change and act as if there is a regime shift (to say regime 2); however, the policy change does not occur, and the behavior reverts back slowly to that given by regime 1.

Thus, an analysis of structural change has to take account of two factors: gradual adjustment to policy changes and expectations (fulfilled or not) about policy changes. In both cases, there is disequilibrium, in the sense that the system does not instantaneously adjust to the new equilibrium position given by, say, regime 2. These two factors in the context of a simple switching regression model will be considered.

### 11.4.1 Partial adjustment to policy changes

The partial adjustment model is the earliest formulation of a disequilibrium model. In the context of policy changes and structural change, one must consider changes in parameters in response to changes in economic policies. If adjustment to the new situation is instantaneous, then one observes a sudden switch and the switching regression model given by equations (11.1) and (11.2) is appropriate with regime 1 describing the system before the policy change and regime 2 describing the system after the policy change. In practice, the adjustment to the new regime is gradual rather than instantaneous because of the *permanent-transitory confusion.* When there is a change in governmental policies economic agents have to guess whether the change is permanent or temporary. Lewis (1988, 1989) and Kaminsky and Peruga (1988), for instance, argue that following the tightening of the US money market in the early 1980s, many economic agents did not immediately believe that the change would persist, but instead learned of the shift rationally.

### 11.4.2 Expectations about policy changes and structural change

If economic agents correctly anticipate future policy changes, then the structural shift would take place before the policy change. On the other hand, agents can keep on expecting a policy change that may not occur. This is known as the "peso problem". In this case agents would put progressively less weight on the significance of a possible policy change, the longer the interval over which the change has not occurred. The peso problem can be modeled by time-varying probabilities of regime shifts or by using what are known as Markov switching models with unknown states of nature. These latter models have become very popular recently in the econometric analysis of structural change.

## 11.5 Markov Switching Models

Goldfeld and Quandt (1973b) suggested a Markov switching regression model in which the regime indicator $I_t$ is a Markov chain with transition probabilities $p_{ij}$ denoting the probability that the system moves from regime $i$ at time $(t-1)$ to regime $j$ at time $t$. Cosslett and Lee (1985) point out that the maximization of the likelihood function presented by Goldfeld and Quandt gives consistent but inefficient estimates. They present the correct likelihood function and an iterative procedure of estimation that gives efficient estimates. The paper by Cosslett and Lee extends the model by Lee and Porter (1984) to the case of serially correlated errors.

These papers form the background for the recent work in the econometric literature on Markov switching models. Hamilton (1988, 1989) suggests that many economic time series can be characterized by Markov switching models. The model, as applied to the foreign exchange market by Engel and Hamilton (1989), is the following: let $x_t$ be the dollar/mark exchange rate; $x_t$ follows two regimes,

$$\begin{aligned} x_t &\sim N(\mu_1, \sigma_1^2) \quad \text{when} \quad s_t = 1 \\ x_t &\sim N(\mu_2, \sigma_2^2) \quad \text{when} \quad s_t = 2\,, \end{aligned}$$

where $s_t$ follows a two-state Markov chain:

$$\begin{aligned} P(s_{t+1} = 1 | s_t = 1) &= p_{11} \\ P(s_{t+1} = 2 | s_t = 1) &= 1 - p_{11} \\ P(s_{t+1} = 1 | s_t = 2) &= 1 - p_{22} \\ P(s_{t+1} = 2 | s_t = 2) &= p_{22}\,. \end{aligned}$$

In this case $s_t = 1$ might denote upward moves so that $\mu_1$ is positive and $s_t = 2$ might denote downward moves so that $\mu_2$ is negative. If $\mu_1$ is positive and large and $p_{11}$ small, upward moves would be short but sharp. If $\mu_2$ is negative and small in absolute value and $p_{22}$ is large, then downward moves would be gradual and drawn out. Other configurations are possible. If $\mu_1$ and $\mu_2$ are opposite in sign and both $p_{11}$ and $p_{22}$ are large, both the upward trend and downward trend would both be long and drawn out. If $p_{11} = 1 - p_{22}$, then we have a random walk.

The observed series $x_t$ comes from a mixture of normal distributions (see Everitt and Hand, 1981), but unlike the case of many mixture distributions the $x_t$ are not independent. The probability that $x_t$ assumes any given value depends on the past values of $x_t$. For instance, if $s_{t-1} = 1$ and $p_{11}$ is high, then $x_t$ is more likely to have been generated from distribution 1, whereas, if $s_{t-1} = 2$ and $p_{22}$ is high, it is more likely to have been generated from distribution 2.

The distribution of $x_t$ depends on the parameter vector

$$\Theta = (\mu_1, \mu_2, \sigma_1, \sigma_2, p_{11}, p_{22})'\,.$$

The states $s_1, s_2, \ldots, s_T$ are unobserved.

The sample likelihood function is found by summing the density $p(x_1, \ldots, x_T, s_1, \ldots, s_T, \Theta)$ over all possible $2^T$ values of $s_1, \ldots, s_T$. The algorithm for the ML estimation of $\Theta$, however, does not require actual calculation of the $2^T$ terms appearing in the sample likelihood function. Details of the algorithm can be found in Hamilton (1989).

Several variants of the Markov switching model have been applied to the foreign exchange by Kaminsky (1989) and to stock market data by Tyssedal and Tjøstheim (1988), Turner *et al.* (1989), and Inman (1989). Inman finds that the stock market can be characterized as shifting between a normal, low beta state and an extreme, high beta state. The advantage of these models is that they allow for discrete jumps in the parameters and capture the non-normality and heteroscedasticity features of the data. They have been found to give better characterizations for many economic time series than models with smoothly varying parameters. However, when Lam (1989) compares the forecasting performance of a generalization of the Hamilton model with that of the Hamilton model, the ARIMA model, and the deterministic trend model for postwar quarterly GNP data, he finds that the results are somewhat mixed with no model dominating the others.

## 11.6 Structural Change with Endogenous Switching

The Markov switching models discussed in Section 11.5 are exogenous switching models. Several practical instances are available to analyze structural change within the context of endogenous switching. As discussed in Maddala (1986), the disequilibrium models and self-selection models fall in the category of endogenous switching models. A case of structural change within the context of self-selection models is presented below.

Suppose that prior to the enactment of a policy, economic agents can choose between two alternatives. The governmental policy eliminates this choice. Structural change in this case involves both changes in the parameters and the elimination of the choice. An example of this is the mandated accounting change in oil and gas exploration discussed in Lys (1984) and Sawyer and Shehata (1989). The accounting change eliminated the full-cost accounting method for oil and gas companies and mandated that they use the successful efforts method only. A company using full-cost (FC) accounting capitalizes on unproductive exploration and drilling costs from one field and writes them off against future income from productive reserves in another field, thus minimizing the earnings impact of unproductive exploration activity. A company using successful effects (SE) accounting, on the other hand, capitalizes only on those expenditures that result in discovery, and immediately expenses the expenditures associated with unproductive efforts. The rule mandating all oil and gas companies to use the SE method of accounting was subsequently reversed, but many other similar rules of accounting changes were not reversed.

In problems like this the analysis of structural change has to take into account the fact that there can be self-selection before the mandated change. A possible formulation of the model would be

$$y = X\beta_1 + u_1 \quad \text{for group 1 adopting the FC method}$$

$$y = X\beta_2 + u_2 \quad \text{for group 2 adopting the SE method.}$$

The group that a firm belongs to is determined by the choice function

$$I^* = Z\gamma + v\,.$$

The firm belongs to group 1 if $I^* \geq 0$ and to group 2 if $I^* < 0$. After the mandated change, all firms belong to group 2. But how do we measure the impact of the mandated change?

First of all, any analysis of parameter changes has to account for the fact that there is self-selection before the mandated change and no self-selection thereafter. Thus, to test for parameter changes, one has to estimate the parameters $\beta_1$ and $\beta_2$ for the two groups taking the self-selection into account. Then we can use the estimated equations to get predictions of $y$ for the two groups using the data after the mandated change and apply the predictive tests for stability.

If these predictive tests for stability show that parameter changes have occurred only for group 1 and not for group 2, then only firms choosing the FC method before the mandated change would have been affected by the new law. If the predictive tests show that parameter changes have occurred for both groups, then changes would have occurred in the oil and gas industry as a whole. In this case, the impact of the mandated change on the firms in group 1 can be obtained only after deducting the industrywide effects. A measure of the average effect of the mandated change is, therefore, given by

$$\begin{bmatrix} \text{average prediction} \\ \text{error for group 1} \end{bmatrix} - \begin{bmatrix} \text{average prediction} \\ \text{error for group 2} \end{bmatrix}.$$

Thus, the prediction errors can be used for both testing for structural change and estimating the effect of the mandated change on the firms that now use the FC accounting method.

In the accounting literature the effect of the mandated accounting changes has always been estimated by using the changes in the rate of return on the securities of the firms. The studies use the capital asset pricing model

$$R_{it} = a_i + b_i R_{mt}\,, \tag{11.7}$$

where $R_{it}$ is the return on stock $i$ in period $t$, and $R_{mt}$ is the return on the market portfolio in period $t$. Equation (11.7) is estimated using weekly data prior to the issuance of the proposal mandating the accounting change. Using the estimated values of $a_i$ and $b_i$, the estimated returns for the post-enactment period are computed. These are compared with the actual returns and the prediction errors computed. These prediction errors for the FC firms and SE firms are then compared.

The problem with this methodology is that no distinction is made between the FC firms and SE firms before the accounting change. The self-selection model captures the fact that the firms have the option of choosing one or the other—whichever maximizes the firm's rate of return. This problem is tackled in Sawyer and Shehata. Furthermore, it is important to consider other explanatory variables besides the market rate of return to explain the rate of return to securing $i$. Finally, in previous studies no statistical tests have been applied to check the significance of the observed effects.

There are alternative ways of formulating the self-selection model. One is the union and wages example of Lee, discussed in Maddala (1983, p. 356). In this formulation the differences in the rate of return for the FC and SE firms occurs as an explicit explanatory variable in the choice equation. Thus, firms choose the method of accounting that gives a higher rate of return for stockholders. Another formulation that includes explicitly the benefits and costs of choosing between the two methods is that outlined in Bjørkland and Moffitt (1987).

It is important to note that the estimation of the model prior to the passage of the law has to be done taking into account the fact that firms have the opportunity to choose between two alternatives. Tests for structural change and measurement of the significance of the effects of the mandated accounting change can be done using the predicted residuals for the period after the accounting change.

## 11.7 Conclusion

Disequilibrium and self-selection models are, as discussed in Maddala (1986), examples of switching regression models with endogenous switching. In recent years two types of switching regression models have received considerable attention in the econometric literature to model structural changes. The first of these is the Markov switching regression model. This model has been applied to analyze long swings in the exchange rates, peso problems in exchange rates, the study of stock market volatility, and the study of non-stationary time series. These models fall in the category of switching regression models with exogenous switching.

The second category of models, switching regression models with endogenous switching, can be fruitfully applied to study structural change in those cases where some laws eliminate the opportunities of self-selection that economic agents have. Examples of this are the mandated accounting changes that require firms to follow only one of two or more possible alternatives. These problems are too numerous and the empirical literature too extensive to be reviewed here. This chapter has outlined the two categories of switching regression models—the Markov switching models and the self-selection models—that have been used extensively in recent econometric literature to study structural changes.

### References

Bjørkland, A. and Moffitt, R. (1987), The estimation of wage gains and welfare gains in self-selection models. *Review of Economics and Statistics*, **69**, 42-49.

Cosslett, S.R. and Lee, L.F. (1985), Serial correlation in discrete variable models. *Journal of Econometrics*, **27**, 79-97.

Engel, C. and Hamilton, J.D. (1989), Long swings in the exchange rate: Are they in the data and do markets know it? Unpublished manuscript, University of Virginia.

Everitt, B.S. and Hand, D.J. (1981), *Finite Mixture Distributions*. London: Chapman Hall.

Goldfeld, S.M. and Quandt, R.E. (1973a), The estimation of structural shifts by switching regressions. *Annals of Economic and Social Measurement*, **2**, 475-485.

Goldfeld, S.M. and Quandt, R.E. (1973b), A Markov model for switching regressions. *Journal of Econometrics*, **1**, 3-16.

Hamilton, J.D. (1988), Rational expectations econometric analysis of changes in regime: An investigation of the term structure of interest rates. *Journal of Economic Dynamics and Contol*, **12**, 385-423.

Hamilton, J.D. (1989), A new approach to the economic analysis of nonstationary time series and the business cycle. *Econometrica*, **57**, 357-384.

Inman, K. (1989), A state switching model of stock market risk premiums. Unpublished manuscript, University of Washington.

Kaminsky, G. (1989), The Peso problem and the behavior of the exchange rate: The Dollar-Pound exchange rate, 1976-87. Working paper, University of California, San Diego.

Kaminsky, G. and Peruga, R. (1988), Credibility crisis: The Dollar in the early eighties. Working paper, University of California, San Diego.

Lam, Pok-sang (1989), The generalized Hamilton model: estimation and comparison with other models of economic time series. Manuscript, Ohio State University.

Lee, L.F., Maddala, G.S., and Trost, R.P. (1979), Testing for structural change by *D*-methods in switching simultaneous equations models. *Proceedings of the American Statistical Association (Business and Economics Section)*, 461-466.

Lee, L.F. and Porter, R. (1984), Switching regression models with imperfect sample separation information—with an application on cartel stability. *Econometrica*, **52**, 391-418.

Lewis, K.K. (1988), The persistence of the Peso problem when policy is noisy. *Journal of International Money and Finance*, **7**, 5-21.

Lewis, K.K. (1989), Changing beliefs and systematic forecast errors with evidence from foreign exchange. *American Economic Review*, **79**, 621-636.

Lys, T. (1984), Mandated accounting changes and debt covenants: The case of oil and gas accounting. *Journal of Accounting and Economics*, **6**, 39-65.

Maddala, G.S. (1983), *Limited Dependent and Qualitative Variables in Econometrics*. New York: Cambridge University Press.

Maddala, G.S. (1986), Disequilibrium, self-selection and switching models, pp. 1633-1688 in Z. Griliches and M.D. Intrilligator (eds.) *Handbook of Econometrics, Volume 3*. Amsterdam: North Holland.

Quandt, R.E. (1988), *The Econometrics of Disequilibrium*. New York: Basil Blackwell.

Sawyer, K.R. and Shehata, M. (1989), Mandated accounting changes and selection bias: The case of oil and gas accounting. Unpublished manuscript, University of Florida.

Turner, C.M., Startz, R., and Nelson, C.R. (1989), A Markov model of heteroscedasticity, risk and learning in the stock market. *Journal of Financial Economics* (forthcoming).

Tyssedal, J.S. and Tjøstheim, K. (1988), An autoregressive model with suddenly changing parameters and an application to stock market prices. *Applied Statistics*, **37**, 353-369.

CHAPTER 12

# Stochastic Trends in Simultaneous Equation Systems

*Andrew C. Harvey and Mariane Streibel*

## Summary

The formulation of an econometric model sometimes requires a trend component in one, or more, of the equations. This is to account for factors, such as technical progress, which are difficult to observe directly. Stochastic trends are more flexible than deterministic trends, and have successfully been used in a number of applications involving single equation dynamic regression. In this chapter the implications of including stochastic trends in simultaneous equation systems are considered. The issue of identifiability is discussed, and full information and limited information maximum likelihood estimation procedures are developed.

## 12.1 Introduction

Stochastic trend components are introduced into econometric equations when the level of the dependent variable cannot be completely explained by observable explanatory variables. Thus in Harvey *et al.* (1986) a stochastic trend played a key role in the specification of an equation relating employment to output where it picked up productivity effects stemming from changes in the capital stock and technical progress. Since technical progress features in many economic relationships, and since it cannot normally be measured, there may be a good case for using stochastic trends in a wide variety of applications. The aim of this chapter is to examine the issues raised when stochastic trends are specified in simultaneous equation models.

Stochastic trends play a central part in structural time series models [see Harvey (1989)]. The essence of such models is that they are formulated in terms of components that have a direct interpretation and are often of interest in themselves. Incorporating a stochastic trend in a regression model yields

$$y_t = \mu_t + \mathbf{x}_t'\boldsymbol{\delta} + \varepsilon_t, \quad t = 1, \ldots, T \tag{12.1}$$

$$\left.\begin{array}{rcl} \mu_t & = & \mu_{t-1} + \beta_{t-1} + \eta_t \\ \beta_t & = & \beta_{t-1} + \zeta_t \end{array}\right\} \tag{12.2}$$

where $\mathbf{x}_t$ is a $k \times 1$ vector of weakly exogenous variables, $\boldsymbol{\delta}$ is a $k \times 1$ vector of regression coefficients, and $\mu_t$ is a stochastic trend. The white noise disturbances, $\eta_t$ and $\zeta_t$, allow the level, $\mu_t$, and the slope, $\beta_t$, to evolve over time provided that their respective variances $\sigma_\eta^2$ and $\sigma_\zeta^2$ are nonzero. The disturbance term $\varepsilon_t$ is also white noise, with variance $\sigma_\varepsilon^2$, and is usually specified to be uncorrelated with $\eta_t$ and $\zeta_t$. The hyperparameters, $\boldsymbol{\psi} = (\sigma_\eta^2, \sigma_\zeta^2, \sigma_\varepsilon^2)'$, can be estimated in the time domain or in the frequency domain. Once this has been done, the model can be used for prediction. In addition the smoothed estimates of $\mu_t$ can be made available for interpretation. Both operations are carried out using filtering techniques based on the state space form.

If $\sigma_\eta^2$ and $\sigma_\zeta^2$ are both zero, the above model collapses to a regression model with a deterministic trend,

$$y_t = \alpha + \beta t + \mathbf{x}_t'\boldsymbol{\delta} + \varepsilon_t\,, \tag{12.3}$$

where $\alpha = \mu_0$. Models of this kind are easy to estimate and appear frequently in applied econometrics. However, if a deterministic trend is assumed when a stochastic one is appropriate, considerable distortion arises and the estimator of $\boldsymbol{\delta}$ will, in general, be inconsistent [see Nelson and Kang (1984)]. Similar distortions will arise when deterministic trends are employed in simultaneous equation systems.

Section 12.2 in this chapter discusses the identifiability of models containing stochastic trends and generalizes the classical rank condition. Maximum likelihood estimation is considered in Section 12.3.

## 12.2 Identifiability

A system of $N$ simultaneous equations with stochastic trends may be written as

$$\boldsymbol{\Gamma}\mathbf{y}_t = \boldsymbol{\mu}_t + \mathbf{D}\mathbf{x}_t + \boldsymbol{\varepsilon}_t\,, \tag{12.4}$$

where $\mathbf{y}_t$ is an $N \times 1$ vector of endogenous variables, $\mathbf{x}_t$ is a $K \times 1$ vector of weakly exogenous variables, and $\boldsymbol{\mu}_t$ is an $N \times 1$ vector of stochastic trends. The set of exogenous variables may be extended to include lagged endogenous variables. The matrices $\boldsymbol{\Gamma}$ and $\mathbf{D}$ are $N \times N$ and $N \times K$, respectively. The specification of $\boldsymbol{\mu}_t$ is just a vector generalization of (12.2) with $\mathrm{Var}(\boldsymbol{\eta}_t)$ and $\mathrm{Var}(\boldsymbol{\zeta}_t)$ both $N \times N$ covariance matrices, and $\boldsymbol{\eta}_t$ and $\boldsymbol{\zeta}_t$ uncorrelated with each other and with the disturbance term $\varepsilon_t$, in all time periods. The disturbance term $\varepsilon_t$ also has an $N \times N$ covariance matrix. Identifiability is therefore concerned with the three covariance matrices as well as with $\boldsymbol{\Gamma}$ and $\mathbf{D}$.

The model in (12.4) is identifiable if no other observationally equivalent structure satisfies the same set of prior constraints. Two structures may be regarded as being observationally equivalent if, in the reduced form, the expected value of $\mathbf{y}_t$ and its covariance matrix are identical for all $t$. Under normality, observational equivalence implies identical distributions. An observationally equivalent structure can be produced by pre-multiplying (12.4) by a nonsingular $N \times N$ matrix $\mathbf{F}$ [see Hsaio (1983) for a full discussion]. If the identity matrix is the only $\mathbf{F}$ matrix that can yield a structure satisfying the constraints on the structure, the model is said to be identifiable.

Suppose that the identifiability of a single equation within the system is to be assessed. Without loss of generality, this equation may be regarded as the first and may be written as

$$y_{1t} = \mu_{1t} + \mathbf{y}_{1t}^{*\prime}\boldsymbol{\gamma}_1 + \mathbf{x}_{1t}'\boldsymbol{\delta}_1 + \varepsilon_{1t}, \qquad (12.5)$$

where $\mathbf{y}_{1t}^*$ and $\boldsymbol{\gamma}_1$ are $(n_1-1)\times 1$ vectors and $\mathbf{x}_{1t}$ and $\boldsymbol{\delta}_1$ are $k_1 \times 1$ vectors. This equation is identifiable if pre-multiplication by the row vector $(1\ \mathbf{f}')$ only yields an equation satisfying the same *a priori* constraints if the $(N-1)\times 1$ vector $\mathbf{f}$ is null. In a classical model with no stochastic trends, a necessary condition for identifiability is the order condition, $K \geq n_1 + k_1 - 1$. A necessary and sufficient condition, the rank condition, requires that the matrix

$$\boldsymbol{\Phi} = [\overline{\boldsymbol{\Gamma}} \vdots \overline{\mathbf{D}}] \qquad (12.6)$$

has full column rank, where $\overline{\boldsymbol{\Gamma}}$ and $\overline{\mathbf{D}}$ are the $(N-1)\times(N-n_1)$ and $(N-1)\times(K-k_1)$ sub-matrices of $\boldsymbol{\Gamma}$ and $\mathbf{D}$ corresponding to coefficients of the variables in equations other than the first, which do not appear in the first equation.

This section seeks to extend the above conditions to take into account the role played by stochastic trends in identifiability. Whether or not stochastic trends play an active role in helping to identify the first equation depends on whether the first equation contains a stochastic trend of the same form as those in the other equations.

### 12.2.1 Random walk trends

The simplest kind of stochastic trend is a random walk, and a good deal of insight into the problem can be obtained by considering this case first.

Suppose, initially, that $N = 2$ and that there are no exogenous variables. The model is then

$$\left.\begin{array}{rcl} y_{1t} & = & \gamma_{12}y_{2t} + \mu_{1t} + \varepsilon_{1t} \\ y_{2t} & = & \gamma_{21}y_{1t} + \mu_{2t} + \varepsilon_{2t} \end{array}\right\}, \quad t = 1,\ldots,T, \qquad (12.7)$$

where

$$\mu_{it} = \mu_{i,t-1} + \eta_{it}\ , \quad t = 1, \ldots, T\ , \tag{12.8}$$

where $\eta_{i0}$ is fixed for $i = 1, 2$. Obviously neither equation is identifiable, but if $\mu_{1t}$ is not stochastic, that is $\mathrm{Var}(\eta_{1t}) = 0$, the first equation is identified since any linear combination involving the second equation would yield a stochastic trend, hence violating the distributional assumptions of the first equation.

It is instructive to rewrite (12.7) so that $\eta_{it}$ is expressed in terms of a deterministic and a stochastic part. Thus

$$\left.\begin{aligned} y_{1t} &= \gamma_{12} y_{2t} + \mu_{10} + \mu_{1t}^{+} + \varepsilon_{1t} \\ y_{2t} &= \gamma_{21} y_{1t} + \mu_{20} + \mu_{2t}^{+} + \varepsilon_{2t} \end{aligned}\right\}\ , \quad t = 1, \ldots, T\ , \tag{12.9}$$

where

$$\mu_{it}^{+} = \mu_{i,t-1}^{+} + \eta_{it}\ , \quad t = 1, \ldots, T \tag{12.10}$$

with $\mu_{i0}^{+} = 0$ for $i = 1, 2$. The exclusion of the stochastic component, $\mu_{1t}^{+}$, from the first equation is similar in its effect to the exclusion of an explanatory variable, insofar as it allows the permanent shifts in the second equation to trace out the first equation.

Some of the implications of identifiability can be illustrated by looking at the relationship between the reduced and structural form parameters. When $\mu_{1t}$ is deterministic, the reduced form of the model is

$$\left.\begin{aligned} y_{1t} &= \pi_{11} + \psi_1 \mu_{2t}^{+} + \mathrm{v}_{1t} \\ y_{2t} &= \pi_{21} + \psi_2 \mu_{2t}^{+} + \mathrm{v}_{2t} \end{aligned}\right\} \tag{12.11}$$

where

$$\begin{aligned} \pi_{11} &= (\mu_{10} + \gamma_{12}\mu_{20})/(1 - \gamma_{12}\gamma_{21}) \\ \pi_{21} &= (\gamma_{21}\mu_{10} + \mu_{20})/(1 - \gamma_{12}\gamma_{21}) \\ \psi_1 &= \gamma_{12}/(1 - \gamma_{12}\gamma_{21}) \\ \psi_2 &= 1/(1 - \gamma_{12}\gamma_{21})\,. \end{aligned}$$

This reduced form contains a common stochastic trend component, $\mu_{2t}^{+}$. Estimators of the parameters $\pi_{11}$, $\pi_{21}$, $\psi_1$, and $\psi_2$ can be computed by ML and unique estimators of the structural parameters $\mu_{10}$ and $\gamma_{12}$ obtained by noting that

$$\gamma_{12} = \psi_1/\psi_2 \tag{12.12}$$

and

$$\mu_{10} = \pi_{11} - \psi_1\pi_{21}/\psi_2\,. \tag{12.13}$$

The first equation is therefore exactly identified. If it were known that $\mu_{10}$ were zero, it would be over-identified as $\gamma_{12}$ could also be estimated from $\pi_{11}/\pi_{12}$. Thus both the deterministic and the stochastic part of a stochastic trend can help in identification, but as will be seen in the general case they do not count in quite the same way.

The assumption that the initial values $\mu_{10}$ and $\mu_{20}$ are fixed is not necessary. The identifiability arguments that follow could all be carried out with respect to a diffuse prior for $\mu_{10}$ and $\mu_{20}$. A general treatment of diffuse priors can be found in de Jong (1988). However, in the context of (12.7) we can proceed by defining a likelihood function conditional on $y_{11}$ and $y_{21}$. The model is then written as

$$\left.\begin{aligned} y_{1t} &= \gamma_{12}y_{2t} + \mu_{11}^* + \mu_{1t}^+ + \varepsilon_{1t}^* \\ y_{2t} &= \gamma_{21}y_{1t} + \mu_{21}^* + \mu_{2t}^+ + \varepsilon_{2t}^* \end{aligned}\right\}, \quad t = 2,\ldots,T, \tag{12.14}$$

where $\mu_{11}^* = y_{11} - \gamma_{12}y_{2t}$, $\mu_{21}^* = y_{21} - \gamma_{21}y_{1t}$, and $\varepsilon_{it}^* = \varepsilon_{it} - \varepsilon_{i1}$, $i = 1,2$. The same rules can be applied as when $\mu_{01}$ and $\mu_{02}$ are fixed.

A final aspect of identifiability of (12.7) concerns restrictions on the covariance matrices of $(\eta_{1t}, \eta_{2t})'$ and $(\varepsilon_{1t}, \varepsilon_{2t})'$. Specifying either of these to be diagonal results in both equations being identifiable. This condition generalizes in a fairly obvious way, but since it would seem difficult to verify in practice we will not pursue it further.

Now consider the identifiability of the first equation in a general model of the form (12.4). If a stochastic trend component does not appear in the first equation, the model may be written as

$$\begin{bmatrix} 1, -\gamma_1' & \mathbf{0}' \\ \boldsymbol{\Gamma}^* & \overline{\boldsymbol{\Gamma}} \end{bmatrix} \mathbf{y}_t = \begin{bmatrix} \boldsymbol{\delta}_1' & \mathbf{0} \\ \mathbf{D}^* & \overline{\mathbf{D}} \end{bmatrix} \mathbf{x}_t + \begin{bmatrix} 0 \\ \overline{\boldsymbol{\mu}}_0 \end{bmatrix} + \begin{bmatrix} \mathbf{0} \\ \overline{\mathbf{S}} \end{bmatrix} \boldsymbol{\mu}_t^+ + \boldsymbol{\varepsilon}_t, \quad t = 1,\ldots,T \tag{12.15}$$

$$\boldsymbol{\mu}_t^+ = \boldsymbol{\mu}_{t-1}^+ + \boldsymbol{\eta}_t^+, \quad \mathrm{Var}(\boldsymbol{\eta}_t^+) = \mathbf{I}, \tag{12.16}$$

where $\boldsymbol{\mu}_t^+$ is an $(N-1)\times 1$ vector, $\boldsymbol{\mu}_0^+ = \mathbf{0}$, and $\overline{\mathbf{S}}$ is a lower triangular $(N-1)\times(N-1)$ matrix. A necessary and sufficient condition for the identifiability of the first equation is that the matrix

$$\boldsymbol{\Phi} = [\ \overline{\boldsymbol{\Gamma}}\ \ \overline{\mathbf{D}}\ \ \overline{\boldsymbol{\mu}}_0\ \ \overline{\mathbf{S}}\ ] \tag{12.17}$$

has rank $N-1$. The result follows because, for any value of $t$, pre-multiplying by $(1\ \mathbf{f}')$ will only give an equation for $y_{1t}$ satisfying the required structural restrictions of $\mathbf{f} = \mathbf{0}$.

As a simple illustration of the above result, suppose that $N = 3$ and that no exogenous or endogenous variables are excluded from the first equation. Then

$$\boldsymbol{\Phi} = \begin{bmatrix} \overline{\mu}_{20} & s_{22} & 0 \\ \overline{\mu}_{30} & s_{32} & s_{33} \end{bmatrix}. \tag{12.18}$$

Hence identifiability is achieved if at least one of the trend components in equations two and three is stochastic; that is $s_{22}$ or $s_{33}$ is strictly positive. On the other hand, if there is no trend component at all in the third equation, all the entries in the bottom row of $\boldsymbol{\Phi}$ are zero and the identifiability condition does not hold.

If the first equation contains a constant term, then

$$\boldsymbol{\Phi} = \begin{bmatrix} s_{22} & 0 \\ s_{32} & s_{33} \end{bmatrix} . \tag{12.19}$$

Hence, for $\boldsymbol{\Phi}$ to be of full rank, both trends must be stochastic and they must not be perfectly correlated—in other words they must not be what Engle and Granger (1987) call cointegrated.

### 12.2.2 Local linear trends

The local linear trend model, (12.2), can be split up into a deterministic and a stochastic part by writing

$$\mu_t = \mu_0 + \beta_0 t + \mu_t^+ + \beta_t^+ , \tag{12.20}$$

where

$$\mu_t^+ = \mu_{t-1}^+ + \eta_t , \quad \mu_0^+ = 0 , \quad t = 1, \ldots, T \tag{12.21}$$

$$\beta_t^+ = 2\beta_{t-1}^+ - \beta_{t-2}^+ + \zeta_{t-1} , \quad \beta_0^+ = \beta_1^+ = 0 , \quad t = 2, \ldots, T . \tag{12.22}$$

If the first equation does not contain a trend component, the model is

$$\begin{bmatrix} 1, -\gamma_1' & \mathbf{0}' \\ \boldsymbol{\Gamma}^* & \overline{\boldsymbol{\Gamma}} \end{bmatrix} \mathbf{y}_t \tag{12.23}$$

$$= \begin{bmatrix} \delta_1' & \mathbf{0} \\ \mathbf{D}^* & \overline{\mathbf{D}} \end{bmatrix} \mathbf{x}_t + \begin{bmatrix} 0 & 0 \\ \overline{\mu}_0 & \overline{\beta}_0 \end{bmatrix} \begin{bmatrix} 1 \\ t \end{bmatrix} + \begin{bmatrix} \mathbf{0}' \\ \overline{\mathbf{S}}_\eta \end{bmatrix} \boldsymbol{\mu}_t^+ + \begin{bmatrix} \mathbf{0}' \\ \overline{\mathbf{S}}_\zeta \end{bmatrix} \boldsymbol{\beta}_t^+ + \varepsilon_t ,$$

where $\boldsymbol{\mu}_t^+$ and $\boldsymbol{\beta}_t^+$ are $(N-1)$ vectors obeying equations of the form (12.21), and $\overline{\mathbf{S}}_\eta$ and $\overline{\mathbf{S}}_\zeta$ are lower triangular matrices of order $N-1$ such that the covariance matrices of the $(N-1) \times 1$ disturbance vectors $\boldsymbol{\mu}_t$ and $\boldsymbol{\zeta}_t$ are identity matrices. Hence

$$\boldsymbol{\Phi} = [\, \overline{\boldsymbol{\Gamma}} \;\; \overline{\mathbf{D}} \;\; \overline{\mu}_0 \;\; \overline{\beta}_0 \;\; \overline{\mathbf{S}}_\eta \;\; \overline{\mathbf{S}}_\zeta \,] . \tag{12.24}$$

The appearance of some kind of trend component in the first equation leads to a modification of this matrix. For example, if the first equation contains a stochastic trend that is a random walk plus drift, then $\overline{\mu}_0$, $\overline{\beta}_0$, and $\overline{\mathbf{S}}_\eta$ disappear from (12.24) and only $\overline{\mathbf{S}}_\zeta$ can help identifiability.

Note that in (12.24), the deterministic components, 1 and $t$, are treated in exactly the same way as the exogenous variables in $\mathbf{x}_t$. The stochastic components, $\boldsymbol{\mu}_t^+$ and $\boldsymbol{\beta}_t^+$, both contribute to identifiability since the first is white noise in first differences while the second is white noise in second differences and so they cannot be confused. Indeed if the model contains no lagged endogenous variables, it is possible to let $\boldsymbol{\eta}_t$ and $\boldsymbol{\zeta}_t$ be stationary, invertible stochastic processes without affecting the identifiability conditions.

### 12.2.3 Seasonality

A stochastic seasonal component, $\gamma_t$, can be incorporated in a model. One specification is to let

$$\sum_{j=0}^{s-1} \gamma_{t-j} = \omega_t \,, \tag{12.25}$$

where $\omega_t$ is white noise. The seasonal component, $\gamma_t$, can be expressed as the sum of a deterministic function of $s-1$ initial values and a stochastic component depending on $\omega_t$. The matrix $\boldsymbol{\Phi}$ can be extended to take into account these factors.

Finally, a contribution to identifiability of the first equation may arise if the disturbance term $\varepsilon_{1t}$ is excluded. The $\boldsymbol{\Phi}$ matrix is then augmented by a matrix $\mathbf{S}_\varepsilon$, such that $\mathbf{S}_\varepsilon'\mathbf{S}_\varepsilon$ is the covariance matrix of the disturbances in the other equations.

## 12.3 Maximum Likelihood (ML) Estimation

The FIML estimator is straightforward in principle. It also forms the basis for the LIML estimator.

### 12.3.1 FIML

Consider the model

$$\boldsymbol{\Gamma}\mathbf{y}_t = \mathbf{Z}_t\boldsymbol{\alpha}_t + \mathbf{D}\mathbf{x}_t + \boldsymbol{\varepsilon}_t \,, \quad t = 1, \ldots, T \,, \tag{12.26}$$

where the stochastic term $\mathbf{Z}_t\boldsymbol{\alpha}_t$ could incorporate trend and seasonal components, with $\boldsymbol{\alpha}_t$ being a state vector obeying a suitably defined transition equation. Assume that all disturbances are normally distributed. To find the likelihood function of the observations, first let $\mathbf{y}_t^+$ denote the stochastic part of the right-hand side of (12.26), that is,

$$\mathbf{y}_t^+ = \boldsymbol{\Gamma}\mathbf{y}_t - \mathbf{D}\mathbf{x}_t = \mathbf{Z}_t\boldsymbol{\alpha}_t + \boldsymbol{\varepsilon}_t \,. \tag{12.27}$$

The distribution of $\mathbf{y}_t^+$, conditional on the information at time $t-1$, is multivariate normal with mean $\tilde{\mathbf{y}}_{t|t-1}^+$ and covariance matrix $\mathbf{F}_t$. Hence the distribution of $\mathbf{y}_t$, conditional on the same information set, is also multivariate normal with mean $\boldsymbol{\Gamma}^{-1}\tilde{\mathbf{y}}_{t|t-1}^+ + \boldsymbol{\Gamma}^{-1}\mathbf{D}\mathbf{x}_t$ and covariance matrix $\boldsymbol{\Gamma}^{-1}\mathbf{F}_t(\boldsymbol{\Gamma}^{-1})'$. The likelihood function of the $T$ sets of observations $\mathbf{y} = (\mathbf{y}_1', \ldots, \mathbf{y}_T')'$ is therefore

$$\begin{aligned} \log L(\mathbf{y}) = & -\frac{TN}{2}\log 2\pi - \frac{1}{2}\sum_t \log|\boldsymbol{\Gamma}^{-1}\mathbf{F}_t(\boldsymbol{\Gamma}^{-1})'| \\ & -\frac{1}{2}\sum_t \left(\mathbf{y}_t - \boldsymbol{\Gamma}^{-1}\tilde{\mathbf{y}}_{t|t-1}^+ - \boldsymbol{\Gamma}^{-1}\mathbf{D}\mathbf{x}_t\right)' \left[\boldsymbol{\Gamma}^{-1}\mathbf{F}_t(\boldsymbol{\Gamma}^{-1})'\right]^{-1} \\ & \times \left(\mathbf{y}_t - \boldsymbol{\Gamma}^{-1}\tilde{\mathbf{y}}_{t|t-1}^+ - \boldsymbol{\Gamma}^{-1}\mathbf{D}\mathbf{x}_t\right) . \end{aligned} \tag{12.28}$$

This can be rewritten as

$$\log L(\mathbf{y}) = -\frac{TN}{2}\log 2\pi + T\log|\boldsymbol{\Gamma}| - \frac{1}{2}\sum_t \log|\mathbf{F}_t| - \frac{1}{2}\sum_t \mathbf{v}_t'\mathbf{F}_t^{-1}\mathbf{v}_t\,, \tag{12.29}$$

where

$$\mathbf{v}_t = \boldsymbol{\Gamma}\mathbf{y}_t - \mathbf{D}\mathbf{x}_t - \mathbf{y}_{t|t-1}^{+}\,.$$

The $\mathbf{v}_t$'s are the innovations and the $\mathbf{F}_t$'s the covariance matrices obtained by running theKalman filter appropriate for the model for $\mathbf{y}_t^+$ on the 'observations' $\boldsymbol{\Gamma}\mathbf{y}_t - \mathbf{D}\mathbf{x}_t$. Hence conditional on all the parameters in the model, the likelihood function is straightforward to evaluate. Difficulties arise because, just as in the application of FIML to a static simultaneous equations system, the presence of the Jacobian term, $\log|\boldsymbol{\Gamma}|$, in the likelihood severely limits the number of parameters that can be concentrated out of the likelihood function. Thus the problem of estimating the model is really the practical one of having to carry out numerical optimization with respect to a large number of unknown parameters.

### 12.3.2 LIML

If interest centers on a single equation, say the first, and there is not enough information to specify restrictions on the remaining equations, a limited information estimation procedure is appropriate. In a classical model the LIML estimator of the parameters in the first equation can be obtained by applying ML to a system consisting of the first (structural) equation and the reduced form for the endogenous variables appearing in that equation. Since the Jacobian of this system is unity, the estimator can be computed by iterating a feasible SURE estimator to convergence [see Pagan (1979)].

Now consider the application of LIML in a system with stochastic trends. For simplicity, these will be assumed to be generated by a multivariate random walk. It will also be assumed that the system contains no lags. The model as a whole is

$$\left.\begin{aligned} \boldsymbol{\Gamma}\mathbf{y}_t &= \boldsymbol{\mu}_t + \mathbf{D}\mathbf{x}_t + \boldsymbol{\varepsilon}_t\,, \quad \mathrm{Var}(\boldsymbol{\varepsilon}_t) = \boldsymbol{\Sigma}_\varepsilon \\ \boldsymbol{\mu}_t &= \boldsymbol{\mu}_{t-1} + \boldsymbol{\eta}_t\,, \quad \mathrm{Var}(\boldsymbol{\eta}_t) = \boldsymbol{\Sigma}_\eta \end{aligned}\right\} \tag{12.30}$$

Hence the reduced form is

$$\mathbf{y}_t = \boldsymbol{\mu}_t^* + \boldsymbol{\Pi}\mathbf{x}_t + \mathbf{v}_t\,, \quad \mathrm{Var}(\mathbf{v}_t) = \boldsymbol{\Sigma}_v = \boldsymbol{\Gamma}^{-1}\boldsymbol{\Sigma}_\varepsilon(\boldsymbol{\Gamma}^{-1})' \tag{12.31}$$

$$\boldsymbol{\mu}_t^* = \boldsymbol{\mu}_{t-1}^* + \boldsymbol{\eta}_t^*\,, \quad \mathrm{Var}(\boldsymbol{\eta}_t^*) = \boldsymbol{\Sigma}_\eta^* = \boldsymbol{\Gamma}^{-1}\boldsymbol{\Sigma}_\eta(\boldsymbol{\Gamma}^{-1})'\,, \tag{12.32}$$

where $\boldsymbol{\mu}_t^* = \boldsymbol{\Gamma}^{-1}\boldsymbol{\mu}_t$. The equation of interest, (12.30), may be written as

$$\left.\begin{aligned} y_{1t} &= \mu_{1t} + \mathbf{y}_{1t}^{*\prime}\gamma_1 + \mathbf{x}_{1t}'\delta_1 + \varepsilon_{1t} \\ \mu_{1t} &= \mu_{1,t-1} + \eta_{1t} \end{aligned}\right\} \tag{12.33}$$

where $\mathbf{y}_{1t}^*$ and $\mathbf{x}_{1t}$ denote the vectors of included endogenous and exogenous variables, respectively. The disturbances $\varepsilon_{1t}$ and $\eta_{1t}$ may be correlated with the corresponding disturbances in the other structural equations. Prior knowledge may suggest the presence of a stochastic trend (12.33), but there is no information on whether or not stochastic trends are present in the other structural equations in the system, and so they must be included for generality.

The reduced form for the endogenous variables included in (12.33) may be written as

$$\left.\begin{aligned} y_{1t}^{*} &= \mu_{1t}^{*} + \boldsymbol{\Pi}_1 \mathbf{x}_t + v_{1t}^{*} \\ \mu_{1t}^{*} &= \mu_{1,t-1}^{*} + \eta_{1t}^{*} \end{aligned}\right\} \tag{12.34}$$

The LIML estimator is obtained by treating (12.33) and (12.34) as though they were the structural form of a system and applying FIML. The complication of the Jacobian term in the likelihood is absent as

$$|\boldsymbol{\Gamma}| = \begin{vmatrix} 1 & -\boldsymbol{\gamma}_1' \\ \mathbf{0} & \mathbf{I} \end{vmatrix} = 1 . \tag{12.35}$$

The model can therefore be regarded as a multivariate structural time series model with explanatory variables, that is,

$$\mathbf{y}_t = \begin{bmatrix} \mathbf{y}_{1t}^{*\prime} & \mathbf{x}_{1t}' & \mathbf{0} \\ \mathbf{0} & \mathbf{0} & \mathbf{I} \otimes \mathbf{x}_t' \end{bmatrix} \begin{bmatrix} \boldsymbol{\gamma}_1 \\ \boldsymbol{\delta}_1 \\ \boldsymbol{\pi}_1 \end{bmatrix} + \begin{bmatrix} \mu_{1t} \\ \boldsymbol{\mu}_{1t}^{*} \end{bmatrix} + \begin{bmatrix} \varepsilon_{1t} \\ \mathbf{v}_{1t}^{*} \end{bmatrix} , \tag{12.36}$$

where $\boldsymbol{\pi}_1 = \mathrm{vec}(\boldsymbol{\Pi}_1')$. Estimation can be carried out in the time domain by generalizing the algorithm of Kohn and Ansley (1985) or in the frequency domain using the approach set out in Fernandez (1989). Whichever procedure is used, it must be iterated to convergence. Because of the endogeneity of $\mathbf{y}_{1t}^{*}$, applying GLS based simply on a consistent estimator of the covariance matrices of $(\varepsilon_{1t}, \mathbf{v}_{1t}^{*\prime})'$ and $(\eta_{1t}, \boldsymbol{\eta}_{1t}^{*\prime})'$ will not necessarily yield consistent estimators of $\boldsymbol{\gamma}_1$ and $\boldsymbol{\delta}_1$ [see Pagan (1979)]. The question of finding initial consistent estimators of $\boldsymbol{\gamma}_1$, $\boldsymbol{\delta}_1$, and $\boldsymbol{\pi}_1$ is considered in Section 12.3.3.

The covariance matrices of $(\varepsilon_{1t}, \mathbf{v}_{1t}^{*\prime})'$ and $(\eta_{1t}, \boldsymbol{\eta}_{1t}^{*\prime})'$ are taken to be unconstrained, unless the specification of the first equation excludes a stochastic trend. In this case, $\mathrm{Var}(\eta_{1t})$ is obviously zero and the parameters to be estimated are all contained within the unconstrained covariance matrix of $\boldsymbol{\eta}_{1t}^{*}$.

The LIML estimator in a simultaneous equation system with vector MA disturbances was derived by Hall and Pagan (1981). Although the system in (12.30) has a vector MA representation when first differences are taken, estimating it by LIML is considerably easier than the LIML estimation of a model with unrestricted MA disturbances. As Hall and Pagan show, the presence of an unrestricted MA disturbance vector means that the reduced form for all endogenous variables in the system, apart from $y_{1t}$, must be used to form the LIML estimator for the first equation. Hence LIML is not a particularly attractive proposition in this case. Furthermore the limited information setup in which all disturbances follow a vector MA, apart from the one in the first equation does not seem to be a very natural one. On the other hand, the specification in (12.33) in which both $\varepsilon_{1t}$ and $\eta_{1t}$ may be correlated with the corresponding disturbances in the other structural equations seems a very reasonable one.

### 12.3.3 Starting values

We now consider the question of obtaining consistent starting values for the LIML estimator of the first equation in (12.30).

If the first equation contains a stochastic trend, consistent estimators of $\gamma_1$ and $\delta_1$ can be obtained by applying 2SLS to first differences (assuming that there are no lagged values of the dependent variable, $y_{1t}$). If there is no stochastic trend in the first equation, consistent estimators may be obtained by OLS provided that the variables in $\mathbf{x}_{1t}$ are integrated of order one [see Stock (1987)].

As regards the reduced form parameters, $\pi_1$, these may be estimated by running OLS on the differenced variables. Such estimators are consistent when lagged dependent variables are absent.

## Acknowledgments

We would like to thank the ESRC for financial support for the project "Multivariate Aspects of Structural Time Series Models", reference number B0 023 2155. The second author would like to thank CNPq for financial support.

## References

de Jong, P. (1988), The likelihood for a state space model. *Biometrika*, **75**, 165–169.

Engle, R.F. and Granger, C.W.J. (1987), Co-integration and error correction: Representation, estimation and testing. *Econometrica*, **55**, 251–276.

Fernandez, F.J. (1989), Estimation and testing of the multivariate exponentially weighted moving average model in the frequency domain. Forthcoming in *Journal of Time Series Analysis*.

Hall, A.D. and Pagan, A.R. (1981), The LIML and related estimators of an equation with moving average disturbances. *International Economic Review*, **22**, 719–730.

Hannan, E.J., Terrell, R.D., and Tuckwell, N. (1970), The seasonal adjustment of economic time series. *International Economic Review*, **11**, 24–52.

Harvey, A.C. (1989), *Forecasting, Structural Time Series Models and the Kalman Filter.* Cambridge: Cambridge University Press.

Harvey, A.C. *et al.* (1986), Stochastic trends in dynamic regression models: An application to the employment-output equation. *Economic Journal*, **96**, 975–985.

Hsaio, C. (1983), Identification, pp. 223–283 in M. Griliches and M.D. Intrilligator (eds.), *Handbook of Econometrics, Volume 1.* Amsterdam: North-Holland.

Kohn, R. and Ansley, C.F. (1985), Efficient estimation and prediction in time series regression models. *Biometrika*, **72**, 694–697.

Nelson, C.R. and Kang, H. (1984), Pitfalls in the use of time as an explanatory variable. *Journal of Business and Economic Statistics*, **2**, 73–82.

Pagan, A.R. (1979), Some consequences of viewing LIML as an iterated Aitken estimator. *Economics Letters*, **3**, 369–372.

Stock, J.H. (1987), Asymptotic properties of least squares estimators of co-integrating vectors. *Econometrica*, **55**, 1035–1056.

CHAPTER 13

# Time-Varying Nonlinear Regression

*Peter M. Robinson*

## Summary

This chapter discusses the estimation of time series models that are possibly nonlinear in parameters, which change smoothly but nonparametrically over time. We describe a time-varying, kernel-based analog of nonlinear least squares and establish consistency and asymptotic normality for the estimates, with allowance for serial dependence of a general kind in the disturbances. These results draw on general theorems for extremum estimates, which can also be applied to more general time-varying models.

## 13.1 Introduction

A time-varying, possibly nonlinear, regression model for a time series $y_t$ is

$$y_t = f(x_t; \theta_t) + u_t \ , \quad t = 1, 2, \dots . \tag{13.1}$$

Here, $x_t$ is a $d$-vector-valued time series that is independent of the zero-mean disturbance $u_t$; neither need be serially independent. The function $f$ is of known form. The $p$-vector $\theta_t$ is allowed to vary with $t$ in an unknown fashion. This model combines a finite-dimensional regression function $f$ with an infinite-dimensional sequence $\{\theta_t\}$, and in the latter sense it might be termed nonparametric. In case $\theta_t$ is *a priori* constant over $t$ we revert to a standard parametric nonlinear regression. With variable $\theta_t$, the model (13.1) extends the linear regression with time-varying parameters considered by Robinson (1989), in which $f(x; \theta) = x'\theta$, and the $u_t$ were assumed to be serially independent. The latter model itself extends a model that has been greatly discussed in the nonparametric function fitting literature, where $f(x;\theta) \equiv \theta$ and the $u_t$ are serially independent [see, e.g., Benedetti (1977)].

A part of the motivation of Robinson (1989) was to allow for the presence of explanatory variables $x_t$; this might be important in an econometric context. For example, we also relaxed the usual assumption of the nonparametric function fitting literature of homoscedesticity of the $u_t$ across time, though these were assumed serially independent. This chapter is concerned with allowing for nonlinearity of $f$ in the time-varying parameters $\theta_t$, as well as serial dependence in the disturbances $u_t$. Even in the simple model $f(x;\theta) \equiv \theta$ the latter possibility has only recently been considered. We describe estimates of the $\theta_t$ and establish their consistency and asymptotic normality. We also indicate that corresponding results can be obtained for more general models involving time-varying parameters.

Robinson (1989) briefly compared this approach to alternative ones in the literature for dealing with time-varying parameters. Perhaps the most attractive of these has involved taking $\theta_t$ to be generated by a stochastic process, depending on finitely many parameters, such as a random walk, reducing the problem to a parametric one. Apart from the possible misspecification introduced when the parametric model for $\theta_t$ is seriously incorrect, which was the main reason for proposing the nonparametric approach for linear models in Robinson (1989), in models nonlinear in $\theta_t$ stochastic modeling of $\theta_t$ is in general not even likely to produce a mathematically tractable model for $y_t$ that is amenable to estimation.

While not necessary for the purposes of computing nonparametric estimates of the $\theta_t$, in deriving statistical properties it is convenient to regard $\theta_t$ as being generated by a function $\theta(t)$ on $[0,1]$,

$$\theta_t = \theta\left(\frac{t}{N}\right) , \quad t = 1,2,\ldots,N ,$$

where $N$ is the amount of data available. As indicated by Robinson (1989), this approach is in the spirit of the nonparametric function fitting literature. The $\theta_t$ are made to depend on sample size due to the fact that estimates will not be consistent unless the number of observations on which they are based increases. Increasing $N$ will not be sufficient for this unless we regard the $\theta_t$ as ordinates of a smooth function on a grid that becomes finer as $N \to \infty$, in view of the local character of the estimates.

## 13.2 Estimation of the $\theta_t$

As in Robinson (1989) we employ a kernel estimation method. We introduce a kernel function $k$ that is bounded, real-valued, integrable, nonnegative, continuous (except possibly at finitely many points), and nonincreasing over the positive real line and nondecreasing over the negative real line, such that

$$\int_{-\infty}^{\infty} k(t)dt = 1$$

[cf., e.g., Benedetti (1977)]. Let

$$k_{\tau t} = k\left(\frac{N\tau - t}{Nh}\right),$$

where $h$ is a positive "bandwidth" number. Let

$$\hat{\theta}(\tau) = \arg\min_{\Theta} Q_\tau(\theta)$$

for some set $\Theta$, where

$$Q_\tau(\theta) = \frac{1}{Nh}\sum_{t=1}^{N} k_{\tau t}\{y_t - f(x_t;\theta)\}^2 .$$

Note that nonnegativity of $k$ implies the same of $Q_\tau(\theta)$, a standard requirement of objective functions.

A simple example of $k$ is

$$k(t) = \begin{cases} \frac{1}{2} & \text{if } |t| \leq 1 \\ 0 & \text{if } |t| > 1 . \end{cases} \tag{13.2}$$

With this $k$, for all large enough $h$, $\hat{\theta}(\tau)$ is the nonlinear least squares estimate of $\theta_0$ in the standard time-invariant nonlinear regression

$$y_t = f(x_t;\theta_0) + u_t .$$

For small enough $h$, $\hat{\theta}_t = \hat{\theta}(t/N)$ using (13.2) is a type of moving nonlinear regression estimate, summands of the objective function being progressively replaced as $t$ increases: we have

$$Q_{t/N}(\theta) = \sum_{|t-s|\leq Nh} \{y_s - f(x_s;\theta)\}^2 .$$

Notice that for given $N$, $h$ must be large enough that $p \leq 2Nh + 1$. Alternative choices of $k$ are discussed by Robinson (1989). Robinson also discusses a cross-validation method of choosing $h$ when $f(x;\theta) = \theta' x$. In principle, this method can be extended to nonlinear models. The discussion for the linear model in Robinson (1989) of rapid computation of all the $\hat{\theta}_t = \hat{\theta}(t/N)$ via the fast Fourier transform can be likewise extended, with respect to computing both the objective function $Q$ and the iterative steps of, say, the Gauss-Newton-type toward $\hat{\theta}(\tau)$. Naive formula for the latter iterates, not using the fast Fourier transform, are

$$\begin{aligned} \hat{\theta}_{j+1}(\tau) &= \hat{\theta}_j(\tau) + \{\sum_{t=1}^{N} k_{\tau t}\, m[x_t;\hat{\theta}_j(\tau)]\, m'[x_t;\hat{\theta}_j(\tau)]\}^{-1} \\ &\quad \times \sum_{t=1}^{N} k_{\tau t}\, m[x_t;\hat{\theta}_j(\tau)]\, \{y_t - f[x_t;\hat{\theta}_j(\tau)]\} \end{aligned}$$

for $j = 1, 2, \ldots$, where $m(x;\theta) = (\partial/\partial\theta) f(x;\theta)$ and we commence from an initial $\hat{\theta}_1(\tau)$.

## 13.3 Consistency of $\hat{\theta}(\tau)$

We find it convenient to begin by mentioning a consistency theorem for quite general extremum estimates. It is one of several variants of such results in the literature [cf., e.g., Malinvaud (1970)] but their application has generally been to standard time-invariant models. Let $S(\varphi)$ be an objective function on $R^q$ depending on $N$ observations such that

$$\hat{\varphi} = \arg\min_{\Phi} S(\varphi)$$

for some set $\Phi$. For a point $\varphi_0 \in \Phi$, let $S(\varphi) - S(\varphi_0) = a(\varphi) + b(\varphi)$, where $a(\varphi)$ is nonstochastic, nonnegative, constant with respect to $N$, and bounded away from 0 outside of a neighborhood of $\varphi_0$, while $\sup_{\Phi} |b(\varphi)| \xrightarrow{p} 0$. Then $\hat{\varphi} \xrightarrow{p} \varphi_0$. The proof of this result follows by noting that, for any $\varepsilon > 0$, with $N = \{\varphi : \varphi \in \Phi, \|\varphi - \varphi_0\| \le \varepsilon\}$ and $\overline{N} = \Phi - N$, we have that $\hat{\varphi} \in \overline{N}$ implies $\inf_{\overline{N}} S(\varphi) \le \inf_N S(\varphi)$, which implies $\inf_{\overline{N}} S(\varphi) \le S(\varphi_0)$. The last event occurs with probability

$$\begin{aligned} P[\inf_{\overline{N}}\{a(\varphi) + b(\varphi)\} \le 0] &\le P[\inf_{\overline{N}} a(\varphi) \le \sup_{\Phi} |b(\varphi)|] \\ &\le P[\sup_{\Phi} |b(\varphi)| \ge \eta] \end{aligned}$$

for some $\eta > 0$, and the last probability tends to 0 as $N \to \infty$ by assumption.

Let us apply this result to $\hat{\theta}(\tau)$, for some arbitrary, fixed $\tau$. We assume $f(x;\theta)$ is a measurable function of $x$ for each $\theta$, and is continuous at $\theta(\tau)$ for each $x$. Let the set $\Theta$ be compact. Let $\Xi$ be the set $\{\theta : \theta(t), 0 \le t \le 1\}$. Let $\theta(t)$ be continuous for $t \in [0,1]$. Let

$$\sup_{\theta \in \Theta \cup \Xi} Ef^2(x_t;\theta) < \infty\,. \tag{13.3}$$

In a neighborhood $N$ of $\theta(\tau)$ let there exist a function $f_*(x_t)$ such that for all $\theta \in N$, $f(x_t;\theta) \le f_*(x_t)$ where $Ef_*(x_t)^2 < \infty$. It follows by dominated convergence that

$$a(\theta) = E\{f(x_t;\theta) - f[x_t;\theta(\tau)]\}^2\,. \tag{13.4}$$

is continuous at $\theta(\tau)$. Condition (13.3) entails some weakening over the corresponding moment condition on $x_t$ of Robinson (1989). For identifiability we assume

$$a(\theta) > 0, \qquad \theta \ne \theta(\tau)\,. \tag{13.5}$$

We assume that $x_t$ is strictly stationary and strongly mixing. The latter assumptions are very strong. Stationarity could easily be relaxed to allow for some sort of non-trending heterogeneity at cost of some extra complexity in the conditions. Trending behavior in explanatory models is generally hard to handle in ordinary nonlinear regression, and even harder in this time-varying version. Incidentally the mixing condition in Robinson (1989) is equivalent to strong mixing. We assume that $u_t$ is also strictly stationary and strongly mixing, independent of $x_s$ for all $t, s$, and

$$E|u_t| < \infty\,.$$

We could somewhat relax strong mixing on $u_t$ if we instead imposed a higher moment condition, such as assuming finite variance. Finally we assume the conditions on $k$ at the beginning of Section 13.2 and

$$h + (Nh)^{-1} \to 0 \quad \text{as } N \to \infty .$$

To prove that

$$\hat{\theta}(\tau) \xrightarrow{p} \theta(\tau),$$

we take $\varphi = \theta$ and $S(\varphi) = Q(\theta)$, then $a(\theta)$ is as stated in (13.4) and, writing

$$\begin{aligned} g_t(\theta) &= f(x_t;\theta) - f[x_t;\theta(\tau)] \\ b(\theta) &= a(\theta)\left(\frac{1}{Nh}\sum_t k_{\tau t} - 1\right) + \frac{1}{Nh}\sum_t k_{\tau t}\{g_t^2(\theta) - a(\theta)\} \\ &\quad + \frac{2}{Nh}\sum_t k_{\tau t}u_t g_t(\theta) + \frac{2}{Nh}\sum_t k_{\tau t} g_t(\theta) g_t(\theta_t) = \sum_{i=1}^{4} b_i(\theta) . \end{aligned}$$

By (13.5) and continuity of $a(\theta)$, it follows that $a(\theta)$ is bounded away from 0 outside a neighborhood of $\theta(\tau)$. By compactness of $\Theta$ and continuity, $\sup_\Theta a(\theta) < \infty$, while

$$\frac{1}{Nh}\sum_t k_{\tau t} \to 1, \quad \text{as} \quad N \to \infty \tag{13.6}$$

slightly extending Lemma 1 of Benedetti (1977). Thus $\sup_\Theta |b_1(\theta)| \xrightarrow{p} 0$. For some $C < \infty$, let $g_t'(\theta) = g_t(\theta)1(g_t(\theta) \le C)$ and $g_t''(\theta) = g_t(\theta) - g_t'(\theta)$, where $1(.)$ is the indicator function. For any fixed $\theta$,

$$E\left|\frac{1}{Nh}\sum_t k_{\tau t}\{g_t''(\theta) - Eg_t''(\theta)\}\right| \le \frac{2}{Nh}\sum_t |k_{\tau t}|E|g_t''(\theta)|$$

which can be made arbitrarily small for $C$ and $N$ large enough, cf. (13.6). Writing

$$\sum_t{}' = \sum_{|\tau N - t| \le BNh}, \quad \sum_t{}'' = \sum_{|\tau N - t| > BNh}$$

for some $B < \infty$, we have

$$E\left|\frac{1}{Nh}\sum_t{}'' k_{\tau t}\{g_t'(\theta) - Eg_t'(\theta)\}\right| \le \frac{2C}{Nh}\sum_t{}'' |k_{\tau t}| \to 0$$

as $N \to \infty$ and $B \to \infty$, as in Robinson (1989), in view of Lemma 1 of Benedetti (1977). Next

$$E\left[\frac{1}{Nh}\sum_t{}' k_{\tau t}\{g_t'(\theta) - Eg_t'(\theta)\}\right]^2 \le \frac{C^2}{(Nh)^2}\left\{\sum_t{}' k_{\tau t}^2 + \sum_{s,t,s\ne t}{}' |k_{\tau t}k_{\tau s}|\alpha_{|t-s|}\right\},$$

where $\alpha_j$ is $x_t$'s $j$-th strong mixing number. This term is of order

$$O\left(\frac{BC^2}{Nh}\right) + O\left(\frac{C^2}{Nh}\sum_{j \leq Nh} \alpha_j\right) \to 0$$

as $N \to \infty$, using monotonicity of the $\alpha_j$. Thus $b_2(\theta) \xrightarrow{p} 0$ for each fixed $\theta$. Uniform convergence follows from compactness of $\Theta$ and a standard equicontinuity argument. Next consider $b_3(\theta)$. We have

$$E\left|\frac{1}{Nh}\sum_t{}'' k_{\tau t} u_t g_t(\theta)\right| \leq \frac{1}{Nh}\sum_t{}'' |k_{\tau t}|\, E|u_t|\, a(\theta)^{1/2} \to 0$$

as $N \to \infty$, $B \to \infty$. For some $D < \infty$, let $u_t' = u_t 1(|u_t| \leq D)$, $u_t'' = u_t - u_t'$. Then

$$E\left|\frac{1}{Nh}\sum_t{}' k_{\tau t}(u_t'' - Eu_t'')\right| \leq \frac{2}{Nh}\sum_t{}' |k_{\tau t}|\, E|u_t''| \to 0$$

as $D \to \infty$, whereas

$$E\left\{\frac{1}{Nh}\sum_t{}' k_{\tau t}(u_t' - Eu_t')g_t(\theta)\right\}^2 \leq \frac{D^2}{(Nh)^2}\sum_t{}'\sum_s{}' \beta_{|t-s|} a(\theta) \to 0$$

as $N \to \infty$, where $\beta_j$ is $u_t$'s mixing number. Thus $b_3(\theta) \xrightarrow{p} 0$ for each $\theta$, and again compactness implies uniformity. By the Schwarz inequality and the previous proof we have $b_4(\theta) \xrightarrow{p} 0$ uniformly on $\Theta$ if

$$\frac{1}{Nh}\sum_t |k_{\tau t}| g_t^2(\theta_t) \xrightarrow{p} 0 .$$

Now

$$E\left\{\frac{1}{Nh}\sum_t{}'' |k_{\tau t}| g_t^2(\theta_t)\right\} = O\left(\frac{1}{Nh}\sum_t{}'' |k_{\tau t}|\right) \to 0$$

as $N$, $B \to \infty$. On the other hand

$$E\left\{\frac{1}{Nh}\sum_t{}' |k_{\tau t}| g_t^2(\theta_t)\right\} = O\left(B \sup_{|t-\tau| \leq Bh} a[\theta(t)]\right) \to 0$$

as $h \to 0$ by continuity of $a$ at $\theta(\tau)$.

## 13.4 Asymptotic Normality of $\hat{\theta}(\tau)$

A central limit theorem for $\hat{\theta}(\tau)$ will be useful to set approximate confidence intervals. Again it is possible to adapt a general theorem for extremum estimates, with implications for more general models than (13.1).

Let the general objective function $S(\varphi)$ of the previous section be twice differentiable and suppose that for a sequence $M$ such that $M \to \infty$ as $N \to \infty$,

$$M^{1/2}\frac{\partial S(\varphi_0)}{\partial \varphi} \xrightarrow{d} N(0, G), \tag{13.7}$$

and that for any sequence $\tilde{\varphi}$ such that $\tilde{\varphi} \xrightarrow{p} \varphi_0$

$$\frac{\partial^2 S(\tilde{\varphi}_0)}{\partial \varphi \partial \varphi'} \xrightarrow{p} H \tag{13.8}$$

as $N \to \infty$, where $H$ is nonsingular and nonstochastic. Then if $\hat{\varphi} \xrightarrow{p} \varphi_0$ and $\varphi_0$ is an interior point of $\Phi$, as $N \to \infty$

$$M^{1/2}(\hat{\varphi} - \varphi_0) \xrightarrow{d} N(0, G^{-1}HG^{-1}).$$

This follows easily from the conditions, because with probability approaching one as $N \to \infty$, $\hat{\varphi}$ satisfies

$$0 = M^{1/2}\frac{\partial S(\hat{\varphi})}{\partial \varphi_i} = M^{1/2}\frac{\partial S(\varphi_0)}{\partial \varphi_i} + H_i(\varphi^i)M^{1/2}(\hat{\varphi} - \varphi_0)$$

where $\varphi_i$ is the $i$-th element of $\varphi$, $H_i$ is the $i$-th row of $H(\varphi) = \partial^2 S(\varphi)/\partial\varphi\partial\varphi'$, and $||\varphi^i - \varphi_0|| \le ||\hat{\varphi} - \varphi_0||$.

So far as $\hat{\theta}(\tau)$ is concerned, we assume that the regularity properties of Section 13.3 are in force, as well as the following. The function $\theta(t)$, $0 \le t \le 1$, satisfies a Lipschitz condition of order $\eta > 0$. Let $\theta(\tau)$ be an interior point of $\Theta$. For some neighborhood $N$ of $\theta(\tau)$, $f(x;\theta)$ is twice continuously differentiable in $\theta$ and

$$m(x;\theta) = \frac{\partial f(x;\theta)}{\partial \theta}, \quad n(x;\theta) = \frac{\partial m(x;\theta)}{\partial \theta'}$$

satisfy

$$||m(x;\theta)|| \le m_*(x), \quad ||n(x;\theta)|| \le n_*(x), \quad \theta \in N$$

where $||.||$ means Euclidean norm and

$$E\{m_*^2(x_t) + n_*^2(x_t)\} < \infty. \tag{13.9}$$

We define $R(\lambda)$ to be a $p \times p$ matrix that is continuous in $\lambda$ from the right, satisfies $R(-\pi) = 0$, has Hermitian nonnegative increments, and satisfies

$$E(m_t m'_{t+j}) = \int_{-\pi}^{\pi} e^{ij\lambda} dR(\lambda),$$

where $m_t = m[x_t; \theta(\tau)]$. The existence of such $R$ follows from stationarity of $x_t$ and (13.9). Let $R(\pi)$ be nonsingular, a local identifiability condition. Let the strong mixing coefficients, $\beta_j$, of $u_t$ satisfy

$$\sum_{i=1}^{\infty} \beta_j^{\delta/(2+\delta)} < \infty$$

for some $\delta > 0$, where also

$$E|u_t|^{2+\delta} < \infty.$$

The conditions imply that $u_t$ has a bounded spectral density, $r(\lambda)$, where

$$r(\lambda) = \frac{1}{2\pi} \sum_{-\infty}^{\infty} \gamma_j e^{-ij\lambda}, \quad -\pi < \lambda \leq \pi$$

and $\gamma_j = E(u_t u_{t+j})$. We assume also that

$$\sum_{j=1}^{\infty} |j\gamma_j| < \infty.$$

Finally, let $k(t)$ have support $(-1, 1)$ and be boundedly differentiable, and let

$$N h^{1+2\eta} \to 0, \quad \text{as } N \to \infty.$$

In fact discontinuities in $k$ at finitely many points, as in (13.2), can be handled. Under these conditions

$$(Nh)^{1/2}\{\hat{\theta}(\tau) - \theta(\tau)\} \xrightarrow{d} N[0, R(\pi)^{-1}\kappa\Sigma R(\pi)^{-1}], \tag{13.10}$$

where

$$\Sigma = 2\pi \int_{-\pi}^{\pi} r(\lambda) dR(\lambda), \quad \kappa = \int_{-\infty}^{\infty} k^2(t) dt.$$

We have to check conditions (13.7) and (13.8). First,

$$(Nh)^{1/2} \frac{\partial Q_\tau[\theta(\tau)]}{\partial \theta} = -\frac{2}{(Nh)^{1/2}} \sum k_{\tau t} u_t m_t - \frac{2}{(Nh)^{1/2}} \sum_t k_{\tau t} g_t(\theta_{t/N}) m_t. \tag{13.11}$$

The last term has variance bounded by

$$\left\{\frac{4}{Nh}\sum_t |k_{\tau t}|\, E\|m_t\|^2\right\}\left\{\sum_t |k_{\tau t}| E g_t^2(\theta_t)\right\}.$$

The first factor is clearly $O(1)$. In view of the domination condition on $m(x;\theta)$ it follows that

$$Eg_t^2(\theta_t) = O(\|\theta_t - \theta(\tau)\|^2) = O(h^{2\eta})$$

for $|t/N - \tau| \le h$. Thus

$$\sum_t |k_{\tau t}|\, Eg_t^2(\theta_t) = O(Nh^{1+2\eta})$$

as $N \to \infty$. This establishes that the second term on the right of (13.11) is $o_p(1)$. Denote the first term $v$. Conditional on $\{x_t\}$

$$[E(vv'|\{x_t\})]^{-1/2} v \xrightarrow{d} N(0, I),$$

where $I$ is the $p \times p$ identity matrix. The proof of this statement follows from an extension to triangular arrays of Theorem 18.5.3 of Ibragimov and Linnik (1971), and the ergodicity of $x_t$. Now

$$E(vv'|\{x_t\}) = \frac{4}{Nh}\sum_t \sum_s k_{\tau t} k_{\tau s} \gamma_{t-s} m_t m_s',$$

which differs from

$$\frac{4}{Nh}\sum_t k_{\tau t}^2 \sum_s \gamma_{t-s} m_t m_s' \tag{13.12}$$

by a quantity whose absolute expectation is bounded by

$$O\left[\frac{1}{(Nh)^2}\sum_t |k_{\tau t}| \sum_s |t-s||\gamma_{t-s}| E\|m_t\|^2\right] = O\left(\frac{1}{Nh}\sum_{j=0}^{\infty} |j\gamma_j|\right) \to 0.$$

For each fixed $j$

$$\frac{1}{Nh}\sum_t k_{\tau t}^2 m_t m_{t+j}' \xrightarrow{p} \kappa E(m_t m_{t+j}')$$

by a proof very similar to that used for $b_2(\theta)$ in Section 13.3. By a standard truncation argument and summability of the $\gamma_j$ it follows that

$$(13.12) \xrightarrow{p} 4\kappa \sum_{-\infty}^{\infty} \gamma_j E(m_t m_{t+j}') = 4\kappa\Sigma.$$

Thus

$$(Nh)^{1/2}\frac{\partial Q[\theta(\tau)]}{\partial\theta}\xrightarrow{d}N(0,4\kappa\Sigma).$$

To verify (13.8) note that

$$\begin{aligned}Q''_\tau(\theta) &= \frac{\partial^2 Q_\tau(\theta)}{\partial\theta\partial\theta'}\\ &= \frac{2}{Nh}\sum_t k_{\tau t}m(x_t;\theta)m'(x_t;\theta)-\frac{2}{Nh}\sum_t k_{\tau t}\{y_t-f(x_t;\theta)\}n(x_t;\theta).\end{aligned}$$

The fact that

$$Q''_\tau[\theta(\tau)]\xrightarrow{p}2R(\pi) \tag{13.13}$$

follows by means of some of the techniques used to deal with the $b_i(\theta)$ in Section 13.2. The condition (13.8) is verified by (13.13), consistency of $\hat{\theta}(\tau)$ and

$$\sup_{\|\theta-\theta(\tau)\|<\varepsilon}|Q''_\tau(\theta)-Q''_\tau[\theta(\tau)]|\xrightarrow{p}0,$$

which follows from dominated convergence as $N\to\infty$, $\varepsilon\to 0$.

With respect to its allowance for disturbance serial correlation, the result (13.10) is an extension of Eicker's (1967) central limit theorem for time-invariant linear regressions, and of subsequent results for other time-invariant models. An estimate of $R(\pi)$ is

$$\frac{1}{N}\sum_t m[x_t;\hat{\theta}(\tau)]\,m'[x_t;\hat{\theta}(\tau)].$$

A possible estimate of $\Sigma$ is

$$\hat{\Sigma}=\sum_{j=-M}^{M}\hat{\Gamma}_j,$$

where for $j\geq 0$

$$\begin{aligned}\hat{\Gamma}_j &= \frac{1}{N}\sum_{t=1}^{N-j}(k_{\tau t}\{y_t-f[x_t;\hat{\theta}(\tau)]\}\,m[x_t;\hat{\theta}(\tau)])\\ &\quad\times(k_{\tau,t+j}\{y_{t+j}-f[x_{t+j};\hat{\theta}(\tau)]\}\,m[x_{t+j};\hat{\theta}(\tau)])'\end{aligned}$$

and $\hat{\Gamma}_{-j}=\hat{\Gamma}'_j$. The consistency of these proposals, with $M$ increasing suitably slowly with $N$, remains to be established. As in the linear case considered by Robinson (1989), $(Nh)^{1/2}\{\hat{\theta}(\tau)-\theta(\tau)\}$ and $(Nh)^{1/2}\{\hat{\theta}(\nu)-\theta(\nu)\}$ can be shown to be asymptotically independent for $\tau\neq\nu$.

## 13.5 Final Comments

This chapter has presented a time-varying version of the methodology and asymptotic statistical theory of nonlinear regression, with allowance for disturbance serial correlation of unknown form. The statistical properties discussed here are relatively basic and among the more useful ones. However, it would no doubt be possible to study the asymptotic theory of the estimates at greater depth and establish a number of the other properties considered in the nonparametric function fitting literature.

The results of Sections 13.3 and 13.4 did not allow the $x_t$ to contain lagged $y_t$, a major drawback in view of some econometric applications. In principle there is no reason why the methodology presented in Section 13.2 cannot be used when there are lagged dependent variables, the difficulty is establishing the asymptotic theory under reasonably attractive and primitive conditions.

The form of the limiting covariance matrix of $\hat{\theta}(\tau)$ in Section 13.4 suggests that efficiency improvements are possible by correcting for the disturbance serial correlation.

In some applications it is likely that only some of the parameters will be allowed to vary with time, the remainder being constant *a priori*. In models where the time-invariant and time-varying parameters can naturally be estimated separately it is easy to see how to proceed, but in general, and in special cases such as where the regressors with time-invariant and time-varying parameters are not orthogonal, matters are more complicated.

The form of our estimate of $\theta(\tau)$ suggests how time-varying versions of standard estimates of a wide variety of models can be constructed. Time-varying modifications of maximum likelihood estimates were indicated by Robinson (1989). Versions of robust $M$-estimates are easily constructed [cf. Härdle and Gasser (1984)], as are the various instrumental variable-type estimates of linear and nonlinear simultaneous equations and transformation models. One simply replaces any sum over the $N$ observations occurring in the objective function by a weighted kernel sum. Many of the techniques of asymptotic theory developed in this chapter for the nonlinear regression model are relevant to a similar analysis of more general models.

### Acknowledgments

This article is based on research funded by the Economic and Social Research Council (ESRC) reference number R000231441.

### References

Benedetti, J.K. (1977), On the nonparametric estimation of regression functions. *Journal of the Royal Statistical Society*, **B-39**, 248-253.

Eicker, F. (1967), Limit theorems for regression with unequal and dependent errors, pp. 59-82 in *Proceedings of the Fifth Berkeley Symposium on Mathematical Statistics and Probability, Volume 1.* Berkeley: University of California Press.

Härdle, W. and Gasser, T. (1984), Robust nonparametric function fitting. *Journal of the Royal Statistical Society*, **B-46**, 42-51.

Ibragimov, I.A. and Linnik, Y.V. (1971), *Independent and Stationary Sequences of Random Variables*. Groningen: Walters-Noordhoff.

Malinvaud, E. (1970), The consistency of nonlinear regressions. *Annals of Mathematical Statistics*, **41**, 956-969.

Robinson, P.M. (1989), Nonparametric estimation of time-varying parameters, pp. 253-264 in P. Hackl (ed.), *Statistical Analysis and Forecasting of Economic Structural Change*. Berlin: Springer-Verlag.

CHAPTER 14

# Stability Analysis Using Kalman Filtering, Scoring, EM, and an Adaptive EM method

*Wolfgang Schneider*


## Summary

This chapter gives a detailed description of the implementation of ML estimation using scoring and EM for the hyperparameters of a particular econometric state-space. Kalman filtering enters these methods in an essential way. The EM method can be turned into an on-line (adaptive) estimation method, which can be conveniently used for speeding up the ML estimation procedure. We apply these techniques to a random walk parameter model of a standard (Goldfeld type) West German money-demand function testing its stability via testing the variances of the random walk for zero. We compare these results to a descriptive stability analysis that uses so-called flexible least squares—a nonstochastic variant of Kalman filtering.


## 14.1 Introduction

The organization of the chapter is as follows: In the first part of the chapter we develop the techniques to be used here in the more general context of a regression relationship with time-varying parameters. To this end we first cast this system into the framework of an econometric state-space. We then apply Kalman filtering to this model to calculate estimates of the time-varying parameters (the states) and to set up the likelihood function for its hyperparameters.

Starting from an elementary exposition of scoring and the EM method we then show in detail how these methods may be applied to the estimation of the hyperparameters of the econometric state-space considered here.

The EM estimators turn out to have a familiar Aitken structure involving the smoothed states and their cross products. Approximating the smoothing solution by the corresponding filtering solution allows us to turn the EM method into a fully recursive estimation method for the state-space hyperparameters ("adaptive filtering" as it is called in the engineering literature). This filter will considerably reduce the storage requirements for intermediate results as compared with the usual EM procedure.

Apart from serving as a tool for estimation within a stochastic framework, the Kalman filter can also be used as a purely descriptive technique assessing the potential gains of parameter variation in terms of reduction in the error variance of the regression equation at the cost of "complicating" the model in terms of additional parameter variance. This interpretation of Kalman filtering has recently been introduced into the econometric literature by Kalaba and Tesfatsion (1986, 1988a, 1988b) under the name of *flexible least squares*. It is closely related to the estimation of time-varying means in time series using certain smoothness priors.

In the second part of the chapter we apply this simple yet powerful tool of exploratory data analysis as a preliminary descriptive stability analysis to the standard West German money-demand function (Goldfeld-type specification). We then compare the outcome with the results obtained from maximum likelihood estimation of individual parameter variances in a random walk model for the parameters of the money-demand function.

It turns out that the simple descriptive technique already reproduces the main features of the Kalman smoothing solution in a state-space containing hyperparameters estimated by maximum likelihood. Looking at the computational search process for these estimates it is also demonstrated that a lot can be gained in terms of likelihood increase per computer time unit by using scoring in combination with one of the EM methods. The latter methods generate large, computationally cheap likelihood increases during their first iteration steps and produce a convenient starting point for a final step of scoring.

## 14.2 Setting up the Econometric State Space

We will consider the following regression equation with time-varying coefficients:

$$y_t = u'_{1t}x_t + u'_{2t}\beta + v_t \tag{14.1}$$

$$x_t = \Phi x_{t-1} + w_t\,, \tag{14.2}$$

where $y_t$ is an observable scalar to be explained as a linear function of observables $u_{it}$, $i = 1, 2$, superimposed by white noise $\{v_t\}$; the coefficient vectors $x_t$ and $\beta$ are of order $k \times 1$ and $l \times 1$, respectively; the coefficients $x_t$ of $u_{1t}$ are time-varying, whereas the coefficients $\beta$ of $u_{2t}$ do not vary over time. The behavior of $x_t$ obeys the law of motion (14.2), which is a first-order autoregressive scheme driven by a white noise process $\{w_t\}$. Also higher-order dynamics may be reduced to this parametric setup [see, e.g., Aoki (1987)].

Observables are available over the time span $t = 1, 2, \ldots, N$; the variables in $u_{it}$ are taken to be nonstochastic. We, however, note that under Gaussian assumptions we can also allow lagged dependent variables among the $u_{it}$ and other stochastic regressors independent of all random variables in (14.3) and still preserve the linear structure of Kalman filtering; for a discussion of stochastic regressors in state-space filtering see Ruskeepää (1985). The stochastic properties of $\{y_t\}$ will be derived from the joint distribution of the random vectors $\{x_0, w_0, \ldots, w_N, v_0, \ldots, v_N\}$. We assume that

The vectors $\{x_0, w_0, \ldots, w_N, v_0, \ldots, v_N\}$ are mutually uncorrelated and form a multivariate normal distribution, where

$$x_0 \sim N(\mu_0, \Sigma_0), \quad v_i \sim N(0, R), \quad w_i \sim N(0, Q), \quad i = 0, 1, \ldots . \tag{14.3}$$

Equations (14.1) and (14.2) and the stochastics (14.3) specify a particular econometric state-space model, (14.1) being the measurement equation and (14.2) being the transition equation of the states $x_t$.

This econometric state-space encompasses a number of interesting models as special cases, e.g., "the return to normality model" [Rosenberg (1973)], the "mixed model" [Sallas and Harville (1981, 1988)], "stochastic polynomial trend models" [Gersch and Kitigawa (1988)]. A stochastic linear trend in the time-varying coefficients $x_t$ translates into a simple random walk model in (14.2). This will also be the parameter dynamics into which we are going to embed a test on the hypothesis which of the parameters are time-varying and which are not. This test will amount to testing the variance components of $Q$. Stated in Bayesian terms: Given the system parameters $(\Phi, Q, \mu_0, \Sigma_0)$, the transition law (14.2) and assumption (14.3) specify a prior distribution on the time-varying coefficients $\{x_t : t = 0, 1, \ldots\}$; whereas the complete system (14.1) and (14.2) yields in conjunction with (14.3) a likelihood function $f(y_1, \ldots, y_N; \theta)$ for the whole set of hyperparameters $\theta = (\beta, \Phi, Q, R; \mu_0, \Sigma_0)$. The Bayesian viewpoint of state-space modeling is taken, e.g., in Sarris (1973), Harrison and Stevens (1976), and Meinhold and Singpurwalla (1983). In the case of a simple random walk, $x_t - x_{t-1} = w_t$, (14.2) defines a "smoothness prior" on deviations from parameter constancy. This interpretation is the one in Gersch and Kitigawa (1988). Viewed from this perspective the estimation of variance components in $Q$ is an instance of an empirical Bayes procedure in the sense of Morris (1983).

The purpose of the statistical analysis of (14.1) and (14.2) will consist in the solution of the following four problems:

(a) Reconstruction of the historical and future path of the time-varying coefficients $\{x_t\colon t = 0, 1, \ldots\}$ by their conditional means $\{E(x_t|y_1, \ldots, y_N; \theta)\}$. Given a quadratic loss function these means are optimal Bayes estimators of $x_t$ based upon the available sample information $y(N) := \{y_1, \ldots, y_N\}$. Depending on whether $t < N$, $t = N$, or $t > N$, they are called *smoothing, filtering*, or *prediction solution*, respectively.

(b) Calculation of a measure of precision for the estimated parameter path, e.g., the covariances $\{\mathrm{Cov}(x_t|y_1, \ldots, y_N; \theta)\}$.

(c) Calculation of an estimate $\hat{\theta}$ for the unknown hyperparameters $\theta$ according to the ML principle.

(d) Derivation of an asymptotic measure of precision for the model parameters thus estimated, e.g., by an appropriate description of the likelihood curvature.

## 14.3 The Kalman Filtering Algorithm

### 14.3.1 Kalman recursions

The Kalman filter algorithm provides a convenient instrument for solving problems (a) and (b). The filter permits one to calculate recursively arbitrary *a posteriori* distributions for the states $x_t$ given any sample information $y(s) = \{y_1, \ldots, y_s\}$ along with predictive distributions for the observables $y_t$ given sample information $y(t-1)$. Under the above assumptions these distributions are normal with parameters:

$$\hat{x}_{t|s} := E[x_t|y(s);\theta], \qquad \Sigma_{t|s} := \mathrm{Cov}[x_t|y(s);\theta] \tag{14.4}$$

$$\hat{y}_t := E[y_t|y(t-1);\theta], \qquad D_t := \mathrm{Cov}[y_t|y(t-1;\theta)]\,. \tag{14.5}$$

The smoothing solutions $(s = N)$ are computed via a series of forward and backward recursions [Anderson and Moore (1979)]:

*Initialization*

$$\hat{x}_{0|0} := \mu_0 \quad \Sigma_{0|0} := \Sigma_0, \tag{14.6}$$

*Forward recursions*

$$\hat{x}_{t|t-1} = \Phi\hat{x}_{t-1|t-1}, \quad \Sigma_{t|t-1} = \Phi\Sigma_{t-1|t-1}\Phi' + Q \tag{14.7}$$

$$\hat{y}_t = u'_{1t}\hat{x}_{t|t-1} + u'_{2t}\beta, \quad D_t = u'_{1t}\Sigma_{t|t-1}u_{1t} + R \tag{14.8}$$

$$\hat{x}_{t|t} = \hat{x}_{t|t-1} + K_t(y_t - \hat{y}_t), \quad \Sigma_{t|t} = (I - K_t u'_{1t})\Sigma_{t|t-1} \tag{14.9}$$

*Backward recursions*

$$\hat{x}_{t|N} = \hat{x}_{t|t} + A_t(\hat{x}_{t+1|N} - \hat{x}_{t+1|t}), \quad \Sigma_{t|N} = \Sigma_{t|t} - A_t(\Sigma_{t+1|t} - \Sigma_{t+1|N})A'_t \tag{14.10}$$

where

$$K_t := \Sigma_{t|t-1}u_{1t}D_t^{-1}, \quad \text{Kalman filter gain} \tag{14.11}$$

$$A_t := \Sigma_{t|t}\Phi'\Sigma_{t+1|t}^{-1}, \quad \text{Kalman smoother gain.} \tag{14.12}$$

There are several ways to derive these recursions:

(a) They can be viewed as an application of the Gram-Schmidt procedure for the computation of the least squares projections $\hat{y}_t = E[y_t|y(t-1)]$ in the space of square integrable random variables $\{y_1, y_2, \ldots\}$. This procedure orthogonalizes the random variables $\{y_1, \ldots, y_N\}$ into the sequence $\{\tilde{y}_1, \ldots, \tilde{y}_N\}$, where $\tilde{y}_t = y_t - \hat{y}_t$, which are mutually uncorrelated [Luenberger (1969, chapter 4)].

(b) The recursions are implied by the standard updating formulas for Gaussian posterior distributions in the present normal setup [Meinhold and Singpurwalla (1983)]. There is a close relationship between the standard formulas for mixed estimation and the Kalman recursions [Cooper (1973)].

(c) One can exploit the duality between the smoothing problem in the Gaussian stochastic environment specified here and the optimal linear regulator problem [Kwakernaak and Sivan (1972)]. The smoothing solution as given in (14.11) and (14.12) is the outcome of maximizing the objective function

$$J[x(N)] \tag{14.13}$$
$$= \sum_{i=1}^{n} R^{-1}(y_t - u'_{1t}x_t - u'_{2t}\beta)^2 + w'_tQ^{-1}w_t + (x_0 - \mu_0)'\Sigma_0^{-1}(x_0 - \mu_0)$$

as a function in $x(N) = \{x_0, x_1, \ldots, x_N\}$ subject to the constraint (14.2). The solution can be found by doing a standard exercise in dynamic programming [see, e.g., Jazwinski (1970, p. 225), Mayne (1966)]. Note that in the present Gaussian setup (14.13) can be interpreted [up to a few constants that do not depend on $x(N)$] as the joint distribution of $y(N)$ and $x(N)$ [see (14.26) below]. The smoothing solution $\{\hat{x}_{t|N}\colon t = 0, 1, \ldots, N\}$ is the posterior mean and mode and thus the maximum of $f[x(N)|y(N)]$, which is the maximum of (14.26) as a function in $x(N)$. This fact is at the heart of the equivalence of penalized least squares techniques, which start from a criterion like (14.13), and state-space smoothing. We will come back to this equivalence during our discussion of flexible least squares.

### 14.3.2 Enhancement of numerical precision

The recursions, (14.7)–(14.10), are known as the *standard covariance form* of the Kalman filter. A disadvantage of this numerical variant consists in the fact that taking differences in (14.10) and (14.9) might produce rounding errors which possibly accumulate as to render the computed covariance matrices no longer positive, (semi)definite. This problem may be avoided by using the so-called square root filter [Kaminski *et al.* (1971), Morf and Kailath (1975)]. There are fully recursive square root versions of the combined filter *and* fixed interval smoothing recursions for the so-called *information* filter [Bierman (1974, 1977), Fraser and Potter (1969), Carraro and Sartore (1986)], which process the inverse of the covariance matrices $\Sigma_{t|s}$. A version of combined filtering and fixed interval smoothing for the *standard* covariance filter can also be found [Schneider (1986, 1988)].

### 14.3.3 Calculation of smoothed cross products of the state

For the implementation of the ML estimation in this econometric state-space we also need the smoothing solution for the cross products $\Sigma_{t-1,t|N}$ of successive states, where

$$\Sigma_{t-1,t|N} := \mathrm{Cov}[x_{t-1}, x_t|y(N); \theta]. \tag{14.14}$$

One way to do this is to expand the state to include $x_{t-1}$ along with $x_t$ [as, e.g., suggested by Watson and Engle (1983, p. 395)]. This, however, unnecessarily blows up the dimension of the filtering problem, since there is a very simple relationship between the cross moments (14.14) and the filtering solution at the horizon $N$. We have

$$\Sigma_{t,t+1|N} = A_t \Sigma_{t+1|N} \,. \tag{14.15}$$

An easy proof using projection arguments in the spirit of Ansley and Kohn (1982) would proceed along the following lines. From the smoothing recursions (14.10) we derive the following difference equation for the errors $\tilde{x}_{t|s} := x_t - \hat{x}_{t|s}$:

$$\tilde{x}_{t|N} = \tilde{x}_{t|t} + A_t(\tilde{x}_{t+1|N} - \tilde{x}_{t+1|t}) \,. \tag{14.16}$$

Post-multiply (14.16) by $x_{t+1}$ on both sides, recalling that $x_{t+1}$ may be decomposed into

$$x_{t+1} = \hat{x}_{t+1|t} + \tilde{x}_{t+1|t} \quad \text{or} \quad x_{t+1} = \hat{x}_{t+1|N} + \tilde{x}_{t+1|N} \,,$$

then take expected values on both sides. Observing that $\tilde{x}_{t|s}$ has zero expectation given $y(s)$ one can deduce result (14.16). This result may also be inferred from Cooley *et al.* (1977) and Rosenberg (1977), who use a more complicated approach. A generalization of the above projection argument to arbitrary covariances of smoothed states can be found in de Jong and MacKinnon (1988).

### 14.3.4 Setting up the likelihood function

The Kalman filter recursions (14.8) provide us with the parameters of the conditional normal densities $f[y_t|y(t-1);\theta]$. Hence the log-likelihood for $\theta$ may be recursively calculated (except for a constant) as:

$$L_N(\theta) = -\frac{1}{2}\sum_{t=1}^{N}\left\{\log[|D_t(\theta)|] + [D_t(\theta)]^{-1}[y_t - \hat{y}_t(\theta)]^2\right\} \,. \tag{14.17}$$

This function is to be maximized in $\theta$. We assume that the solution can be found by differentiation within the interior of the parameter space, i.e., the solution $\hat{\theta}_N$ may be characterized as $\nabla L_N(\hat{\theta}_N) = 0$ and $\nabla^2 L_N(\hat{\theta}_N) < 0$.

### 14.3.5 Scoring

The scoring method [Rao (1973)] for the computation of $\hat{\theta}_N$ is an iterative search procedure consisting in the following iterations:

$$\hat{\theta}_N^{(i+1)} = \hat{\theta}_N^{(i)} + \alpha[\hat{\mathbf{I}}_N(\hat{\theta}_N^{(i)})]^{-1}\nabla L_N[\hat{\theta}_N^{(i)}] \,. \tag{14.18}$$

The parameter $\alpha$ denotes an appropriate step length [optimized as, e.g., in Lasdon (1970, pp. 11-13)] and $[\hat{\mathbf{I}}_N(\hat{\theta}_N^{(i)})]$, an estimate of the information matrix $\mathbf{I}_N(\theta) = E[-\nabla^2 L_N(\theta)]$

evaluated at $\theta = \hat{\theta}_N^{(i)}$. Following the rules of matrix differentiation [see, e.g., Magnus and Neudecker (1988)], we compute the derivative of (14.18) with respect to the $i$-th component of $\theta$ as

$$\nabla_i L_N(\theta) = -\frac{1}{2}\sum_{t=1}^{N} \left[D_t^{-1}\nabla_i D_t(I - D_t^{-1}\tilde{y}_t^2) - 2\, D_t^{-1}\tilde{y}_t\nabla_i\hat{y}_t\right] , \tag{14.19}$$

where $\{\tilde{y}_t\} := \{y_t - \hat{y}_t\}$ is the so-called *innovation sequence* of the Kalman filter. In a correctly specified model this series is (Gaussian) white noise with mean zero and variance $D_t$ [as defined in (14.8)]. Exploiting the moments of the innovation sequence, one can deduce the $(i,j)$-th element of the information matrix as [Gupta and Mehra (1974), Pagan (1980)]

$$[\mathbf{I}_N(\theta)]_{ij} = \frac{1}{2}\sum_{t=1}^{N} \left[D_t^{-1}\nabla_i D_t D_t^{-1}\nabla_j D_t + 2\, E(\nabla_i\hat{y}_t D_t^{-1}\nabla_j\hat{y}_t)\right] . \tag{14.20}$$

Neglecting in (14.20) the expectations operator, one derives an estimator for the information matrix, which only depends on first-order derivatives of the moments $\hat{y}_t$ and $D_t$. These derivatives are taken numerically during two filtering runs in a small neighborhood of the current iteration solution $\hat{\theta}_N^{(i)}$. The scoring method is a modified Newton method (the modification consists in substituting the Hessian matrix for its estimated expectation). Near the likelihood maximum scoring has quadratic convergence properties, but far-off the maximum this method may generate misleading search vectors due to a bad approximation of the Hessian matrix. As an alternative one may use the EM method, which converges only linearly near the likelihood maximum [Dempster *et al.* (1977)], but which—as experience shows—also generates satisfying increases in likelihood far-off the likelihood maximum. In addition the computational burden of the EM method is far less than that of scoring. In the framework of model (14.1) and (14.2) the EM method amounts to the solution of a standard least squares regression problem.

## 14.4 EM method

### 14.4.1 The basic algorithm

The general situation where the EM method is applicable is the following: There is a joint distribution $f_{X,Y}(x, y; \theta)$ of "observables Y" and "latent variables X". We are looking for an ML estimate of $\theta$ based upon the likelihood derived from the marginal distribution $f_Y(y;\theta)$. The EM method starts from a decomposition of the log-likelihood into two auxiliary functions. The definition of conditional densities implies

$$L_N(\theta) := \log f_Y(y;\theta) = \log f_{X,Y}(x, y;\theta) - \log f_{X|Y}(x|y;\theta) .$$

If one takes expectations on both sides with respect to the distribution $f_{X|Y}(x|y;\bar{\theta})$ one arrives at

$$\log f_Y(y;\theta) = E[\log f_{X,Y}(x,y;\theta)|y,\bar{\theta}] - E[\log f_{X|Y}(x|y;\theta)|y,\bar{\theta}] \tag{14.21}$$

or briefly in obvious notation

$$\log f_Y(y;\theta) = A_{X,Y}(\theta,\bar{\theta}) - A_{X|Y}(\theta|\bar{\theta})\,. \tag{14.22}$$

Starting from an iteration solution $\theta^{(i)}$ the next solution $\theta^{(i+1)}$ is constructed in two steps:

(1) *Expectation step:* form the auxiliary function $A_{X,Y}[\theta,\theta^{(i)}]$ (14.23)

(2) *Maximization step:* find $\theta^{(i+1)}$ such that for all $\theta \in \Theta$ (parameter space)

$$A_{X,Y}[\theta^{(i+1)},\theta^{(i)}] \geq A_{X,Y}[\theta,\theta^{(i)}]\,. \tag{14.24}$$

For all iteration sequences $\{\theta^{(i)}\}$ constructed according to (14.23) and (14.24) the corresponding likelihood sequence $\{L[\theta^{(i)}]\}$ is nonincreasing. It will converge to a stationary point of the likelihood function under suitable continuity and differentiability conditions on $A_{X,Y}(\theta_1,\theta_2)$ [see Dempster *et al.* (1977), Wu (1983)]. The EM method does not guarantee convergence to the global likelihood maximum. In addition, one cannot exclude convergence to a saddle point; the situation has to be checked numerically by an analysis of the Hessian. The great advantage of the EM method lies in the fact that usually the maximization problem of the auxiliary function $A_{X,Y}(\theta_1,\theta_2)$ is far simpler than direct maximization of $L(\theta)$. This is also the case in state-space models [Watson and Engle (1983), Shumway and Stoffer (1982)]. In this framework the latent variables $X$ may be identified with the state vectors $X = (X_0, X_1, \ldots, X_N)$, and the observables with the output vectors $Y = (Y_1, \ldots, Y_N)$. The joint distribution of $X$ and $Y$ is determined by the stochastic specification (14.3) as

$$\begin{aligned}\log f_{X,Y}(x,y;\theta) &= \log f_{X(N),Y(N)}[x(N),y(N);\theta] \\ &= \log f_{Y(N)|X(N)}[y(N)|x(N);\theta] + \log f_{X(N)}[x(N);\theta]\,.\end{aligned} \tag{14.25}$$

Exploiting the special Markovian structure of this econometric state-space we have

$$\begin{aligned}\log f_{X,Y}(x,y;\theta) &= \sum_{t=1}^{N} [\log f(y_t|x_t;\theta) + \log f(x_t|x_{t-1};\theta)] + \log f(x_0;\theta) \\ &= \sum_{t=1}^{N} \left[\log \mathcal{N}(y_t : u'_{1t}x_t + u'_{2t}\beta; R) + \log \mathcal{N}(x_t : \Phi x_{t-1}; Q)\right] + \log \mathcal{N}(x_0 : \mu_0; \Sigma_0)\,.\end{aligned} \tag{14.26}$$

*Expectation step*

The auxiliary function $A_{X,Y}(\theta,\theta^{(i)})$ is the expected value of (14.26) with respect to the conditional distribution $f_{X(N)|Y(N)}[x(N)|y(N);\theta^{(i)}]$ given the last iteration solution $\theta^{(i)}$. It will be convenient to decompose the function $A_{X,Y}$ into the partial sums

$$A_{X,Y}[\theta,\theta^{(i)}] = E[\log f_{X,Y}(x,y;\theta)|y;\theta^{(i)}] = \sum_{j=1}^{3} A_j[\theta_j;\theta^{(i)}]$$

where

$$A_1[\theta_1,\theta^{(i)}] = \sum_{t=1}^{N} E_{X_t|Y(N)}[\log \mathcal{N}(y_t : u'_{1t}x_t + u'_{2t}\beta; R)|y(N);\theta^{(i)}] \tag{14.27}$$
$$= \text{const} - \frac{N}{2}\log|R| - \frac{1}{2}\sum_{t=1}^{N}\{R^{-1}E[v_t^2|y(N);\theta^{(i)}]\}$$

$$A_2[\theta_2,\theta^{(i)}] = \sum_{t=1}^{N} E_{X_t,X(t-1)|Y(N)}[\log \mathcal{N}(x_t : \Phi x_{t-1}; Q)|y(N);\theta^{(i)}] \tag{14.28}$$
$$= \text{const} - \frac{N}{2}\log|Q| - \frac{1}{2}\sum_{t=1}^{N}\text{trace}\{Q^{-1}E[w_t w_t'|y(N);\theta^{(i)}]\}$$

$$A_3[\theta_3,\theta^{(i)}] = E_{X_0|Y(N)}\{\log \mathcal{N}[x_0 : \mu_0; \Sigma_0|y(N);\theta^{(i)}]\} \tag{14.29}$$
$$= \text{const} - \frac{1}{2}\log|\Sigma_0| - \frac{1}{2}\text{trace}\{\Sigma_0^{-1}E[w_0 w_0'|y(N);\theta^{(i)}]\}$$

The cross moments of $\{v_t\}$ and $\{w_t\}$ in (14.27)–(14.29) can be deduced from the Kalman filter recursions (14.6)–(14.10) using model specification $\theta^{(i)}$. For this purpose the following decompositions of the noise vectors turn out to be convenient:

$$v_t = y_t - u'_{1t}x_t - u'_{2t}\beta = \hat{v}_{t|N} - u'_{1t}\tilde{x}_{t|N}, \qquad t = 1,\ldots \tag{14.30}$$
$$w_t = x_t - \Phi x_{t-1} = \hat{w}_{t|N} + [\tilde{x}_{t|N} - \Phi\tilde{x}_{t-1|N}], \qquad t = 1,\ldots \tag{14.31}$$
$$w_0 = x_0 - \mu_0 = \hat{v}_{0|N} + \tilde{x}_{0|N}\,, \tag{14.32}$$

where $\hat{v}_{t|N}$ and $\hat{w}_{t|N}$ are the smoothed system errors given specification $\theta^{(i)}$, i.e.,

$$\hat{v}_{t|N} := y_t - u'_{1t}\hat{x}_{t|N} - u'_{2t}\beta, \quad t = 1,2,\ldots \tag{14.33}$$
$$\hat{w}_{0|N} := \hat{x}_{0|N} - \mu_0; \qquad \hat{w}_{t|N} := \hat{x}_{t|N} - \Phi\hat{x}_{t-1|N}, \qquad t = 1,2,\ldots. \tag{14.34}$$

The prediction errors $\tilde{x}_{t|N} := x_t - \hat{x}_{t|N}$, where $\hat{x}_{t|N} = E[x_t|y(N);\theta^{(i)}]$, have zero expectation given $y(N)$ and $\theta^{(i)}$. Since $\hat{w}_{t|N}$ and $\hat{v}_{t|N}$ are linear functions in $y(N)$, we have

$$\left.\begin{array}{l} E[\hat{v}_{t|N}\tilde{x}'_{s|N}|y(N);\theta^{(i)}] = 0 \\ E[\hat{w}_{t|N}\tilde{x}'_{s|N}|y(N);\theta^{(i)}] = 0 \end{array}\right. \quad \text{for arbitrary } s,t = 0,1,\ldots. \tag{14.35}$$

Observing these orthogonality conditions in (14.30)–(14.32), we can complete the expectation step as

$$E[v_t^2|y(N);\,\theta^{(i)}] = \hat{v}_{t|N}^2 + u'_{1t}\Sigma_{t|N}u_{1t} \tag{14.36}$$
$$E[w_t w_t'|y(N);\,\theta^{(i)}] = \hat{w}_{t|N}\hat{w}'_{t|N} + \Sigma_{t|N} - \Phi\Sigma_{t-1,t|N} - \Sigma'_{t-1,t|N}\Phi' + \Phi\Sigma_{t-1|N}\Phi' \tag{14.37}$$
$$E[w_0 w_0'|y(N);\theta^{(i)}] = \hat{w}_{0|N}\hat{w}'_{0|N} + \Sigma_{0|N}\,. \tag{14.38}$$

*Maximization step*

Obviously we can maximize (14.26) by maximizing each component $A_j$ separately. Each maximization corresponds to a well-known regression problem. All critical points may be found by differentiation, but their actual computation may involve the (iterative) solution of nonlinear systems. We compute the derivatives with respect to the components of $\theta$ as

$$\frac{\partial A_1}{\partial \beta} = \sum_{t=1}^{N} \left[u_{2t}(y_t - u'_{1t}\hat{x}_{t|N} - u'_{2t}\beta)\right] R^{-1} \tag{14.39}$$

$$\frac{\partial A_2}{\partial \Phi'} = \sum_{t=1}^{N} \left[M_{t-1,t|N} - M_{t-1|N}\Phi'\right] Q^{-1} \tag{14.40}$$

$$\frac{\partial A_1}{\partial R} = -\frac{N}{2} + \frac{1}{2}\sum_{t=1}^{N} R^{-2} E[v_t^2|y(N);\theta^{(i)}] \tag{14.41}$$

$$\frac{\partial A_2}{\partial Q} = -\frac{N}{2} + \frac{1}{2}\sum_{t=1}^{N} Q^{-1} E[w_t w'_t|y(N);\theta^{(i)}]Q^{-1} \tag{14.42}$$

$$\frac{\partial A_3}{\partial \mu'_0} = (\hat{x}_{0|N} - \mu_0)'\Sigma_0^{-1} \tag{14.43}$$

$$\frac{\partial A_3}{\partial \Sigma_0} = -\frac{1}{2}\Sigma_0^{-1}\left[M_{0|N} - \hat{x}_{0|N}\mu'_0 - \mu_0\hat{x}'_{0|N} + \mu_0\mu'_0\right]\Sigma_0^{-1}, \tag{14.44}$$

where $M_{t-1|N}$ and $M_{t-1,t|N}$ are the smoothed cross moments of the states, which are [using (14.15) and (14.12)]

$$M_{t-1|N} := E[x_{t-1}x'_{t-1}|y(N);\theta^{(i)}] = \Sigma_{t-1|N} + \hat{x}_{t-1|N}\hat{x}'_{t-1|N} \tag{14.45}$$

$$M_{t-1,t|N} := E[x_{t-1}x'_t|y(N);\theta^{(i)}] = \Sigma_{t-1|t-1}\Phi'\Sigma_{t|t-1}^{-1}\Sigma_{t|N} + \hat{x}_{t-1|N}\hat{x}'_{t|N}\,. \tag{14.46}$$

Collecting all nonzero elements of $\Phi$ in the vector $\varphi$, such that $vec(\Phi') = C\varphi$, hence $\frac{\partial A_2}{\partial \varphi} = C'vec(\frac{\partial A_2}{\partial \Phi'})$, where $C$ is an appropriate selection matrix, we arrive at the final set of equations determining the EM solution $(\hat{\beta}, \hat{\varphi}, \hat{R}, \hat{Q}, \hat{\mu}_0, \hat{\Sigma}_0)$ at step $(i+1)$ given model specification $\theta^{(i)}$ [assumed in the computation of all smoothed moments used below and indicated by the superindex $(i)$]

$$\sum_{t=1}^{N} u_{2t}\cdot(y_t - u'_{1t}\hat{x}_{t|N}) = (\sum_{t=1}^{N} u_{2t}u'_{2t})\cdot\hat{\beta} \tag{14.47}$$

$$[C'(\hat{Q}^{-1}\otimes I_k)vec(\sum_{t=1}^{N} M_{t-1,t|N})] = [C'(\hat{Q}^{-1}\otimes\sum_{t=1}^{N} M_{t-1|N})C]\cdot\hat{\varphi} \tag{14.48}$$

$$\hat{\mu}_0 = \hat{x}_{0|N}\,, \quad \hat{\Sigma}_0 = \Sigma_{0|N} \tag{14.49}$$

$$\hat{R} = \frac{1}{N}\sum_{t=1}^{N} E[v_t^2|y(N),\theta^{(i)}] \tag{14.50}$$

$$\hat{Q} = \frac{1}{N}\sum_{t=1}^{N} E[w_t w'_t|y(N),\theta^{(i)}]\,. \tag{14.51}$$

The solution for $\mu_0$, $\Sigma_0$ is trivial, since it coincides with the smoothing solution. The solution for $\beta$ and $R$ are the standard OLS estimate formulas for a single regression equation. The normal equations for $\varphi$ and $Q$ are nonlinear in these parameters, and we can exploit the usual estimation techniques available for systems of seemingly unrelated regression equations [Zellner (1962)], i.e., two-step or iterated Aitken estimators. The two-step Aitken estimator consists in updating $\varphi$ given the last iteration solution for $Q$ and then substituting these updates for (14.51) to generate a new solution of $Q$, i.e.,

$$\hat{\beta}^{(i+1)} = \left[\sum_{t=1}^{N} u_{2t}u_{2t}'\right]^{-1} \left\{\sum_{t=1}^{N}[y_t - u_{1t}'\hat{x}_{t|N}^{(i)}]u_{2t}\right\} \tag{14.52}$$

$$\hat{\varphi}^{(i+1)} = \left\{C'\left[[\hat{Q}^{(i)}]^{-1} \otimes \sum_{t=1}^{N} M_{t-1|N}^{(i)}\right]C\right\}^{-1} \times C'\{[\hat{Q}^{(i)}]^{-1} \otimes I_k\}\, vec\left[\sum_{t=1}^{N} M_{t-1,t|N}^{(i)}\right] \tag{14.53}$$

$$\hat{R}^{(i+1)} = \frac{1}{N}\sum_{t=1}^{N}[\hat{v}_{t|N}^{(i)\,2} + u_{1t}'\Sigma_{t|N}^{(i)}u_{1t}] \tag{14.54}$$

$$\hat{Q}^{(i+1)} = \frac{1}{N}\sum_{t=1}^{N}[\hat{w}_{t|N}^{(i)}w_{t|N}^{(i)\prime} + \mathcal{W}_{t|N}^{(i)}], \tag{14.55}$$

where

$$\hat{v}_{t|N}^{(i)} := y_t - u_{1t}'\hat{x}_{t|N}^{(i)} - u_{2t}'\hat{\beta}^{(i+1)} \tag{14.56}$$

$$\hat{w}_{t|N}^{(i)} := \hat{x}_{t|N}^{(i)} - \hat{\Phi}^{(i+1)}\hat{x}_{t-1|N}^{(i)} \tag{14.57}$$

$$\mathcal{W}_{t|N}^{(i)} := \Sigma_{t|N}^{(i)} + \hat{\Phi}^{(i+1)}\Sigma_{t-1|N}^{(i)}\hat{\Phi}^{(i+1)\prime} - \hat{\Phi}^{(i+1)}\Sigma_{t-1,t|N}^{(i)} - \Sigma_{t-1,t|N}^{(i)\prime}\hat{\Phi}^{(i+1)\prime}. \tag{14.58}$$

Note that $\mathcal{W}_{t|N}^{(i)}$, as defined in (14.58), should be positive definite. It is a quadratic form in the smoothed covariance matrix of $(x_{t-1}', x_t')'$. We have

$$\begin{aligned}\mathcal{W}_{t|N}^{(i)} &:= \mathrm{Cov}[\tilde{x}_{t|N} - \Phi\tilde{x}_{t-1|N}|y(N);\theta^{(i)}] \\ &= \begin{bmatrix} \hat{\Phi}^{(i+1)} & -I \end{bmatrix} \begin{bmatrix} \Sigma_{t-1|N}^{(i)} & \Sigma_{t-1,t|N}^{(i)} \\ \Sigma_{t-1,t|N}^{(i)\prime} & \Sigma_{t|N}^{(i)} \end{bmatrix} \begin{bmatrix} \hat{\Phi}^{(i+1)\prime} \\ -I \end{bmatrix}.\end{aligned} \tag{14.59}$$

The iterated Aitken estimator for $\varphi$ and $Q$ is computed iterating between (14.53) and (14.55). The convergence of these (sub)iterations are guaranteed if the matrix series $\{\hat{Q}\}$, obtained during these subiterations, can be uniformly bounded from above and below by positive definite matrices [Oberhofer and Kmenta (1974)].

The EM iterations will generate increases in the likelihood in each step. If there are no ridges in the likelihood, they will converge to a stationary point of the likelihood

[Wu (1983)]. Since convergence to a saddle point or to the global maximum is not guaranteed a final step of scoring is required to check the curvature of the likelihood at that point. The Hessian or an approximation of it is also needed for generating confidence intervals for the complete set of hyperparameters $\theta$.

### 14.4.2 Adaptive EM estimation

There is an obvious way to turn the EM algorithm into an on-line estimation technique for all hyperparameters. This is simply done by approximating the *smoothing* solutions $\hat{x}_{t|N}$, $\Sigma_{t|N}$ by the corresponding *filtering* solutions available at time $t$. Substituting $t$ for $N$ in all subindices of (14.52)–(14.58) denoting the information set being conditioned upon, we arrive at a series of equation systems that can be built up recursively during one filtering run. Making the relevant substitutions we have for $N = \bar{N}, \bar{N}+1, \ldots$ (where $\bar{N}$ is set in such a way that invertibility in the formulas below is assured)

$$\hat{\beta}^{(N+1)} = \left[\sum_{t=1}^{N} u_{2t}u_{2t}'\right]^{-1} \left\{\sum_{t=1}^{N} u_{2t}[y_t - u_{1t}'\hat{x}_{t|t}^{(t)}]\right\} \tag{14.60}$$

$$\hat{\varphi}^{(N+1)} = \left\{C'\left[[\hat{Q}^{(N)}]^{-1} \otimes \sum_{t=1}^{N} M_{t-1|t}^{(t)}\right] C\right\}^{-1} \times C'\{[\hat{Q}^{(N)}]^{-1} \otimes I_k\}\, vec \left[\sum_{t=1}^{N} M_{t-1,t|t}^{(t)}\right] \tag{14.61}$$

$$\hat{R}^{(N+1)} = \frac{1}{N}\sum_{t=1}^{N}[\hat{v}_{t|t}^{(t)2} + u_{1t}'\Sigma_{t|t}^{(t)}u_{1t}] \tag{14.62}$$

$$\hat{Q}^{(N+1)} = \frac{1}{N}\sum_{t=1}^{N}[\hat{w}_{t|t}^{(t)}w_{t|t}^{(t)\prime} + \mathcal{W}_{t|t}^{(t)}], \tag{14.63}$$

where

$$\hat{v}_{t|t}^{(t)} := y_t - u_{1t}'\hat{x}_{t|t}^{(t)} - u_{2t}'\hat{\beta}^{(t+1)} \tag{14.64}$$

$$\hat{w}_{t|t}^{(t)} := \hat{x}_{t|t}^{(t)} - \hat{\Phi}^{(t+1)}\hat{x}_{t-1|t}^{(t)} \tag{14.65}$$

$$\mathcal{W}_{t|t}^{(t)} := \Sigma_{t|t}^{(t)} + \hat{\Phi}^{(t+1)}\Sigma_{t-1|t}^{(t)}\hat{\Phi}^{(t+1)\prime} - \hat{\Phi}^{(t+1)}\Sigma_{t-1,t|t}^{(i)} - \Sigma_{t-1,t|t}^{(i)}{}'\hat{\Phi}^{(t+1)\prime}. \tag{14.66}$$

All estimators can be computed in a fully recursive fashion for $N = \bar{N}, \bar{N}+1, \ldots$ using the Kalman filter recursions (14.7)–(14.9) as well as those for the so-called *one-step back smoother*, which is just a special version of (14.10) (substituting $t$ for $N$) [Anderson and Moore (1979, pp. 187-190)]:

$$\hat{x}_{t-1|t} = \hat{x}_{t-1|t-1} + A_{t-1}(\hat{x}_{t|t} - \hat{x}_{t|t-1}) \tag{14.67}$$

$$\Sigma_{t-1|t} = \Sigma_{t-1|t-1} + A_{t-1}(\Sigma_{t|t} - \Sigma_{t|t-1})A_{t-1}', \tag{14.68}$$

where

$$A_{t-1} \;:=\; \Sigma_{t-1|t-1}\Phi'\Sigma_{t|t-1}^{-1}.$$

The superscript $t$ for the state moments in (14.60)–(14.63) indicates that the most recent model specification $\theta^{(t)}$ is used for the recursions (14.7)–(14.9) and (14.67) and (14.68). Approximating the usual EM method in this way substantially reduces the storage requirements for intermediate results.

Some comments on the character of the variance estimators follow. The decomposition (14.30)–(14.32) is also valid for $N = t$, whence we have

$$E[v_t^2|y(t)] \;=\; \hat{v}_{t|t}^{2}{}' + u_{1t}'\Sigma_{t|t}u_{1t} \tag{14.69}$$

$$E[w_t w_t'|y(t)] \;=\; \hat{w}_{t|t}\hat{w}_{t|t}' + \mathcal{W}_{t|t}\,, \tag{14.70}$$

where

$$\mathcal{W}_{t|t} \;=\; \Sigma_{t|t} - \Phi\Sigma_{t-1,t|t} - \Sigma_{t-1,t|t}'\Phi' + \Phi\Sigma_{t-1|t}\Phi'\,. \tag{14.71}$$

Taking expected values on both sides of (14.69)–(14.71) we get in a correctly specified state-space:

$$R \;=\; E\left(\frac{1}{N}\sum_{t=1}^{N}\hat{v}_{t|t}^{2}\right) + \frac{1}{N}\sum_{t=1}^{N}u_{1t}'\Sigma_{t|t}u_{1t} \tag{14.72}$$

$$Q \;=\; E\left(\frac{1}{N}\sum_{t=1}^{N}\hat{w}_{t|t}\hat{w}_{t|t}'\right) + \frac{1}{N}\sum_{t=1}^{N}\mathcal{W}_{t|t}\,. \tag{14.73}$$

Hence the variance estimator (14.62) and (14.63) may be interpreted as a special "method of moments estimator", where $E[\frac{1}{N}\sum_{t=1}^{N}\hat{v}_{t|t}^2]$ and $E[\frac{1}{N}\sum_{t=1}^{N}\hat{w}_{t|t}\hat{w}_{t|t}']$ are substituted by the corresponding observed values and where all moments have been calculated as the filtering solution based on some prior [or recursively updated as in (14.60) and (14.61)] values for the state-space hyperparameters. This estimator is very closely related to an adaptive filter proposed by Louv (1984), whose approach amounts to the solution of a slightly modified version of (14.72) and (14.73), namely:

$$\frac{1}{N}\sum_{t=1}^{N}(I - u_{1t}'\Sigma_{t|t}u_{1t}R_0^{-1})R \;=\; \frac{1}{N}\sum_{t=1}^{N}\hat{v}_{t|t}^{2} \tag{14.74}$$

$$\frac{1}{N}\sum_{t=1}^{N}(I - \mathcal{W}_{t|t}Q_0^{-1})Q \;=\; \frac{1}{N}\sum_{t=1}^{N}\hat{w}_{t|t}\hat{w}_{t|t}'\,, \tag{14.75}$$

where $Q_0$ and $R_0$ are prior values for $Q$ and $R$, "not too different" from the true specification. Louv actually only equates the diagonal elements of the matrix equation (14.74) and (14.75) and computes the filtering solutions for a fixed prior specification, where $\Phi$ is known and $\beta$ is zero. This procedure may be interpreted as an approximation to Rao's (1971) *minimum norm quadratic unbiased estimators* (MINQUE) for the variance components of $R$ and $Q$ in the stochastic setup of state-space model (14.1) and (14.2) [for details see Schneider (1989)].

## 14.5 A Descriptive Interpretation of Kalman Filtering

In three recent papers Kalaba and Tesfatsion (1986, 1988a, 1988b) have suggested an interesting descriptive reinterpretation of Kalman filtering.

As pointed out in the Introduction the model assumptions (14.1)–(14.3) amount to specifying a model for the observables $y := \{y_1, y_2, \ldots, y_N\}$ given the unobservable states $x := \{x_0, x_1, \ldots, x_N\}$ and a prior distribution for the unobservable states $x$, thus yielding the common distribution for $x$ and $y$:

$$\log f_{X,Y}(x, y; \theta) = \sum_{t=1}^{N} [\log f(y_t|x_t; \theta) + \log f(x_t|x_{t-1}; \theta)] + \log f(x_0; \theta), \qquad (14.76)$$

which under the normal setup used here yields (14.26).

Kalman filtering can be viewed as a computational tool for calculating the marginal distribution of the observables $y$ in a recursive fashion according to

$$\log f_Y(y; \theta) = \sum_{t=1}^{N} \log f(y_t|y_{t-1}; \theta), \qquad (14.77)$$

which under assumption (14.3) reduces to (14.17) (except for a constant). The essence of the argument in Kalaba and Tesfatsion (1988a, 1988b) is that going from (14.76) to (14.77) amounts to a forced scalarization of what should better be regarded as a multi-criterion objective function consisting of conceptually distinct types of model specification error, namely:

*Measurement specification error*

$$c_M(x) := \sum_{t=1}^{N} v_t^2, \qquad \text{where} \qquad v_t = y_t - u_{1t}'x_t - u_{2t}'\beta. \qquad (14.78)$$

*Dynamic specification error*

$$c_D(x) := \sum_{t=1}^{N} w_t'Q^{-1}w_t, \qquad \text{where} \qquad w_t = x_t - \Phi x_{t-1}. \qquad (14.79)$$

*Initialization specification error*

$$c_I(x) := (x_0 - \mu_0)'\Sigma_0^{-1}(x_0 - \mu_0). \qquad (14.80)$$

These three sources of specification error are collapsed into one aggregate cost measure given by the log likelihood (14.17). This scalarization (computationally done by Kalman filtering) is achieved using the stochastic framework (14.3) parameterized by $\theta$.

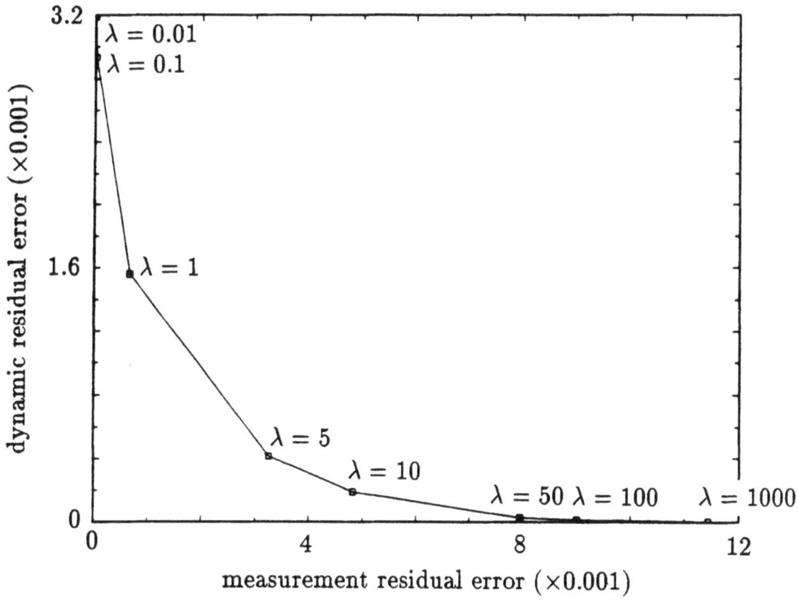


*Figure 14.1:* Residual efficiency frontier [FLS of (14.87)].

If, however, we are very uncertain about the stochastic parameterization $\theta$ (in particular, not knowing which coefficients are constant and which are not), then we might want to start out with a purely descriptive stability analysis using a simplified version of (14.78)–(14.80), namely:

$$c_M(x) = \sum_{t=1}^{N}(y_t - u_t'x_t)^2 \tag{14.81}$$

$$c_D(x) = \sum_{t=1}^{N}(x_t - x_{t-1})'\bar{Q}^{-1}(x_t - x_{t-1}) \tag{14.82}$$

$$c_I(x) = 0\,, \tag{14.83}$$

where (14.81) mirrors our belief in an approximately linear regression model, (14.82) our preference for a simple (i.e., constant) model, and (14.83) our complete ignorance of a plausible initial state. The matrix $\bar{Q}$ contains prior weights, which we attach to deviations from constancy in particular coefficients.

Our goal consists now in finding a time path $\{x_t\colon\ t = 0, 1, \ldots, N\}$, which vector minimizes both error types simultaneously. Obviously there is a trade-off between $c_M$ and $c_D$: The more we allow $x$ to vary (i.e., the higher the deviations from the preferred model of constancy) the better will be the fit in the measurement equation and vice versa.

This trade-off becomes explicit in the so-called *residual efficiency frontier*, which is the set of all points $\{c_M(x), c_D(x)\}$ , where it is not possible to decrease both $c_M$ and $c_D$. This

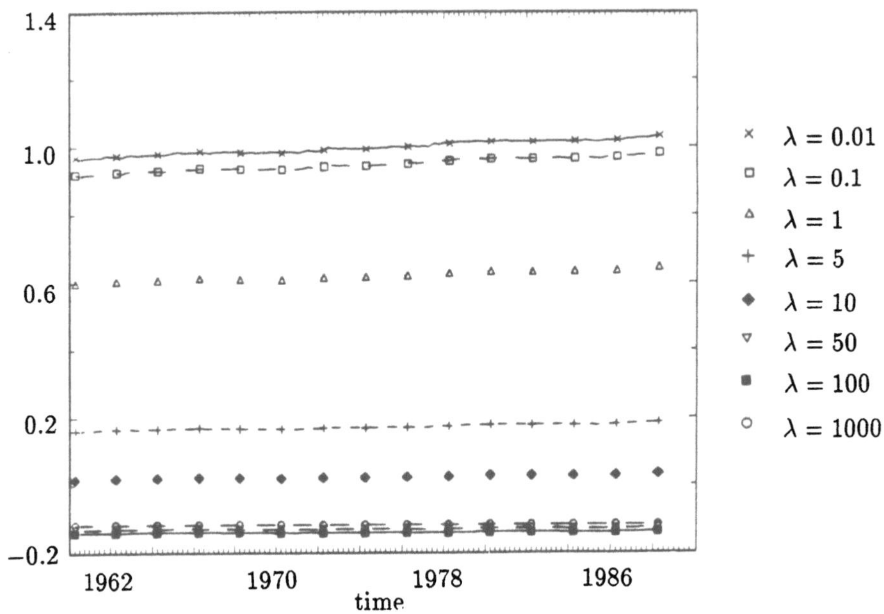


*Figure 14.2:* Time path of the constant [FLS of (14.87)].

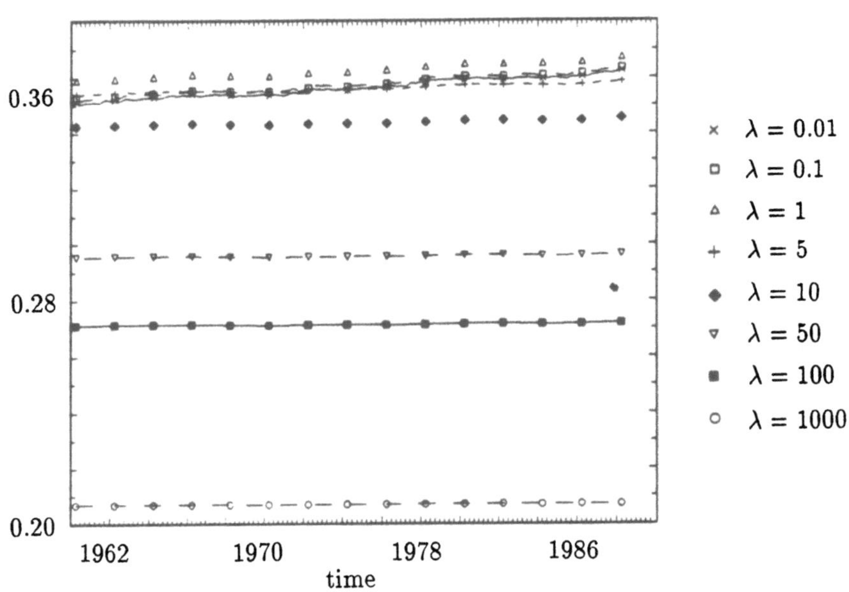


*Figure 14.3:* Time path of the real income coefficient [FLS of (14.87)].

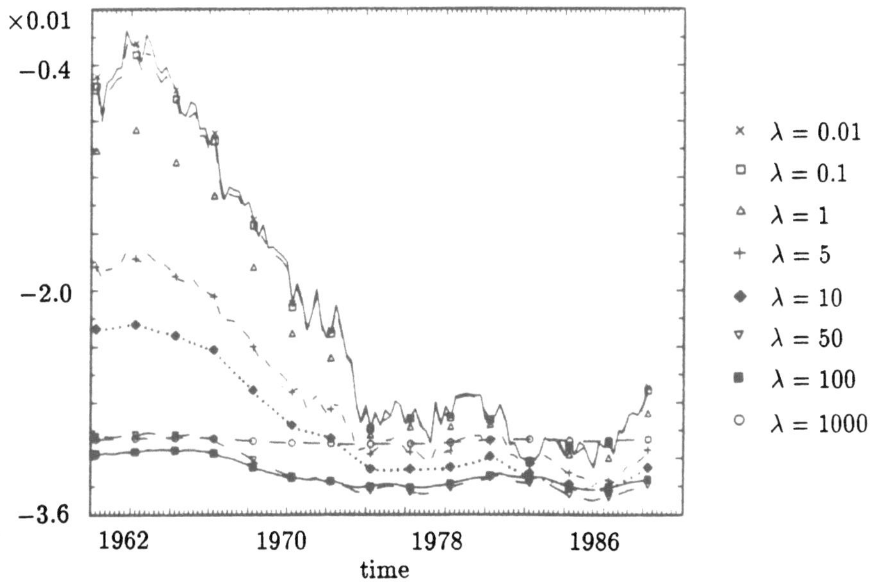


*Figure 14.4:* Time path of the short-term interest coefficient [FLS of (14.87)].

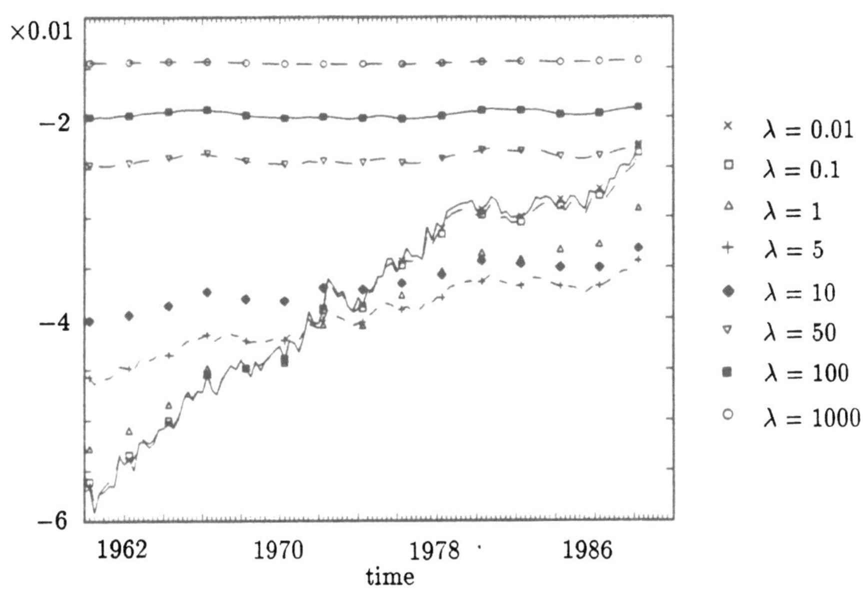


*Figure 14.5:* Time path of the long-term interest coefficient [FLS of (14.87)].

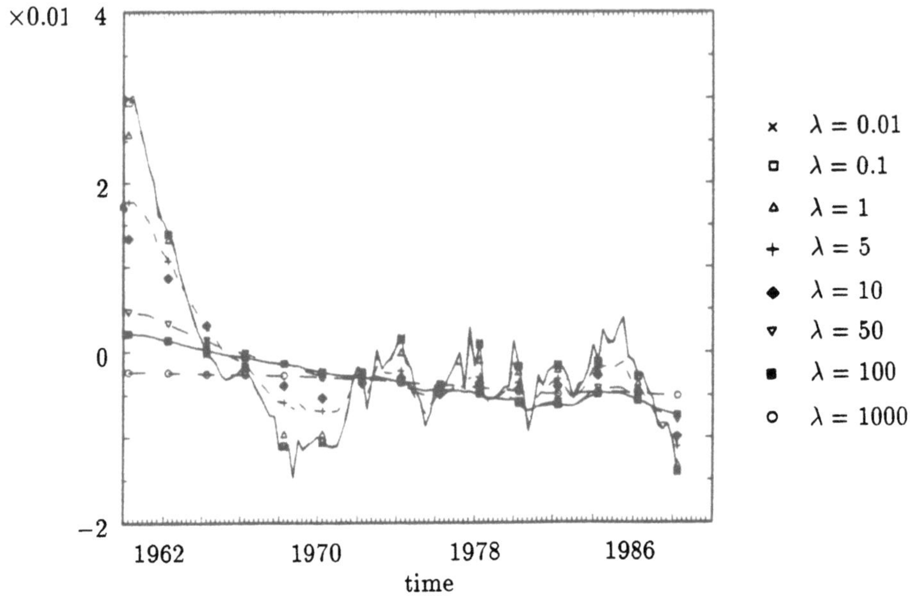


*Figure 14.6:* Time path of the swap rate coefficient [FLS of (14.87)].

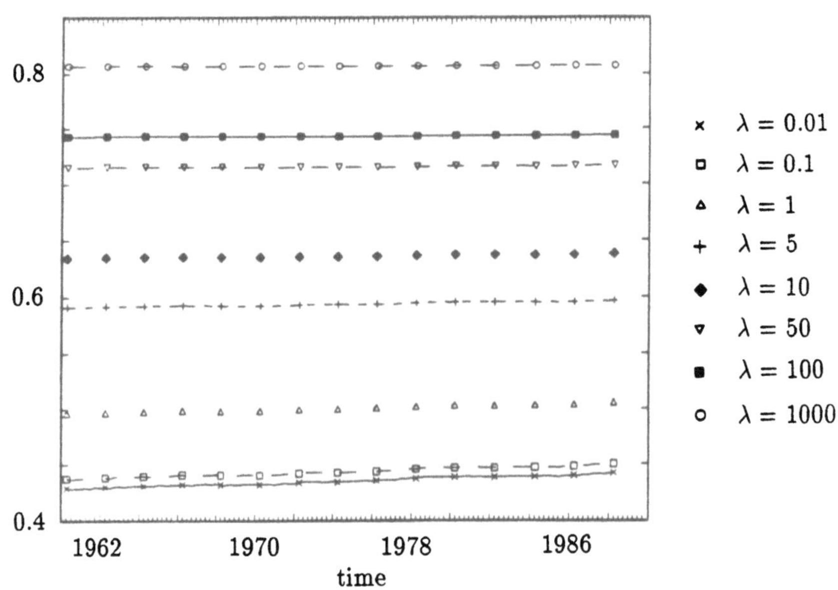


*Figure 14.7:* Time path of the lagged money coefficient [FLS of (14.87)].

set is a convex curve in the first quadrant of the Euclidean $R^2$ plane (see *Figure 14.1*). It can be traced out by maximization of the cost aggregate

$$c = \lambda c_D(x) + c_M(x) \tag{14.84}$$

and letting $\lambda$ parametrically vary in the interval $[0, \infty)$. The maximum is achieved if $x := \{x_0, x_1, \ldots, x_N\}$ is set to the Kalman smoothing solution $\{\hat{x}_{t|N}\colon\ t = 0, 1, \ldots, N\}$ computed from (14.7)–(14.10) using

$$R = \lambda, \quad Q = \bar{Q}, \quad \beta = 0, \quad \Phi = I, \tag{14.85}$$

and initializing $\Sigma_0$ as a very large number. Alternatively one can use the information variant of the Kalman filter and set the inverse of $\Sigma_0$ formally to zero, thus suppressing the initialization cost term. The equivalence between the FLS solution and the Kalman smoother using (14.85) can be inferred from comparing (14.81)–(14.83) and (14.84) with (14.13).

Kohn and Ansley (1988) also showed the equivalence between certain Bayesian smoothness priors in state-space smoothing and optimal smoothing for function estimation using penalized least squares, i.e., minimizing a criterion of the form

$$\sum_{t=1}^{N}[y_t - f(t)]^2 + \lambda \sum_{t=1}^{N}[\Delta^k f(t)]^2 \tag{14.86}$$

of which the FLS criterion is just a special case ($\Delta$ is the difference operator).

Having computed the FLS solution for several values of $\lambda$ the researcher has at his or her disposal a collection of time paths $\{\hat{x}_{t|N}(\lambda)\colon\ t = 0, 1, \ldots, N\}$ with associated cost vectors $\{c_M(\lambda), c_D(\lambda)\}$.

If $\lambda$ approaches infinity, the OLS solution is eventually obtained. The lower $\lambda$ is specified the more time variation we allow—$\lambda$ may be interpreted as a kind of smoothing parameter controlling the "resolution" of parameter variation (analogous to the size of a spectral window). In a stochastic setting an increase in $\lambda$ for fixed $Q$ can be interpreted as a decrease in the ratio of transition noise to measurement noise. Multiplication of (14.13) by a scalar does not change the corresponding smoothing solution for $x(N)$; it only depends on the ratio $(Q/R)$. The graphs in *Figures 14.1* to *14.7* visually demonstrate the effect of decreasing this ratio.

There still remains the problem of how to specify the weights in $\bar{Q}$. A good case can be made for $\bar{Q}$ to be set to $(\sum_{t=1}^{N} u_t u_t')^{-1}$, since for this particular choice the residual efficiency frontier remains invariant to the units of measurement for $u_t$, otherwise one implicitly weighs the components of dynamic cost by the units of measurement in the corresponding regressor.

A judgment on stability must now resort to descriptive criteria, e.g., the issue can be made dependent on how favorable the trade-off between $c_M$ and $c_D$ is, or dependent on the range of $\lambda$, over which certain characteristic features of the parameter paths persist. To this end it will be convenient to normalize $\lambda$ to the interval $[0,1]$, e.g., by a reparameterization of $\lambda$ as $\delta = \frac{\lambda}{\lambda+1}$. The particular choice of the normalization entails some arbitrariness, however.

Confidence intervals for the parameter path or significance tests on individual parameter variances can of course not be made. This requires a stochastic model for the parameter transition process.

In Section 14.6 I will apply flexible least squares as a preliminary descriptive stability test to a standard form of the West German money-demand function and will check how the ML estimation via Kalman filtering will pick an "optimal" time path for the regression coefficients, when the descriptive cost functions $c_M$ and $c_D$ are reinterpreted as elements of a likelihood function generated from a random walk model for the regression coefficients.

## 14.6 Stability Analysis for a Standard Demand Function for West German *M1*

Standard West German money-demand functions have already been subjected to a series of stability tests [see, e.g., Buscher (1984), Gaab and Seitz (1988)]. The results have been somewhat mixed depending on the specification used, the particular stability test employed and the estimation period. We will use a slightly extended version of the so-called Goldfeld specification [Goldfeld (1976)] for West German *M1*, namely,

$$y_t = \sum_{i=1}^{6} u_{it}' x_t + v_t \,, \tag{14.87}$$

where

$y_t$ = log of real *M1* (using the implicit GNP deflator with 1980=100, quarterly averages, seasonally adjusted)

$u_{1t}$ = 1

$u_{2t}$ = log of real GNP (quarterly data, seasonally adjusted)

$u_{3t}$ = log of three month money-market rate at Frankfurt (quarterly averages)

$u_{4t}$ = log of yield on fully taxed newly issued bonds (quarterly averages)

$u_{5t}$ = 90-day swap rate (DM/US \$), quarterly averages

$u_{6t}$ = lagged dependent variable

$t$ = 1960(II)–1988(II) (quarters, 113 observations)

The data were derived from monthly issues of the *Deutsche Bundesbank*.

One problem with the CUSUM tests and variants thereof [Brown *et al.* (1975)] and the Cooley-Prescott test [Cooley and Prescott (1973)] is the fact that they constitute what might be called a global stability test. They do not allow one to identify the particular source of instability once global instability is detected. Garbade (1977) pointed out this in a simulation study comparing CUSUM, recursive residuals, Cooley and Prescott tests, and ML estimation of random walk variances. This is the advantage of flexible least squares (FLS) and ML estimation of individual coefficient variances-methods, which analyze individual time paths.

The graphs in *Figures 14.1* to *14.7* illustrate the results of FLS when applied to (14.87). The main features to be deduced are that seemingly only the coefficients for the interest rates and the swap rate exhibit a distinct time-varying behavior. Especially the behavior of the short-term interest rate is remarkable: an apparent stabilization from 1974 onward coincides with the date when the German central bank officially switched from an interest-rate target regime to a money target regime. Some marked troughs in the swap-rate coefficient occurred when there was massive speculation against the US dollar. There are also slight upward movements in the coefficients for the constant, real income and the lagged dependent variable, but these increases are small relative to the size of the coefficients. The level of these latter coefficients change substantially at resolutions from $\lambda = 5$ onward.

The pattern of the individual FLS path remains persistent for all $\lambda < 10$ (corresponding to $0 < \delta = \frac{\lambda}{\lambda+1} < .9$). If we pick that degree of parameter resolution, where the trade-off between measurement residual error and dynamic residual error substantially deteriorates (see *Figure 14.1*), we should choose some $\lambda$ between 1 and 5.

We want to check now how maximum likelihood estimation picks an "optimal" time path from the above collections, when the parameter variances are estimated individually. We assume a random walk model for the regression coefficients

$$x_{t+1} = x_t + w_t, \quad \mathrm{Cov}(w_t) = Q, \quad Q \text{ being diagonal}. \tag{14.88}$$

We specified the initial state as the OLS solution of the regression coefficients calculated from data over the time span 1960(II)–1970(I). Starting from several points in the parameter space for $\theta$ we arrived at the following ML estimates for the random walk variances. The estimated standard deviations taken from the estimated information matrix and the corresponding $t$-ratios are also supplied. Apparently the likelihood had several local maxima, but a common feature of those local maxima was the "stability ordering" given in *Table 14.1*.

Little, however, is known about the sampling distributions of these ML estimators. For particular econometric state-spaces (e.g., stationary ARMA parameter processes and nonstochastic regressors) consistency and asymptotic normality results are known [Pagan (1980)]. Also the asymptotic normality results in Pagan (1980) are only obtained if the true parameter vector is an interior point of the parameter space. In the case of parameter constancy the true variances are on the boundary of the parameter space (namely, zero).

It is to be expected that some part of the measurement error will be transformed into parameter variation in a constant coefficient model. A simulation study was undertaken to find out to what extent this phenomenon might have taken place here. A simulation study using 500 replications of a constant coefficient model (14.87), which assumed the OLS

*Table 14.1:* Estimates of the variances in $Q$ and estimated precision.

| Parameter | Estimate $\times 10^{-6}$ | Std. Dev. $\times 10^{-6}$ | $t$-ratio |
|---|---|---|---|
| Constant | 10.38 | 244.59 | 0.05 |
| Real income | 0.18 | 16.88 | 0.01 |
| Short-term interest | 5.48 | 11.76 | 0.45 |
| Long-term interest | 4.49 | 15.07 | 0.32 |
| Swap rate | 8.83 | 8.30 | 0.79 |
| Adjustment parameter | 1.73 | 16.41 | 0.12 |

*Table 14.2:* Percentage points of the ML estimators.

| Percentage points $\times 10^{-6}$ | 90.0 | 95.0 | 97.5 | 99.0 |
|---|---|---|---|---|
| Constant | 12.3 | 12.7 | 17.4 | 17.7 |
| Real income | 1.05 | 1.34 | 1.53 | 1.86 |
| Short-term interest | 5.02 | 5.09 | 5.24 | 5.45 |
| Long-term interest | 5.08 | 5.46 | 5.58 | 5.68 |
| Swap rate | 5.84 | 6.23 | 6.37 | 6.45 |
| Adjustment parameter | 2.34 | 2.51 | 2.76 | 2.77 |

estimates from 1960(II) to 1988(II) to be true, turned up with the (thus bootstrapped) percentage points of the ML estimators for the random walk variances shown in *Table 14.2*.

According to these results the movements in the coefficients for the short-term interest rate and the swap rate are significantly identified to be time-varying (within this particular model of parameter dynamics) at a type I error level of 1% (not taking into account the sampling variation in the estimates of the percentage points, though). We note that the simulated distributions of the variance component estimators resemble much more highly skewed chi-square distributions heavily concentrated at zero than a symmetric normal distribution, which at best turns the $t$-ratios quoted in *Table 14.1* into "stability ordering indices". This is also in line with asymptotic results from Garbade (1977) and Nabeya and Tanaka (1988), whose test statistic for parameter constancy essentially is an estimate of $(Q/R)$ for a single time-varying regression coefficient [corresponding to $\frac{1}{\lambda}$ in (14.84)].

The low estimates of the variance components are to be expected with the particular parameter transition model used here, since the random walk model spreads out the time variation over the entire sample period. The FLS paths of the short-term interest rate at levels of high parameter resolution do suggest that also a step change might have occurred in 1974 due to a regime shift of the West German central bank. A similar case for a step change can be made for the introduction date of flexible exchange rates; note the movement of the swap-rate coefficient around that time. So a different type of parameter dynamics might also produce an adequate model for the data.

Another possible explanation of the parameter variation observed here is the different degree of capital market integration in the 1960s as opposed to the 1970s: During the 1970s new financial instruments have been developed and this to a greater extent in the short-term range than in the long-term range. The 1970s have opened a broader menu of possibilities to exploit the opportunity cost captured here as a short-term interest rate—a fact, which makes the increased sensitivity to this opportunity cost variable appear very plausible.

An obvious test for the adequacy of the model estimated here consists of course in an analysis of residuals. Conveniently, Kalman filtering provides us with the innovation sequence $\{\tilde{y}_t := y_t - \hat{y}_t\}$. Normalizing this sequence with the estimated standard deviations in $D_t$ [see equation (14.8)] should result in a sequence, which behaves like a Gaussian white noise process, if the model is correctly specified. A test on the correlations and the periodogram of this sequence could not reject the hypothesis of a white noise process at the 5% level. Also the first four moments of the normalized innovation sequence matched those of a Gaussian process fairly well (using a simple 95% confidence interval for the first four sample moments of the estimated innovations $\tilde{y}_t$).

*Table 14.3* illustrates a particular search process in the likelihood space (the starting values being the OLS measurement variance and $Q$ being set to one-hundredth of that variance). ML estimation was not able to generate good search vectors at these starting points, whereas both EM and adaptive EM generated large likelihood increases initially and propelled the search procedure to points from where scoring could converge in one step.

Furthermore, the EM methods were in this particular case an order of magnitude faster than scoring (the time difference grows exponentially with the number of parameters to be estimated). Adaptive EM requires 70% of the time needed for a full EM cycle and saves the need to store the whole set of filtering solutions that would be required for EM (which makes adaptive EM especially attractive for memory-scarce environments as in a PC).

It seems that during the first iteration cycles adaptive EM generates larger likelihood increases than full EM but then slows down considerably in comparison with EM, which leads to a slightly higher number of iterations required for convergence of adaptive EM. This experience was also confirmed in other simulation experiments [Schneider (1989)].

Inserting the maximum likelihood estimates for $Q$ and $R$ into the state-space permits us then to compute the corresponding estimates of the *a posteriori* distributions

$$f[x_t|y(N);\hat{\theta}_{ML}] = \mathcal{N}[x_t : \hat{x}_{t|N}(\hat{\theta}_{ML}), \Sigma(\hat{\theta}_{ML})]\,. \tag{14.89}$$

The graphs of *Figures 14.8* to *14.13* show the mean and a two-$\sigma$-band from the above distributions (14.89). They essentially exhibit the same features as the FLS solution in the range $1 < \lambda < 5$.

We finally note that the specific behavior in the short-term interest-rate coefficient observed here does not seem to depend on the particular type of money-demand specification

*Table 14.3:* Increases of log likelihood for EM and adaptive EM.

| Iteration | EM method | Adap. EM method |
|---|---|---|
| 1 | 89.31 | 105.77 |
| 2 | 36.75 | 39.68 |
| 3 | 20.06 | 22.93 |
| 4 | 13.75 | 14.65 |
| 5 | 10.40 | 9.07 |
| 6 | 8.07 | 9.08 |
| 7 | 6.27 | 2.98 |
| 8 | 4.85 | 1.61 |
| 9 | 3.71 | 0.86 |
| 10 | 2.83 | 0.46 |
| ⋮ | ⋮ | ⋮ |
| Convergence in step | 16 | 18 |
| At log-likelihood | 434.60 | 434.50 |
| Final scoring increase | 0.44 | 0.41 |

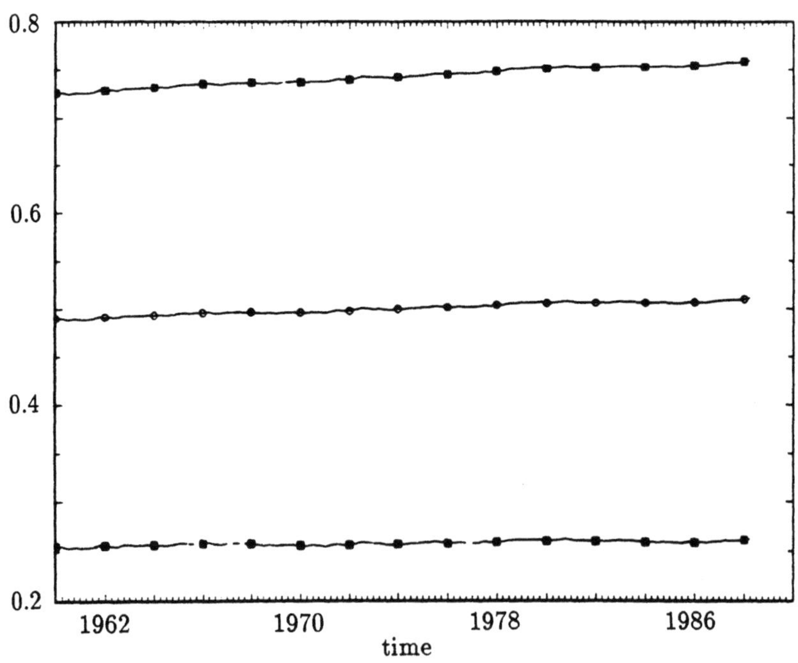


*Figure 14.8:* Time path of the constant [Kalman smoothing of (13.87)].

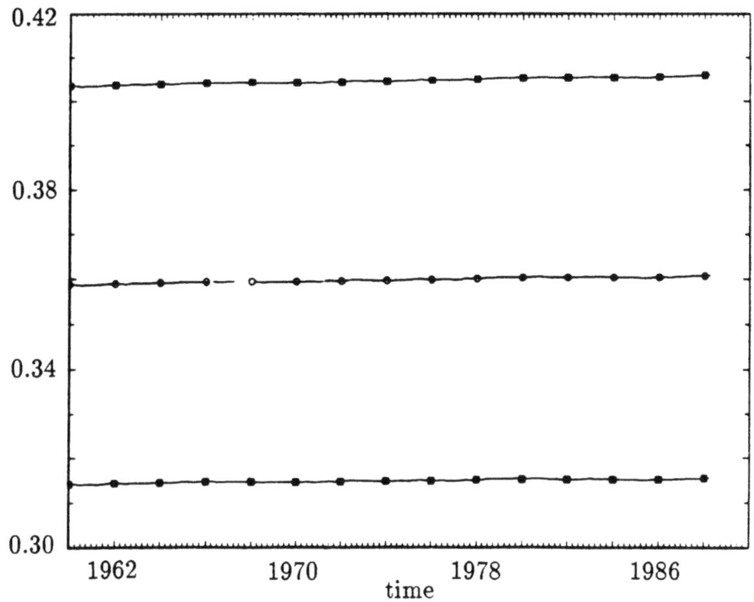


*Figure 14.9:* Time path of the real income coefficient [Kalman smoothing of (13.87)].

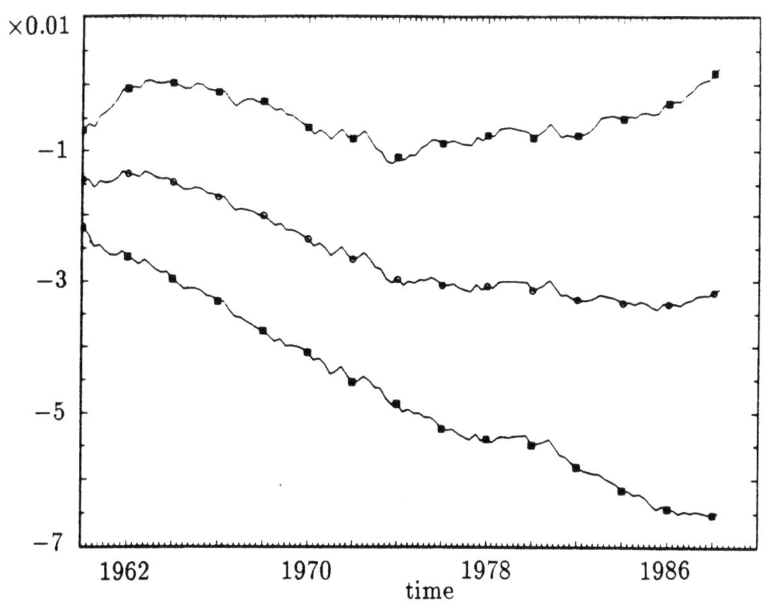


*Figure 14.10:* Time path of the short-term interest coefficient [Kalman smoothing of (13.87)].

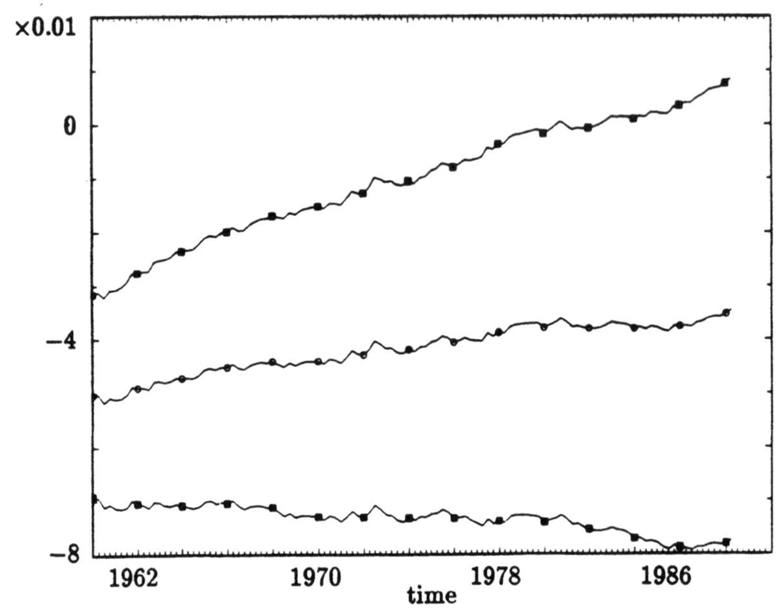


*Figure 14.11:* Time path of the long-term interest coefficient [Kalman smoothing of (13.87)].

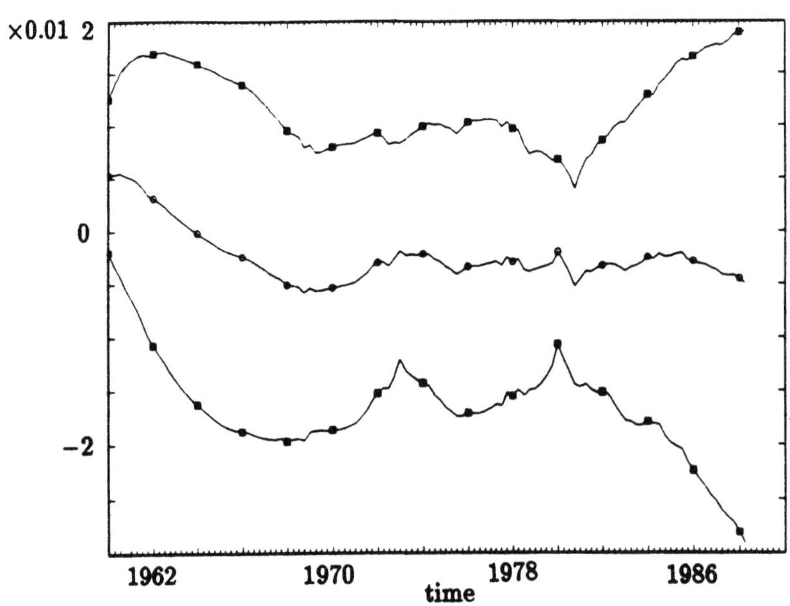


*Figure 14.12:* Time path of the swap-rate coefficient [Kalman smoothing of (13.87)].

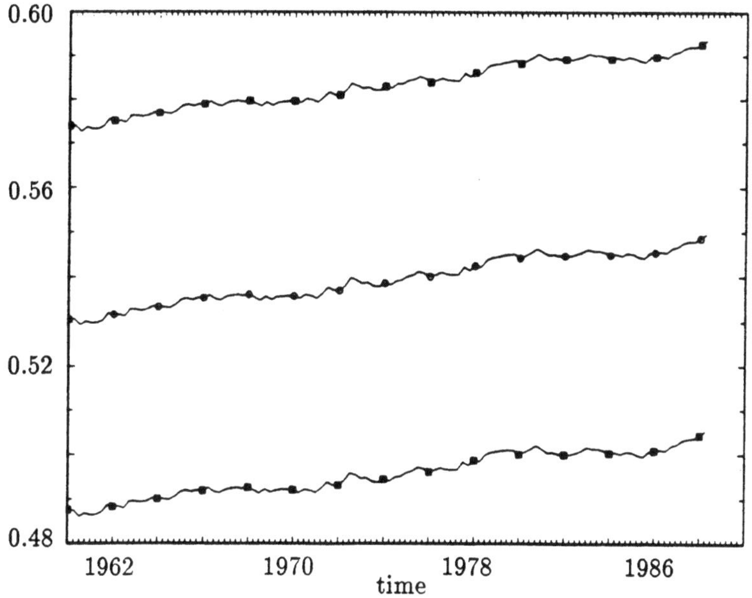


*Figure 14.13:* Time path of the lagged money coefficient [Kalman smoothing of (13.87)].

used; the pattern persists also in a more elaborate error correction model of money demand, which Gaab and Seitz (1988) found to be an adequate description of West German data. Defining $u_{3t}$ to be the short-term interest rate (taking no logs), an FLS analysis of the model

$$\Delta y_t = u_{1t}x_{1t} + \Delta u_{2t}x_{2t} + \Delta u_{3t}x_{3t} + u_{2,t-1}x_{4t} + u_{3,t-1}x_{5t} + y_{t-1}x_{6t} + v_t \qquad (14.90)$$

yielded FLS time paths for $x_{5t}$ as shown in *Figure 14.14*.

## 14.7 Conclusion

Summing up we detected instability of a standard *M1* function for West Germany, and we attribute that instability to variability in the coefficients for interest rates and the swap rate. A very simple excercise in exploratory data analysis (flexible least squares) was able to detect that phenomenon and could be confirmed in a formal hypothesis-testing setup within a random walk model for the parameters.

The movement in these coefficients may reflect the adjustment to regime changes of monetary policy in the 1970s. Increasing sensitivity with respect to short-term interest rates and decreasing sensitivity with respect to long-term interest rates may also be a plausible feature when economic agents become increasingly aware of the risks of money holding and increasingly reluctant to engage in long-term nominal commitments when exposed to prolonged inflationary experiences as happened in the 1970s and in the beginning

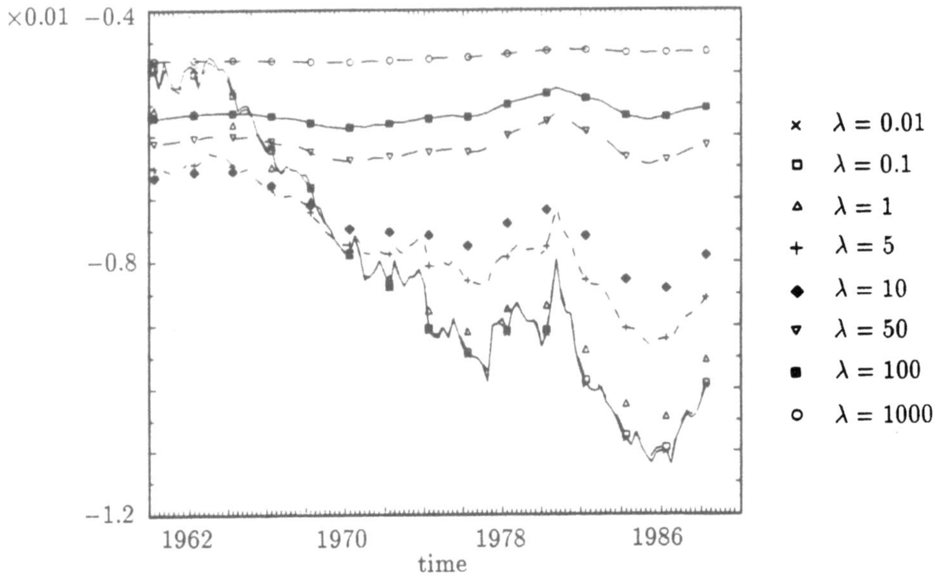


*Figure 14.14:* Time path of the lagged short-term interest coefficient [FLS of (13.90)].

of the 1980s. The observed parameter variation is also compatible with a higher degree of financial innovation in short-term capital markets than in long-term capital markets.

As far as the technical question of computational procedures is concerned, the relative performance of EM and adaptive EM suggests that for convergence enhancement the faster adaptive EM technique may be used.

## Acknowledgments

I would like to thank Ludwig Fahrmeir, Heinz Kaufmann, and Walter Krämer for their comments on this chapter.

## References

Anderson, B.D.O. and Moore, J.B. (1979), *Optimal Filtering.* Englewood Cliffs: Prentice-Hall.

Ansley, C.F. and Kohn, R. (1982), A geometrical derivation of the fixed interval smoothing algorithm. *Biometrika*, **69**, 486-487.

Aoki, M. (1987), *State Space Modeling of Time Series.* Berlin: Springer-Verlag.

Bierman, G.J. (1974), Sequential square root filtering and smoothing of discrete linear systems. *Automatica*, **10**, 147-158.

Bierman, G.J. (1977), *Factorization Methods for Discrete Sequential Estimation.* New York: Academic Press.

Brown, R.L., Durbin, J., and Evans, J.M. (1975), Techniques for testing the constancy of regession relationships over time. *Journal of the Royal Statistical Society*, **B-37**, 149-192.

Buscher, H.S. (1984), The stability of the West German demand for money. *Weltwirtschaftliches Archiv*, **120**, 256-277.

Carraro, C. and Sartore, D. (1986), Square root iterative filter: Theory and applications to econometric models. Working Paper 8611, University of Venice.

Cooley, T.F. and Prescott, E.C. (1973), Varying parameter regression: A theory and some applications. *Annals of Economic and Social Measurement*, **2**, 463-473.

Cooley, T.F., Rosenberg, B., and Wall, K.D. (1977), A note on the optimal smoothing for time varying coefficient problems. *Annals of Economic and Social Measurement*, **6**, 453-462.

Cooper, J.P. (1973), Time-varying regression coefficients: A mixed estimation approach and operational limitations of the general Markov structure. *Annals of Economic and Social Measurement*, **4**, 525-530.

de Jong, P. and MacKinnon, M.J. (1988), Covariances for smoothed estimates in state-space models. *Biometrika*, **75**, 601-602.

Dempster, A.P., Laird, N.M., and Rubin, D.B. (1977), Maximum-likelihood from incomplete data via the EM-algorithm. *Journal of the Royal Statistical Society*, **B-39**, 1-38.

Duncan, D.B. and Horn, S.D. (1972), Linear recursive estimation from the viewpoint of regression analysis. *Journal of the American Statistical Association*, **67**, 815-821.

Fraser, D.C. and Potter, J.E. (1969), The optimum linear smoother as a combination of two optimum linear filters. *IEEE Transactions on Automation and Control*, **AC-14**, 387-390.

Gaab, W. and Seitz, H. (1988), Zur Frage der Stabilität der Geldnachfrage $M1$, pp. 93-122 in W. Franz, W. Gaab, and J. Wolters (eds.), *Theoretische und angewandte Wirtschaftsforschung*. Berlin: Springer-Verlag.

Garbade, K. (1977), Two methods for examining the stability of regression coefficients. *Journal of the American Statistical Association*, **72**, 54-63.

Gersch, W. and Kitigawa, G. (1988), Smoothness priors in time series, pp. 431-476 in C. Spall (ed.), *Bayesian Analysis of Time Series and Dynamic Models*. New York: Marcel Dekker.

Goldfeld, S.M. (1976), The case of the missing money. *Brookings Papers on Economic Activity*, **3**, 683-730.

Gupta, N.K. and Mehra, R.K. (1974), Computational aspects of maximum likelihood estimation and reduction of sensitivity function calculations. *IEEE Transactions on Automation and Control*, **AC-19**, 774-783.

Harrison, P.J. and Stevens, C.F. (1976), Bayesian forecasting (with discussion). *Journal of the Royal Statistical Society*, **B-38**, 205-247.

Jazwinski, A.H. (1970), *Stochastic Processes and Filtering Theory*. New York: Academic Press.

Kalaba, R. and Tesfatsion, L. (1986), Time-varying regression via flexible least squares. Modelling Research Group Working Paper 8633, University of Southern California, Los Angeles.

Kalaba, R. and Tesfatsion, L. (1988a), U.S. money demand instability: A flexible least squares approach. Modelling Research Group Working Paper 8809, University of Southern California, Los Angeles.

Kalaba, R. and Tesfatsion, L. (1988b), An organizing principle for dynamic estimation.

Modelling Research Group Working Paper 8818, University of Southern California, Los Angeles.

Kaminski, P.G., Bryson, A.E., and Schmidt, S.F. (1971), Discrete square root filtering: A survey of current techniques. *IEEE Transactions on Automation and Control*, **AC-16**, 727-736.

Kohn, R. and Ansley, C.F. (1988), Equivalence between Bayesian smoothness priors and optimal smoothing for function estimation, pp. 393-430 in C. Spall (ed.), *Bayesian Analysis of Time Series and Dynamic Models*. New York: Marcel Dekker.

Kwakernaak, H. and Sivan, R. (1972), *Linear Optimal Control Systems*. New York: John Wiley.

Lasdon, L.S. (1970), *Optimization Theory for Large Systems*. New York: John Wiley.

Louv, W.C. (1984), Adaptive filtering. *Technometrics*, **26**, 399-409.

Luenberger, D.G. (1969), *Optimization by Vector Space Methods*. New York: John Wiley.

Magnus, J.R. and Neudecker, H. (1988), *Matrix Differential Calculus*. New York: John Wiley.

Mayne, D.Q. (1966), A solution for the smoothing problem for linear dynamic systems. *Automatica*, **4**, 73-92.

Meinhold, R.J. and Singpurwalla, N. (1983), Understanding the Kalman filter. *The American Statistician*, **37**, 123-127.

Morf, M. and Kailath, T. (1975), Square-root algorithms for least-squares estimation. *IEEE Transactions on Automation and Control*, **AC-20**, 487-497.

Morris, C.R. (1983), Parametric empirical Bayes inference: Theory and applications. *Journal of the American Statistical Association*, **78**, 47-55.

Nabeya, S. and Tanaka, K. (1988), Asymptotic theory of a test for the constancy of regression coefficients against the random walk alternative. *The Annals of Statistics*, **16**, 218-235.

Oberhofer, W. and Kmenta, J. (1974), A general procedure for obtaining maximum likelihood estimates via generalized regression models. *Econometrica*, **42**, 579-590.

Pagan, A. (1980), Some identification and estimation results for regression models with stochastically varying coefficients. *Journal of Econometrics*, **13**, 341-363.

Rao, C.R. (1971), Estimation of variance and covariance components: MINQUE theory. *Journal of Multivariate Analysis*, **1**, 257-277.

Rao, C.R. (1973), *Linear Statistical Inference and Its Applications*. New York: John Wiley.

Rosenberg, B. (1973), A survey of stochastic parameter regression. *Annals of Economic and Social Measurement*, **2**, 381-398.

Rosenberg, B. (1977), Estimation of error covariance in regession with sequentially varying parameters. *Annals of Economic and Social Measurement*, **6**, 457-462.

Ruskeepää, H. (1985), Conditionally Gaussian distributions and an application to Kalman filtering with stochastic regressors. *Communications in Statistics, Theory and Methods*, **A-14**, 2919-2942.

Sallas, W.M. and Harville, D.A. (1981), Best linear recursive estimation for mixed linear models. *Journal of the American Statistical Association*, **76**, 860-869.

Sallas, W.M. and Harville, D.A. (1988), Noninformative priors and restricted maximum likelihood estimation in the Kalman filter, pp. 477-508 in C. Spall, (ed.), *Bayesian Analysis of Time Series and Dynamic Models*. New York: Marcel Dekker.

Sarris, A.H. (1973), Kalman filter models: A Bayesian approach to estimation of time varying regression coefficients. *Annals of Economic and Social Measurement*, **2**, 501-523.

Schneider, W. (1986), Der Kalmanfilter als Instrument zur Diagnose und Schätzung variabler Parameter in ökonometrischen Modellen, volume 27 in K.-A. Schäffer, P. Schönfeld, and W. Wetzel (eds.), *Arbeiten zur Angewandten Statistik*. Heidelberg: Physica.

Schneider, W. (1988), Analytical uses of Kalman filtering in econometrics: A survey. *Statistical Papers*, **29**, 3-33.

Schneider, W. (1989), Maximum likelihood estimation of seemingly unrelated regression equations with time varying coefficients by Kalman filtering: A comparison of scoring, EM- and an adaptive EM-method. Working paper, University of Kiel.

Shumway, R.H. and Stoffer, D.S. (1982), An approach to time series smoothing and forecasting using the EM-algorithm. *Journal of Time Series Analysis*, **3**, 253-264.

Watson, M.W. and Engle, R.F. (1983), Alternative algorithms for the estimation of dynamic factor, mimic and varying coefficient regression models. *Journal of Econometrics*, **23**, 385-400.

Wu, C.F.J. (1983), On the convergence properties of the EM-algorithm. *The Annals of Statistics*, **11**, 95-103.

Zellner, A. (1962), An efficient method of estimating seemingly unrelated regressions and tests for aggregation bias. *Journal of the American Statistical Association*, **57**, 348-368.

# Part III

# Forecasting in the Presence of Structural Change

CHAPTER 15

# A Note on Bayesian Forecast Combination Procedures

*Francis X. Diebold*


## Summary

The properties of Bayesian composite forecasts are studied. It is argued, and illustrated with an example, that the asymptotic performance of such composite forecasts depends on the validity of a maintained assumption, namely, that one of the models among those whose forecasts are combined is the true data-generating process. The implications of this phenomenon are explored.


## 15.1 Introduction

Scientific knowledge obtained from research in one area often proves useful in others. Such has been the case with the Bayesian theory of econometric model selection, as developed by Geisel (1970, 1974) and Zellner (1971, 1972, 1979, 1984), which has generated insights useful not only for model selection but also for prediction. In particular, it is now known that under certain conditions the posterior probabilities associated with various forecasting models may be used as weights in forming a linear composite forecast, and that the resulting composite forecast is optimal, in the sense of minimizing posterior expected loss.

In this chapter, I focus on one of those "certain conditions"—in particular, the assumption that one of the models among those whose forecasts are combined is the true data-generating process (DGP)—and I explore the effects of its failure on the performance of Bayesian composite forecasts. I argue that, if the assumption is satisfied, such Bayesian composite forecasts will have certain desirable asymptotic properties relative to their classical counterparts, but that the result is reversed if the assumption is not satisfied.

In Section 15.2, I give an explicit derivation of the Bayesian composite forecast, and I show that (under certain conditions) it minimizes posterior expected loss. I explore the linkage between the maintained assumption of truth of one of the models and the resulting good performance of the Bayesian composite forecast. In Section 15.3, I illustrate the argument with a simple example involving the combination of forecasts from two linear regression models. In Section 15.4, I conclude with a summary and directions for future research.

## 15.2 Bayesian Model Selection and Forecast Combination

The Bayesian solution to the model selection problem (under symmetric loss) is well known: It is optimal to choose the model with highest posterior probability. Zellner (1972, 1984), Zellner *et al.* (1989), and others have suggested that the posterior probabilities may be used fruitfully not only for model selection, but also for forecast combination. Forming a composite forecast with weights equal to the posterior probabilities seems reasonable, and the case for doing so is easily formalized.

Consider two models, $M_1$ and $M_2$, with associated posterior probabilities $p_1$ and $p_2$, respectively, where $p_1+p_2 = 1$. [The generalization to the case of more than two competing models is immediate.] Then posterior expected loss is given by

$$E(y-\hat{y})^2 = p_1[E(y-\hat{y})^2|M_1] + p_2[E(y-\hat{y})^2|M_2], \tag{15.1}$$

where $\hat{y}$ is any point forecast. Let $\bar{y}_1$ ($\bar{y}_2$) be the mean of the predictive probability density function for $M_1$ ($M_2$). Then we can write

$$\begin{aligned} E[(y-\hat{y})^2|M_i] &= E\{[(y-\bar{y}_i)-(\hat{y}-\bar{y}_i)]^2|M_i\} \\ &= E\{[(y-\bar{y}_i)^2+(\hat{y}-\bar{y}_i)^2-2(y-\bar{y}_i)(\hat{y}-\bar{y}_i)]|M_i\} \\ &= E[(y-\bar{y}_i)^2|M_i]+(\hat{y}-\bar{y}_i)^2 \\ &= C_i+(\hat{y}-\bar{y}_i)^2, \quad i=1,2, \end{aligned} \tag{15.2}$$

where $C_i = E[(y-\bar{y}_i)^2|M_i]$. But then

$$\begin{aligned} E(y-\hat{y})^2 &= p_1[C_1+(\hat{y}-\bar{y}_1)^2]+p_2[C_2+(\hat{y}-\bar{y}_2)^2] \\ &= C+p_1(\hat{y}-\bar{y}_1)^2+p_2(\hat{y}-\bar{y}_2)^2, \end{aligned} \tag{15.3}$$

where $C \equiv p_1C_1+p_2C_2$. The first-order condition for minimization of (15.3) with respect to $\hat{y}$ is

$$2\hat{y}^* - 2(p_1\bar{y}_1+p_2\bar{y}_2) = 0 \tag{15.4}$$

or

$$\hat{y}^* = p_1\bar{y}_1+p_2\bar{y}_2\,. \tag{15.5}$$

On what does this standard Bayesian result depend? Most important is the assumption that the posterior probabilities associated with the models being combined sum to 1. This is equivalent to the assumption (often made explicitly) that one of the models is true. To see this, note that $p_1 + p_2 = 1$ is equivalent to $p_1 + p_2 - P(M_1 \cap M_2) = 1$, because $P(M_1 \cap M_2) = 0$ so long as $M_1 \neq M_2$. But $p_1 + p_2 - P(M_1 \cap M_2) = 1$ is equivalent to $P(M_1 \cup M_2) = 1$. The motivation for the assumption seems to be the idea that it should be possible to write down an exhaustive listing of candidate models, one of which must (by construction) be the true DGP. In practice, of course, the forecasts of only a small number of models are combined; enunciation of an exhaustive set of candidate models for the true DGP is always infeasible and probably impossible. In short, it seems difficult to take seriously the assumption that the true DGP is among the candidate models whose forecasts are being combined.

To better understand the effects of the assumption, let us first suppose that it *is* true. Without loss of generality, assume that $M_1$ is true. Then, if the Bayesian model selection procedure is consistent, $p_1$ will approach 1 as sample size ($T$) gets large. The implication of consistency of the Bayesian model selection procedure for Bayesian forecast combination, of course, is that progressively more weight is placed on $M_1$ as $T$ gets large; in the limit, $M_1$ receives unit weight and $M_2$ receives zero weight. In other words, the Bayesian model selection and Bayesian forecast combination procedures coincide asymptotically. This result is natural and desirable, *if* the true DGP is among the models whose forecasts are combined.

But what happens when the true DGP is *not* among those whose forecasts are combined, as is likely to be the case in practice? Is there any harm in maintaining the assumption of truth of one of the models, in order to make the Bayesian analysis operational? Recent work on estimation and testing in misspecified models [e.g., White (1982), Gourieroux *et al.* (1984), Vuong (1989)] furnishes a useful perspective on this question. We now know that, under general conditions, an estimator of the parameter of a misspecified model will converge to a pseudo-true value, that is, to a parameter configuration closest (within the misspecified class) to the true DGP. Furthermore, the metric defining "closeness" is induced by the estimation procedure.

Now, if the true DGP is not among the models entertained, it is of course impossible for any model selection procedure—Bayesian or otherwise—to be consistent for the true model. But, as the discussion above indicates, we might expect the model selected by the Bayesian procedure to be consistent for *something*, namely, the model *closest* to the true DGP. Without loss of generality, assume that $M_1$ is closer. Then it is reasonable to expect that $p_1$ will converge to 1, as was the case when $M_1$ was true. For model selection, such a property is very useful—it is often desired to determine which among a set of models provides the best approximation to the true DGP. The implications for forecast combination, however, appear less desirable. Consistency of the Bayesian model selection procedure for the closest model implies that Bayesian composite forecasts will asymptotically place unit weight on $M_1$ and zero weight on $M_2$. But $M_1$ and $M_2$ are *both* false models; the fact that $M_1$ is closer to the true DGP does not mean that the information contained in $M_2$ cannot be usefully combined with that in $M_1$ to produce a superior composite forecast. (Contrast this with the case where $M_1$ is in fact the true DGP.) This insight, of course, is the entire motivation for forecast combination [see Clemen (1990)].

In summary, then, it would appear that Bayesian composite forecasts will perform well in large samples, placing all weight on the forecast of the true model, *if* the true model is among those whose forecasts are combined. Otherwise, it would appear that Bayesian composite forecasts will perform poorly in large samples, placing all weight on the forecast of one false model, and thereby discarding the potentially valuable information contained in other false models. In the next section, the truth of these conjectures in the context of simple linear regression is illustrated.

## 15.3 Combination of Forecasts from Regression Models

Consider a simple comparison of two regression models, as in Zellner (1971, pp. 306-312),

$$M_1: \quad y_t = X_{1t}\beta_1 + \mu_{1t}, \qquad t = 1, \ldots, T \tag{15.6}$$

$$M_2: \quad y_t = X_{2t}\beta_2 + \mu_{2t}, \qquad t = 1, \ldots, T \tag{15.7}$$

one of which is the true model, where $X_1$ and $X_2$ are nonstochastic matrices of maximum and equal column rank, $\beta_1$ and $\beta_2$ contain no common elements, and the disturbances of the true model are i.i.d. Gaussian with zero mean and constant variance. Then, in a Bayesian analysis with diffuse priors over parameters and models, the posterior odds for $M_1$ versus $M_2$ are given by

$$\frac{p_1}{p_2} = \left[\frac{s_2}{s_1}\right]^T, \tag{15.8}$$

where $s_i$ is the square root of the usual unbiased estimator of $\sigma_i^2$, $i = 1, 2$. [The result also holds for informative-prior Bayesian analyses if $T$ is large and certain other regularity conditions are satisfied.]

Consider now the implications of the earlier-derived Bayesian forecast combination procedure. Rearranging (15.8) yields

$$p_2 = \left[\frac{s_1}{s_2}\right]^T p_1 \tag{15.9}$$

or, because $p_1 + p_2 = 1$ by assumption,

$$p_1 + \left[\frac{s_1}{s_2}\right]^T p_1 = 1. \tag{15.10}$$

Thus,

$$p_1 = \frac{s_2^T}{s_1^T + s_2^T}. \tag{15.11}$$

Therefore, $p_1$ depends only on the ratio $s_1/s_2$, which is emphasized by writing

$$p_1 = \frac{1}{1 + \left[\frac{s_1}{s_2}\right]^T}. \tag{15.12}$$

Note that

$$\lim_{T\to\infty} p_1 = \begin{cases} 0 & \text{if} \quad \sigma_1 > \sigma_2 \\ 1/2 & \text{if} \quad \sigma_1 = \sigma_2 \\ 1 & \text{if} \quad \sigma_1 < \sigma_2 . \end{cases} \tag{15.13}$$

Now compare the weight arising from the Bayesian forecast combination procedure (15.5) with the weight arising from the classical variance-covariance forecast combination procedure. [By classical variance-covariance combining weight I mean the weight that minimizes combined prediction error variance, as developed by Bates and Granger (1969) and discussed in Granger and Newbold (1987).] As is well known, the classical forecast combination is

$$\hat{y} = \phi^* f_1 + (1 - \phi^*) f_2 , \tag{15.14}$$

where $f_1$ and $f_2$ are forecasts (possibly but not necessarily the means of predictive probability density functions),

$$\phi^* = \frac{1 - s_{12}/s_2^2}{1 + s_1^2/s_2^2 - 2s_{12}/s_2^2} \tag{15.15}$$

and $s_{12}$ is the usual estimator of the covariance of the forecast errors associated with $M_1$ and $M_2$. If $s_{12} = 0$, the classical weight is

$$\phi^* = \frac{1}{1 + \left[\frac{s_1}{s_2}\right]^2}. \tag{15.16}$$

While the Bayesian and classical combining weights are very similar, several interesting differences are apparent. For example, the Bayesian weights are required to be convex, while the classical weights need not be. The convexity restriction is not necessarily beneficial. A forecast with a negative weight can play the same useful role in producing a combined forecast as an asset sold short plays in producing the return on a portfolio. Convexity of the Bayesian weights follows immediately from the definition of probability and the assumption that one of the two models is true. In addition, the Bayesian weights do not exploit covariance information, while the classical weights (in general) do. Presumably this again reflects the assumption that one, and only one, of the models is true.

These differences are of minor importance, however, compared with those associated with the nature of dependence on sample size. Both the classical and Bayesian weights change implicitly with sample size, as the underlying estimators of the relevant variances (and, in the classical case, covariances) converge to their pseudo-true values. The Bayesian weight changes explicitly with sample size, however, as is made clear by the appearance of $T$ in (15.12).

To understand the significance of the role played by sample size in the construction of the Bayesian combining weight, it will again prove useful to segment the discussion into two parts, according to the truth of the assumption that one of the two models is the true DGP. Suppose first that the assumption is true, and with no loss in generality assume that $M_1$ is true; then by (15.12) and (15.13) the Bayesian weight placed on $M_1$ converges to unity, while that placed on $M_2$ converges to zero. (The truth of $M_1$ implies that it has a smaller disturbance variance.) As argued earlier, it is desirable that this should happen, and it is made possible by virtue of the validity of the assumption that one of the two models is true. The desirability follows from the fact that the true model encompasses all rivals. [For a discussion of encompassing in its relation to forecast combination see Diebold (1990).]

Suppose now that neither $M_1$ nor $M_2$ is the true DGP. Suppose also, without loss of generality, that $M_1$ is closer than $M_2$ to the true DGP, in the sense that $\text{plim}(s_1) < \text{plim}(s_2)$. As before, the Bayesian weight placed on $M_1$ converges to unity, while that placed on $M_2$ converges to zero. Such convergence is no longer desirable, however, because asymptotically all weight is placed on a false model, $M_1$. The essential insight of forecast combination, of course, is that loss may be reduced by pooling the information contained in false models, all of which represent different approximations to reality.

## 15.4 Concluding Remarks

I have conjectured that the asymptotic performance of Bayesian composite forecasts is intimately linked to the truth of a maintained assumption, namely, that the true DGP is among the models whose forecasts are combined. The conjecture was verified in the context of a particular linear regression example. The argument points to the desirability of exploring Bayesian approaches to forecast combination that do not assume the truth of one of the underlying models. Is such a problem well-posed? If so, how would such an analysis proceed? [The difficulty is related to the fact that the calculations for posterior expected loss, (15.1)–(15.3), are apparently not meaningful unless one of the models is assumed true.] What relationship would the resulting combining weights have to the Bayesian and classical weights discussed in this chapter?

It is worth noting that, regardless of the answers to the questions posed above, Bayesian insights will likely contribute in other ways to the advancement of forecasting methodology and forecast combination methodology. Shrinkage techniques, for example, have been used advantageously by Zellner and Hong (1987) and Zellner *et al.* (1988, 1989) to forecast international growth rates and turning points, and by Diebold and Pauly (1990) to shrink classical composite forecasts toward a prior location, such as the sample mean.

Finally, I am happy to report on concurrent and independent work by Zellner (1989) who has recently initiated development of Bayesian methods for combining forecasts from sets of models whose posterior probabilities do not sum to unity. I hope that my paper will stimulate additional work along similar lines.

## Acknowledgments

I gratefully acknowledge financial support from the National Science Foundation and the University of Pennsylvania Research Foundation. John Geweke, Bruce Mizrach, and especially Arnold Zellner provided useful comments that improved the content and exposition of this chapter. All remaining errors, omissions, and inaccuracies are mine.

## References

Bates, J.M. and Granger, C.W.J. (1969), The combination of forecasts. *Operations Research Quarterly*, **20**, 451-468.

Clemen, R.T. (1990), Combining forecasts: A review and annotated bibliography (with discussion). Forthcoming in *International Journal of Forecasting.*

Diebold, F.X. (1990), Forecast combination and encompassing: Reconciliation of two divergent literatures. Forthcoming in *International Journal of Forecasting.*

Diebold, F.X. and Pauly P. (1990), The use of prior information in forecast combination. Forthcoming in *International Journal of Forecasting.*

Geisel, M.S. (1970), Comparing and choosing among parametric statistical models: A Bayesian analysis with macroeconomic applications. Ph.D. dissertation, University of Chicago.

Geisel, M.S. (1974), Bayesian comparisons of simple macroeconomic models, pp. 227-256 in S. Feinberg and A. Zellner (eds.), *Studies in Bayesian Econometrics and Statistics.* Amsterdam: North-Holland.

Gourieroux, C., Monfort A., and Trognon, A. (1984), Pseudo maximum likelihood method theory. *Econometrica*, **52**, 681-700.

Granger, C.W.J. and Newbold, P. (1987), *Forecasting Economic Time Series*, second edition. New York: Academic Press.

Vuong, Q.H. (1989), Likelihood ratio tests for model selection and nonnested hypotheses. *Econometrica*, **57**, 307-334.

White, H. (1982), Maximum likelihood estimation of misspecified models. *Econometrica*, **50**, 1-25.

Zellner, A. (1971), *An Introduction to Bayesian Inference in Econometrics.* New York: John Wiley.

Zellner, A. (1972), A note on optimal prediction and control with two alternative regression models. Unpublished manuscript, Graduate School of Business, University of Chicago.

Zellner, A. (1979), Posterior odds ratios for regression hypotheses: General considerations and some specific results. Invited paper presented to the Econometric Society; reprinted in Zellner (1984).

Zellner, A. (1984), *Basic Issues in Econometrics.* Chicago: University of Chicago Press.

Zellner, A. (1989), Bayesian and non-Bayesian methods for combining models and forecasts. Unpublished manuscript, Graduate School of Business, University of Chicago.

Zellner, A. and Hong, C. (1987), Forecasting international growth rates using Bayesian shrinkage and other procedures. *Journal of Econometrics*, **40**, 183-202.

Zellner, A. and Hong, C. (1988), Bayesian methods for forecasting turning points in economic time series: Sensitivity of forecasts to asymmetry of loss structures, forthcoming in K. Lahiri and G. Moore (eds.), *Leading Economic Indicators: New Approaches and Forecasting Records.* Cambridge: Cambridge University Press.

Zellner, A., Hong, C., and Gulati, G.M. (1988), Turning points in economic time series, loss structures and Bayesian forecasting, forthcoming in J. Hodges, S. Press, and A. Zellner (eds.), *Bayesian Likelihood Methods in Statistics and Econometrics: Essays in Honor of George A. Barnard.* Amsterdam: North-Holland.

Zellner, A., Hong, C., and Min, C. (1989), Forecasting turning points in international output growth rates using Bayesian exponentially weighted autoregression, time-varying parameter, and pooling techniques. Manuscript, Graduate School of Business, University of Chicago.

CHAPTER 16

# A New Approach to Statistical Forecasting

*Spyros Makridakis*

## Summary

Available approaches to statistical forecasting suffer from several deficiencies that can render their predictions for real-world economic/business series inappropriate. In this chapter I illustrate such deficiencies, with real-life data and propose an approach that corrects their negative impact. The proposed approach is based on three premises: First, model selection is based not on historical information but rather on accuracy measures computed from out-of-sample data. Second, *two* types of model selection are done on out-of-sample—the first chooses the best model from those available *within* a single method, while the second selects the best *among* four methods run in parallel. Third, the within-method or among-methods model selection is done *for each forecasting horizon separately*, making it possible to have different models or methods or both to predict each of the $m$ horizons. In addition to being theoretically appealing, this new approach outperforms the best method of the M-competition by a large margin when tested empirically.

## 16.1 Introduction

The last ten years have become a learning ground for those working in the field of forecasting. Large-scale empirical studies [Ahlers and Lakonishok (1983), Makridakis and Hibon (1979), Makridakis *et al.* (1982), Zarnowitz (1984)] have provided us with invaluable information to judge the accuracy of statistical methods and to help us better understand their advantages and limitations. Few will disagree that the findings of these studies are fundamentally changing the field of forecasting. The fact that relatively simple methods were found to be as accurate as complex or sophisticated ones, and the conclusion that combining methods by a simple arithmetic average did better than the individual methods being combined, brought a great deal of disappointment. The initial disillusion, however, has been replaced by a sense of realism. The alternative of abandoning statistical forecasting does not seem attractive, since the accuracy of judgmental forecasts has been found to be even worse than those of statistical methods [Dawes (1979), Dawes (1986), Goldberg (1970), Hogarth and Makridakis (1981)]. Furthermore, statistical forecasting is, usually, substantially less expensive. It became evident, therefore, that the problems facing statistical forecasting had to be understood and new, imaginative ways of correcting them found.

In this chapter I use several real-world economic/business series to illustrate the deficiencies surrounding the prevalent approach to statistical forecasting and to discuss the reasons for such deficiencies. Second, I propose and empirically test a new approach to deal with these problems. The results obtained are superior, by a large margin, to the best method of the Makridakis or M-competition [Makridakis *et al.* (1982)].

## 16.2 Deficiencies of Statistical Forecasting

The most positive outcome of the empirical findings has been the realization *and* acceptance by a majority of those working in the field of the disappointing conclusion that major problems beset the area of statistical forecasting. Problem awareness has brought a gradual but also fundamental change of attitudes. It is now accepted that the prevalent approach to statistical forecasting cannot adequately deal with many real-world series [Armstrong (1986), Clemen and Winkler (1986), Mahmoud (1984), Makridakis *et al.* (1982), Makridakis (1986, 1987), Gardner and Makridakis (1988)]. In this section I summarize the two major deficiencies of the prevalent approach, and provide examples to illustrate these deficiencies.

### 16.2.1 Model fitting versus forecasting

In the prevalent approach to statistical forecasting, a model of a method (or methodology) is fitted to all available data. The choice of the method (or methodology) is a matter of personal preferences with some guidelines drawn from previous empirical studies. Once a method (or methodology) has been selected, the specific model that *best fits* the available historical data *for one-period-ahead* forecasts is selected and used to predict for the future (post-sample). This is done by making at period $t$, $m$ forecasts: $X_{t+i}$, $i = 1, 2, \ldots, m$. "Best fit" commonly means the model that minimizes the Mean Square Error (MSE), the Mean Absolute Percentage Error (MAPE), Mean Absolute Deviation (MAD), Median, Akaike's information criterion, or some analogous function. Theoretically, models that minimize $m$-period-ahead forecasting errors (by making at period $t$ forecast $X^m_{t+m}$, aimed at $m$-periods later) also exist, but in practice their usage is limited. This is due to three reasons: (a) the theoretical advantages of these models over those making one-period-ahead forecasts are not clear; (b) empirical evidence has shown no real differences in post-sample forecasting accuracy [Meese and Geweke (1984), Makridakis *et al.* (1982), Andersen and Carbone (1983)] between one- and multi-step-ahead forecasts; and (c) available software rarely includes options for multiple-period forecasts.

In some methods such as ARIMA or regression, model errors or the random disturbances need to be independent, constant, and normally distributed. In other methods (exponential smoothing, Bayesian forecasting) there is no restriction about the disturbances although it is desirable that they be random, constant, and normally distributed. None of the methods can know the post-sample forecasting errors, which are assumed to possess properties analogous to those of the model's disturbances, although this is not ordinarily true [Makridakis and Winkler (1988)]. Two assumptions are implicit in the prevalent approach to model selection. First, it is assumed that the model that "best" fits the available data will also be the best model to predict beyond these data (post-sample). Second, it is assumed that the model that "best" forecasts for one-period ahead, will also be best for predicting two-, three-, ..., $m$-periods ahead. Both assumptions, however, do *not* hold true for many real-world economic/business series. *Figure 16.1*, for instance, shows the ranking of five methods for one-, two-, ..., 18 forecasting horizons at different chronological time periods. That is, the best (denoted by 1), second best (denoted by 2), ..., worst (denoted by 5) method at each time period was found for one-, two-, ..., and 18-period-ahead forecasts. Such rankings are not consistent at different time periods *or* forecasting horizons. The series in *Figure 16.1* is typical of those of the M-competition.

The same conclusion can be drawn by computing the rank correlation between how well the methods in the M-competition fitted past data versus how well they forecast beyond these data. Such rank correlations were small to start with (about 0.20) and dropped to zero after forecasting horizons of longer than four periods [Makridakis (1986)].

The implications of the fact that the model that best fits the available data might *not* be the best model for post-sample forecasting have not been adequately considered [Priestley (1979)]. Even during the 1970s this possibility was not mentioned in the most popular forecasting or econometric textbooks [Box and Jenkins (1970), Johnston (1972)]. Furthermore, no serious effort has been made to validate the ability of the selected model to forecast accurately for out-of-sample periods. [A practice followed by those using regression is to split the data into two halves and validate that the regression coefficients of the two parts are not statistically different. Although such validation allows us to test for nonrandom changes in the regression coefficients, it cannot tell us how well the model will predict out-of-sample data.] This is partly because *all* data are being used to develop the "best" model, and partly due to the belief (originated in natural/physical sciences) that a "true best" model exists, and that such a model could be correctly identified and used for forecasting. In as much as most series used in the social sciences are short, measurement errors abound, and controlled experimentation is not possible, the basic premise that the "best" model fitted to past data exists and can be identified, and that such a model is also the best model to forecast beyond these data is invalid.

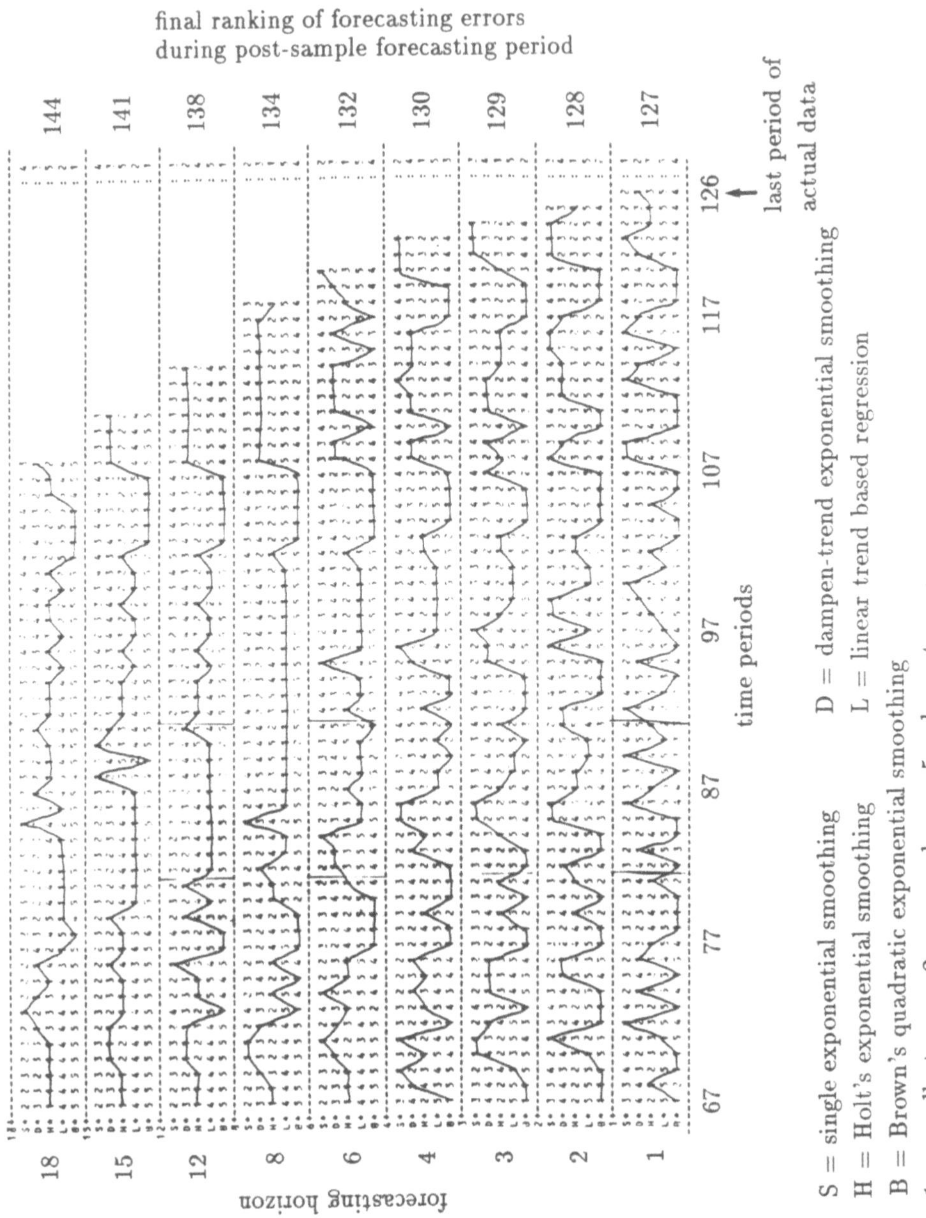


*Figure 16.1:* Ranking according to forecasting errors (five methods at different time periods and for several forecasting horizons).

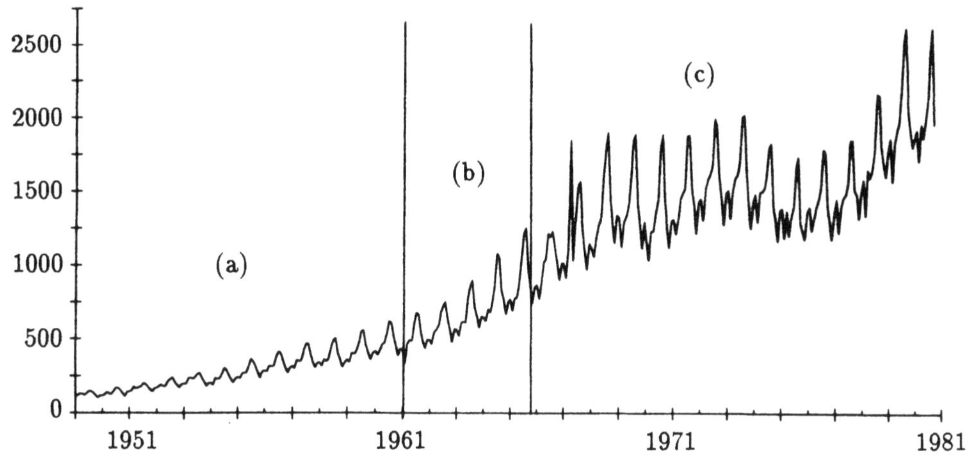


*Figure 16.2:* International airline passengers, 1949–1981.

### 16.2.2 Constant versus changing patterns/relationships

*Figure 16.2* shows monthly international airline passengers (in thousands) between 1949 and 1981. The data are divided into three parts. Part (a) consists of 144 observations (1949-1961), which are the airline data widely used in the forecasting literature since the early 1960s [Box and Jenkins (1970), Brown (1963)]. Part (b) includes data from 1961-1967, while the data in part (c) are after 1967. There is an obvious change in the pattern of international airline passengers after 1967 [part (c)]. Both the exponential trend and the seasonal fluctuations are different from those existing before 1967. The prevalent approach to statistical forecasting assumes constancy of patterns or relationships or both. Such an assumption permits the use of the best model fitted to available data to forecast beyond these data. Unfortunately, however, constancy is not a realistic assumption [Makridakis (1981, 1986)] as far as business or economic data are concerned, which raises a major issue about the validity of the prevalent approach to statistical forecasting.

Since constancy of pattern/relationships is a prerequisite of the prevalent approach, "nice" series such as the airline data [part (a) of *Figure 16.2*] had to be found to test new forecasting methods and to illustrate their alleged "superior" performance. Furthermore, it was considered normal to test a new method on a single series (such as the airline data) and then generalize that the same accuracy level would hold for any other series. Having followed this practice myself [Makridakis and Wheelwright (1977)], I can now say that from a methodological point of view it is ludicrous to consider it possible to generalize from the past to the future and from a single series to all series. As part (c) of *Figure 16.2* shows this is not even possible for what seemed to be the perfect series back in the 1960s.

In addition to patterns, relationships can and do change. *Figure 16.3* is the scatter diagram between paper orders in France and pulp prices. At least four relationships can be identified (A, B, C, and D), as well as two cases (E and F) where pulp price increases

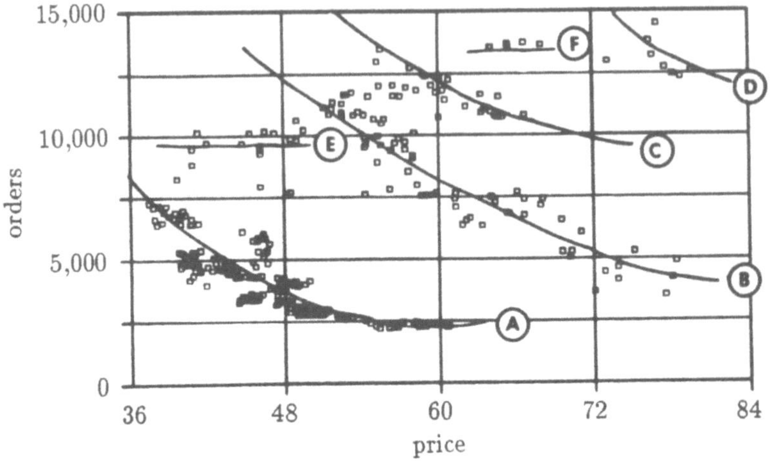


*Figure 16.3:* Orders and pulp prices in constant US $.

did not affect (reduce) orders. Assuming that the relationship between price and orders is constant is not realistic and would result in inaccurate forecasts. Econometricians might argue that the factors causing the relationship between paper orders and prices to shift could be found (if no factors can be found they include dummy variables). Although in some cases this might be possible, it cannot help to forecast more accurately (although the $R^2$ of the model fit could be high) since the majority of the factors causing the relationship to change are exogenous and, therefore, unpredictable themselves.

Although most forecasting methods would provide equally good forecasts for data series when there are no changes in established patterns or relationships or both, the forecasts and their accuracy will vary substantially when changes in patterns or relationships or both occur. It is necessary, therefore, to understand how various methods forecast when such changes do take place, since this is the key to understanding the deficiencies of available methods and to becoming capable of forecasting in the real world when constancy of patterns/relationships cannot be assured. Although changes of relationships also need to be considered, the remainder of this chapter concentrates on the effects of pattern changes on forecasting.

*Figures 16.4, 16.5*, and *16.6* (the data of the three figures have been deseasonalized to better illustrate pattern changes and their consequences) show three kinds of pattern change during forecasting. In *Figure 16.4* the exponential growth trend changed into an abrupt decline. There was nothing in the past data to indicate that such a change was forthcoming. It was impossible, therefore, to have anticipated a pattern change without exogenous judgmental knowledge. All methods, except for single exponential smoothing, forecast a continuation of the established trend (single exponential smoothing always forecasts horizontally at the most recent smoothed data level). In *Figure 16.4* exponential smoothing performs the best, since all methods (except linear trend regression) forecast

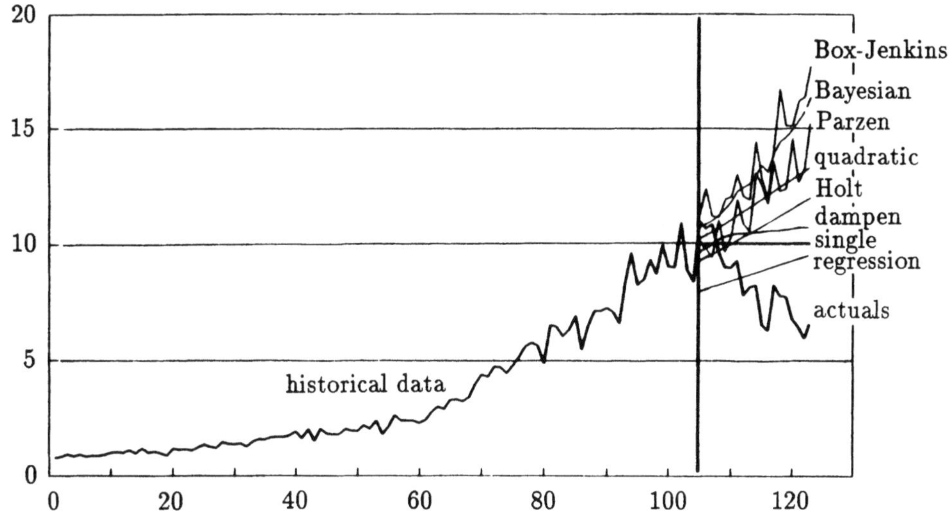


*Figure 16.4:* A monthly series (MNM61) of the M-competition (with an unexpected pattern change while forecasting) and the forecasts of eight methods.

by extrapolating the established exponential trend.

The data of *Figure 16.5* start increasing at period 34 and do so until period 39. The figures then decrease for two consecutive periods. Two methods (Box-Jenkins and quadratic smoothing) ignore the latest two-period decline and forecast a continuation of the recent increase from periods 34 to 39.Bayesian forecasting assumes that the decline in period 39 and 40 is not random, and forecasts by extrapolating the downward trend implicitly assuming the latest decline to be permanent. By so doing, the Bayesian procedure produces forecasts that beat all other methods. Linear trend regression ignores all fluctuations around the trend line, assuming them to be random, and extrapolates the trend to arrive at linearly growing forecasts. The forecasts of the other methods are between those of regression and Box-Jenkins. Interestingly, the forecasts of single exponential smoothing are fairly accurate although they ignore both the initial increase (periods 34 to 39) in the actual data and the subsequent two-period decline (see *Figure 16.5*). The series in *Figure 16.5*, contrary to that of the series in *Figure 16.4*, has in the past indicated that it *might* decline after several periods of continuous increase. Such a decline has happened twice in the past. One could therefore have anticipated that a similar decline might occur in the future and have forecasted in this light.

The data of *Figure 16.6* reach a trough at period 96, then increase (with small interruptions) until period 120 at which point they start declining until period 125. Finally, there is a single increase at period 126. Bayesian forecasting, although doing best with the data of *Figure 16.5*, does the worst with those of *Figure 16.6*. It assumes a growing trend, thus providing increasing forecasts. Quadratic exponential smoothing, which did

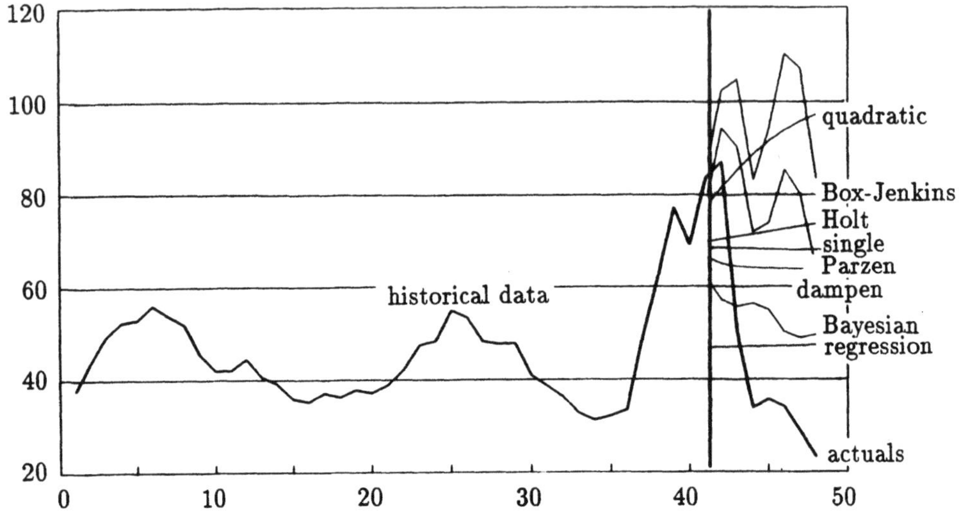


*Figure 16.5:* A quarterly series (QND37) of the M-competition (with a pattern change just before forecasting) and the forecasts of eight methods.

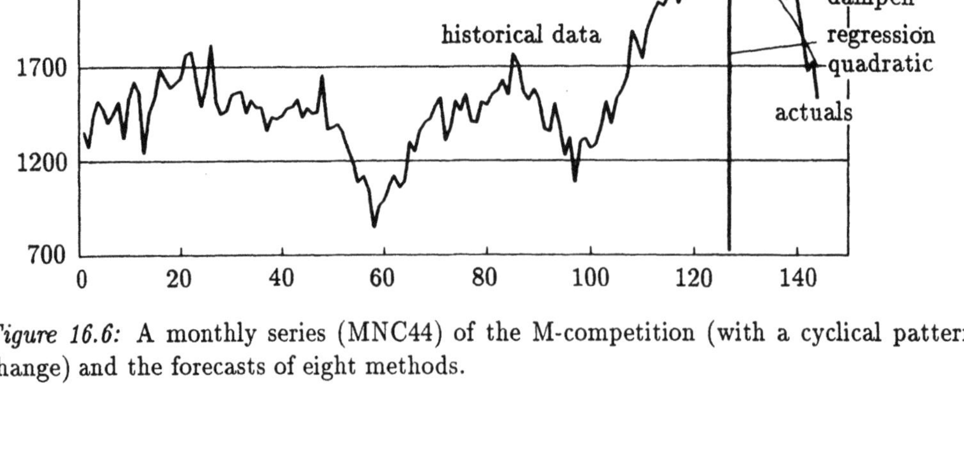


*Figure 16.6:* A monthly series (MNC44) of the M-competition (with a cyclical pattern change) and the forecasts of eight methods.

the worst with the data of *Figure 16.5*, now does the best by ignoring the increase in the last period and forecasting a continuing decline from periods 120 onward. The forecasts of the other methods are in between those of quadratic smoothing and Bayesian forecasting. The series of *Figure 16.6* is similar to that of *Figure 16.5* in that several declines after persistent increases, similar to the latest one, have occurred in the past. It is not unreasonable, therefore, to anticipate that similar declines might occur in the future during forecasting (although the exact timing might not be predictable).

Three observations are worth making at this point. First, the forecasts of the various methods are all over the graph when the data pattern changes (see *Figures 16.4, 16.5*, and *16.6*) during the forecast period (this is one reason why combining various methods by simple arithmetic averaging does well). Second, the accuracy of the methods depends upon whether the latest nonrandom change in pattern is temporary or permanent [Makridakis (1986)]. Some methods, such as Bayesian forecasting, are reactive in extrapolating recent nonrandom changes in the data pattern by assuming them to be permanent. Other methods are slower in identifying and extrapolating the continuation of nonrandom changes in the data. Linear trend regression, for instance, ignores all changes around the long-term trend, while single exponential smoothing assumes a no-change situation. Third, single exponential smoothing seems to do well, not because it can predict pattern changes, but rather because its forecasts are robust, staying in the middle of the data and usually being in the middle of the forecasts of the various methods when patterns change. This seems to be a good strategy, at least for the short term, since empirically the accuracy of single exponential smoothing for one-period-ahead forecasts was found to be the best of all methods in the M-competition.

## 16.3 The Proposed Approach

For any forecasting approach to be realistic and practically relevant it must avoid the two major problems facing the prevalent approach to statistical forecasting—that is, selection based upon how well a model fits historical data for one-step-ahead forecasts, and assuming constancy of patterns/relationships. In addition, it needs to incorporate what we have learned from empirical studies (see *Table 16.1* for a summary), and it must permit one to test forecasting performance on out-of-sample data [see also Fildes (1986)].

Initially, the desired characteristics of the new approach might seem contradictory. Any time series model, for instance, must be based on past data. At the same time, it seems that the future might be different from the past. Furthermore, all data should be used to develop the forecasting model (otherwise some information might be lost), while at the same time it is advocated that out-of-sample testing needs to be done. These seemingly contradictory requirements can be simultaneously achieved if we are willing to reconceptualize our approach to statistical modeling and forecasting.

*Figure 16.7* shows an approach to model selection, which, instead of using all $n$ data points to develop a forecasting model, only $s < n$ data points are initially used and $m$ forecasts are made. Since actual data exist beyond $s$, the *actual* forecasting accuracy

*Table 16.1:* Major empirical evidence and its implications.

| Major findings | Empirical evidence | Implications |
|---|---|---|
| 1. Simple methods | Simple, automatic and inexpensive methods give realistic forecasts. | Use simple methods to a greater extent unless specific reasons exist that can be substantiated by concrete empirical evidence. For instance, use exponential smoothing methods. |
| 2. Seasonality | Seasonality can be accurately predicted no matter what approach is being used. | Deseasonalize the data to develop a model and forecast. Then reseasonalize forecasts. |
| 3. Combining | Combining different methods (by a simple arithmetic average) improves forecasting accuracy and reduces the variance of errors. | No matter what the approach utilized, use several methods and combine their forecasts. Choose methods in such a way as their forecasts will be as complementary (therefore independent) as possible. |
| 4. Short versus long term | Some models are more accurate for the short term (e.g., single exponential smoothing) others are more accurate for the long term (e.g., long-term ARARMA models). | In addition to traditional methods also use an AR($p$) model where the length of $p$ is large. Such AR($p$) (called long memory) is appropriate for capturing and extrapolating the long-term trend. |
| 5. Dampening the trend | Dampening the trend improves forecasting accuracy. | Dampen the trend extrapolation using a dampen-trend exponential smoothing model. |

of the model can be tested *for each of the* $m$ forecasts. Accuracy measures (such as MAPE, MSE, or median) for one, two, ..., $m$-period-ahead forecasts can, therefore, be found. Subsequently, one more data point can be used, $m$ forecasts made, and their actual forecasting accuracy recorded. The process can be done, each time using one more data point, until all observations except one have been used. This type of testing (sliding simulation) can be called out-of-sample and is shown schematically in *Figure 16.7.*

Once the sliding simulation has been completed, $K$ (where $K = n-s$, and where $n$ is the number of data and $s$ the start of the simulation) one-step-ahead accuracy measures, $K-1$ two-step-ahead accuracy measures, ..., $K - m + 1$ $m$-step-ahead accuracy measures are available (see *Figure 16.7*). The average of these measures can subsequently be computed for each of the $m$ forecasting horizons (see *Figure 16.7*). Consequently, model selection can be based on actual out-of-sample forecasting performance without any loss of information, since in the final analysis *all* data have been used for making the final forecast. Such a type of model selection is fundamentally different from the prevalent approach in two respects. First, model selection is *based* on forecasts of *out-of-sample* data. Second, *forecasting performance is measured*, in addition to one-period-ahead, *for two-, three-, ..., and m-step-ahead forecasts*. Once a model has been selected for each of $m$ forecasting horizons based on its out-of-sample performance, it can then be used to predict for the future—that is, for making post-sample forecasts, for the specific period(s) it has been found to be the "best".

### 16.3.1 Among-methods model selection

Several authors [e.g., see Jenkins (1982)] have pointed out that combining forecasts makes no sense from a theoretical point of view. Yet, the empirical evidence showing a "consensus" forecast to outperform the individual methods being combined is indisputable [Clemen and Winkler (1986), Gupta and Wilton (1987), Mahmoud (1984), Zarnowitz (1984)]. Reasons contributing to the more accurate performance of combining over individual methods can be deduced from *Figures 16.4, 16.5*, and *16.6*. In *Figures 16.4, 16.5*, and *16.6*, the forecasts of the different methods vary widely, thus making their average robust and closer to the center of the unpredictably changing pattern. This average, therefore, provides not only more accurate forecasts, but also forecasts with smaller variance [Makridakis and Winkler (1983)].

There is another fundamental reason why combining works well with real-data series, one that relates to our concept of what constitutes the "best" model to represent reality. In the frictionless physical/natural sciences a best model might exist. This is hardly the case, however, in the business/economic fields where the "best" model will be different from series to series (e.g., see *Figures 16.4, 16.5*, and *16.6*), and where the best can vary at each period and with each forecasting horizon (see *Figure 16.1*). Under such circumstances, combining different methods or models or both provides a satisfactory solution that seems to forecast more accuratly than the individual methods being combined.

Although additional research might be required to decide which methods to include, four nonseasonal methods can be used in parallel, with the ultimate selection to be made from the "best" among them. The reasons for proposing these four methods are the following:

1. Empirical evidence has shown that these methods are accurate.
2. They are simple.
3. Their forecasts are intuitive.
4. They can produce forecasts and confidence intervals in an automatic, push-button manner.
5. They are complementary, specifically (a) single exponential smoothing assumes that changes cannot be predicted; (b) Holt's exponential smoothing extrapolates a linear trend (weighting recent data more heavily); (c) dampen-trend exponential smoothing [Gardner and McKenzie (1985)] is similar to Holt's except, as its name implies, it dampens the trend for longer forecasting horizons; and (d) a long-term trend [a long nonstationary auto-regressive (AR) model similar to the long-memory ARAR models proposed by Parzen (1982)] model that captures the long-term tendency of a series.

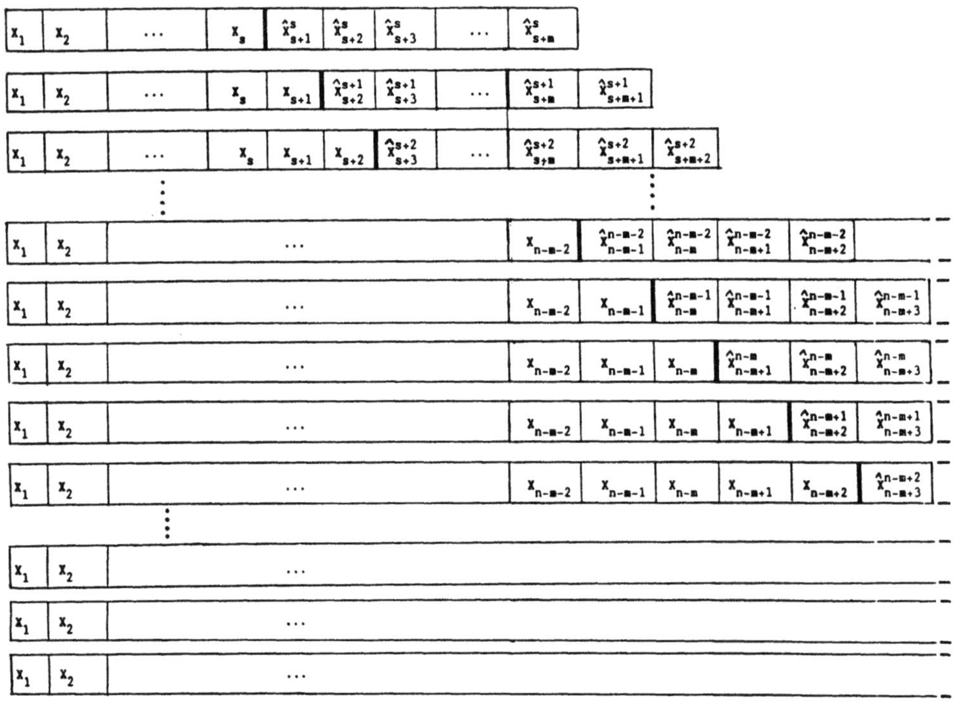

$X_t$ = actual data at $t$

$X^t_{t+i}$ = forecast $X_{t+i}$ made at period $t$ ($t = s, s+1, \ldots, n-1$, $i = 1, 2, \min[m, n-t]$)

$e^k_t = X_t - X^k_t$ = $k$-step-ahead forecasting error made made at period $t$ ($k = 1, 2, \ldots, m$)

*Figure 16.7:* A forecasting simulation of six forecasts using the jackknife approach assuming selection criterion is MAPE.

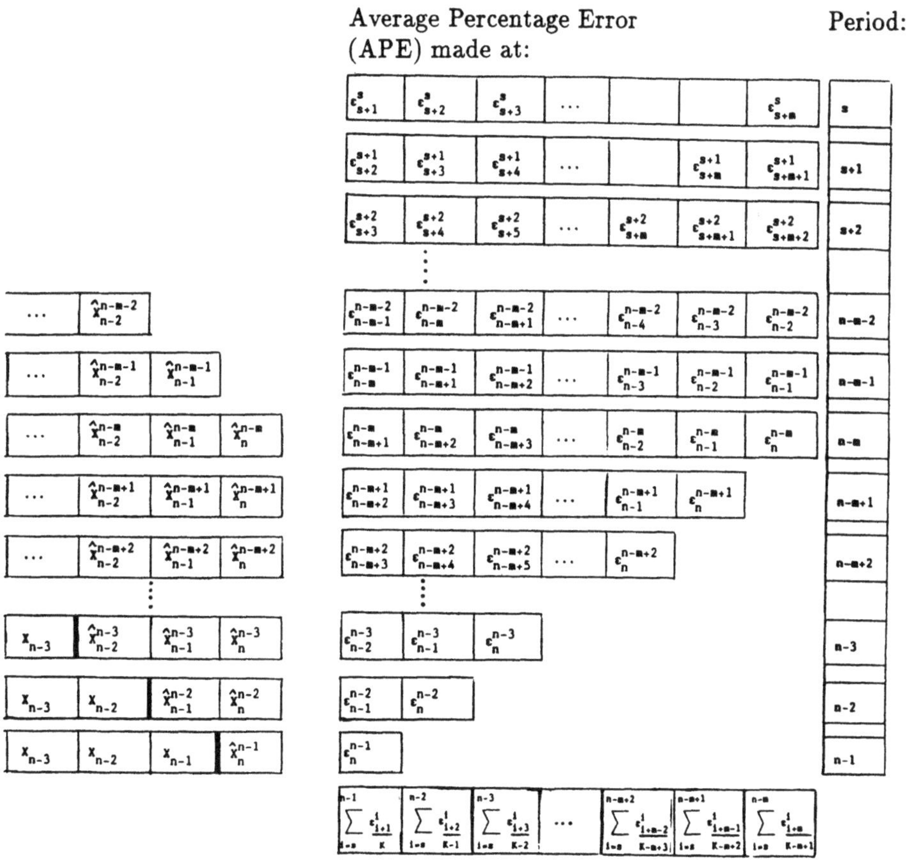


$\varepsilon_t^k = \phi(e_t^k)$ = error measure [e.g., $\varepsilon_t^k = e_t^k / X_t$, i.e., the APE]

$K = n - s$ = number of sliding simulations

average of the APE

*Figure 16.7:* Continued.

During the sliding simulation the aforementioned four methods are run in parallel, using an optimal model from each method that minimizes some error-selection criterion. Although *within*-method model selection is discussed below, for the moment assume that the "best" model for each method is selected by minimizing the MSE at each period of the simulation. Thus, the model that minimizes the one-step-ahead model-fitting errors for the first $s$ data points is selected and $m$ forecasts are made. Then, the first $s + 1$ data points are used, the "best" model is found, and $m$ forecasts are made; and so on until *all* data points except one have been used. The process allows us to compute $m$ forecasts (based on the "best" model as defined by the prevalent approach to statistical forecasting) *at each period* of the simulation, compute the square error, the absolute percentage error, or other accuracy measures, at each step of the simulation. Once the simulation has been completed the average of the error function used *for each of m forecasting horizons can be found* (see *Figure 16.7*). There will be $K$ one-period-ahead forecasts, $K - 1$ two-period-ahead forecasts, ..., $K - m + 1$ $m$-period-ahead forecasts (see *Figure 16.7*) whose out-of-sample average accuracy can be computed.

Unlike the prevalent approach, a different method can be selected for each series and each of the $m$ forecasting horizons, depending upon the out-of-sample performance of each method during the sliding simulation. Furthermore, confidence intervals based on actual forecasting errors can be constructed [see Williams and Goodman (1971)]. In addition, information (i.e., standard errors) can be obtained about the sampling variation of the accuracy measure we are concerned with (e.g., the MSE or MAPE), since the individual values of such measures are known through the sliding simulation. Knowing the empirical sampling distribution can provide us with invaluable information not available through the prevalent approach to statistical forecasting (or statistical modeling in general) which can permit us to select the "best" method among the four using criteria other than the smallest MSE, MAPE, or median.

*Table 16.2* shows the forecasting performance of the proposed approach, together with the results of the most accurate and important methods of the M-competition as well as the accuracy of dampen exponential smoothing [Gardner and McKenzie (1985)]. The model selection for each method was done using the prevalent approach of choosing the model that *minimizes the one-step-ahead fitting errors*. The first part of *Table 16.2* shows the individual accuracy of the four methods as well as the accuracy of their combined forecasts. The second part lists the accuracy of the remaining most accurate and important methods of the M-competition.

The third part of *Table 16.2* shows the accuracy when the method among the four with the smallest MSE or MAPE was selected to forecast for $m$-periods ahead using the procedure of the prevalent approach if more than one method is involved (i.e., select the method, among the four, with the smallest one-step-ahead MSE or MAPE when the best model is fitted to the past data). The MSE and MAPE of such selection are worse than that of combining the four methods by a simple arithmetic average. Notice, however, that the MSE and MAPE *for the model fitting* are better than the corresponding ones of the four individual methods. The evidence in *Table 16.2* demonstrates that the strategy of selecting the method that minimizes the MSE or MAPE when a model is fitted to past

*Table 16.2:* Among-methods model selection using different selection criteria. Average MAPE; all data (111).

| | Fitting | Forecasting horizons | | | | | | | | Average of forecasting horizons | | | | |
|---|---|---|---|---|---|---|---|---|---|---|---|---|---|---|
| Method | model | 1 | 2 | 3 | 4 | 6 | 8 | 12 | 18 | 1-4 | 1-6 | 1-8 | 1-12 | 1-18 |
| (1) | 8.6 | 7.8 | 10.8 | 13.1 | 14.5 | 17.2 | 16.5 | 13.6 | 30.1 | 11.6 | 13.2 | 14.1 | 14.0 | 16.8 |
| (2) | 10.1 | 7.8 | 10.2 | 12.4 | 14.4 | 16.8 | 18.1 | 14.0 | 30.6 | 11.2 | 12.9 | 14.2 | 14.3 | 17.2 |
| (3) | 8.6 | 7.9 | 10.5 | 13.2 | 15.1 | 19.0 | 23.1 | 16.5 | 35.2 | 11.7 | 13.8 | 16.1 | 16.4 | 19.7 |
| (4) | 6.8 | 9.6 | 8.6 | 10.3 | 12.2 | 14.1 | 14.7 | 14.7 | 24.5 | 10.2 | 11.4 | 12.3 | 12.7 | 14.3 |
| (5) | 8.2 | 7.7 | 9.2 | 10.8 | 12.9 | 15.2 | 14.8 | 13.8 | 26.6 | 10.1 | 11.7 | 12.6 | 13.2 | 14.2 |
| (6) | 10.8 | 9.8 | 11.3 | 13.7 | 15.1 | 18.8 | 23.3 | 16.2 | 33.9 | 12.5 | 14.3 | 16.3 | 16.2 | 19.0 |
| (7) | 13.3 | 10.3 | 12.8 | 13.6 | 14.4 | 17.1 | 19.2 | 16.1 | 30.6 | 12.8 | 14.1 | 15.2 | 15.0 | 17.6 |
| (8) | 0.0 | 10.3 | 10.7 | 11.4 | 14.5 | 17.1 | 18.9 | 16.4 | 34.2 | 11.7 | 13.4 | 14.8 | 15.1 | 18.0 |
| (9) | 12.3 | 11.6 | 12.8 | 14.5 | 15.3 | 17.6 | 18.9 | 17.0 | 28.6 | 13.5 | 14.7 | 15.5 | 15.6 | 18.6 |
| (10) | 8.9 | 10.6 | 10.7 | 10.7 | 13.5 | 14.7 | 16.0 | 13.7 | 26.5 | 11.4 | 12.4 | 13.3 | 13.4 | 15.4 |
| (11) | 6.7 | 8.4 | 8.3 | 11.2 | 13.8 | 16.0 | 17.8 | 17.0 | 34.6 | 10.4 | 12.0 | 13.5 | 14.1 | 18.0 |
| (12) | 6.1 | 8.4 | 8.9 | 11.9 | 15.0 | 16.7 | 19.5 | 15.4 | 31.6 | 11.0 | 12.7 | 14.3 | 14.4 | 17.5 |
| (13) | n.a | 7.6 | 8.5 | 10.0 | 12.3 | 14.2 | 16.1 | 14.2 | 20.4 | 9.6 | 10.9 | 12.2 | 12.5 | 13.8 |
| (14) | n.a | 7.0 | 8.4 | 9.6 | 12.3 | 14.6 | 16.4 | 14.2 | 20.5 | 9.3 | 10.9 | 12.1 | 12.5 | 13.8 |
| (15) | n.a | 7.4 | 8.8 | 10.0 | 12.1 | 14.1 | 15.7 | 13.4 | 20.6 | 9.6 | 10.9 | 12.2 | 12.3 | 13.9 |
| (16) | n.a | 7.3 | 8.6 | 9.8 | 12.6 | 14.9 | 16.7 | 12.7 | 20.5 | 9.6 | 11.1 | 12.6 | 12.7 | 14.1 |
| (17) | n.a | 7.2 | 8.6 | 9.7 | 13.0 | 14.8 | 17.9 | 12.6 | 21.1 | 9.6 | 11.2 | 12.8 | 12.9 | 14.4 |
| (18) | n.a | 7.3 | 8.8 | 9.6 | 11.8 | 14.2 | 15.6 | 13.3 | 20.7 | 9.4 | 10.8 | 12.0 | 12.2 | 13.8 |
| (19) | n.a | 7.1 | 8.3 | 9.8 | 12.6 | 14.9 | 16.4 | 12.8 | 20.5 | 9.4 | 11.0 | 12.5 | 12.6 | 14.0 |
| (20) | n.a | 7.1 | 8.3 | 9.7 | 13.0 | 14.9 | 17.8 | 12.6 | 21.1 | 9.5 | 11.1 | 12.7 | 12.9 | 14.3 |
| (21) | n.a | 7.2 | 8.4 | 9.3 | 12.2 | 14.3 | 15.9 | 14.2 | 20.4 | 9.3 | 10.7 | 11.9 | 12.3 | 13.7 |
| (22) | n.a | 7.7 | 8.7 | 10.0 | 12.7 | 13.8 | 15.3 | 14.4 | 20.7 | 9.8 | 11.0 | 12.2 | 12.5 | 13.7 |
| (23) | n.a | 7.5 | 8.2 | 9.8 | 12.8 | 13.9 | 15.3 | 13.7 | 20.9 | 9.6 | 10.9 | 11.9 | 12.3 | 13.4 |
| (24) | n.a | 7.5 | 8.4 | 9.8 | 12.3 | 13.6 | 14.4 | 14.0 | 20.5 | 9.5 | 10.7 | 11.6 | 12.0 | 13.2 |

The methods are: (1) single exponential smoothing; (2) dampen-trend exponential smoothing; (3) Holt's linear exponential smoothing; (4) long term memory AR($p$) model; (5) above four methods combined; (6) automatic AEP filter; (7) Bayesian forecasting; (8) Box-Jenkins' ARIMA models; (9) Lewandowski's FORSYS; (10) Parzens' ARARMA models; (11) method with best MSE model fit; (12) method with best MAPE model fit; (13) best MAPE for out-of-sample; (14) best MSE for out-of-sample; (15) combine two best MAPE out-of-sample; (16) combine three best MAPE out-of-sample; (17) combine four best MAPE out-of-sample; (18) combine two best MSE out-of-sample; (19) combine three best MSE out-of-sample; (20) combine four best MSE out-of-sample; (21) combine best MSE and MAPE out-of-sample; (22) combine best MSE, MAPE, and rank out-of-sample; (23) reference; (24) best in confidence interval.

data does not improve post-sample forecasting accuracy.

The remainder of *Table 16.2* shows the accuracy of the four methods when the selection among the four methods was done with out-of-sample information. Several selection criteria for choosing one of the four methods for each series *and* for each of the $m$ forecasting horizons were tried. The effect of combining the best two, three, or four methods is also shown.

There is a large improvement if, instead of using the MSE or MAPE of model fit, the method with the best MSE, MAPE *for each of the m-forecasting horizons* is selected *to forecast for that specific horizon.* (The improvement in MAPE using this selection scheme beats the best method in the M-competition—Parzen's ARARMA models—by more than ARARMA models did better than single exponential smoothing.) The results seem equally accurate for short- as well as long-forecasting horizons. Moreover, combining the forecasts of the methods with the best MAPE or MSE for each forecasting horizon *does* not improve the results. This is contrary to previous findings [Makridakis and Winkler (1983), Russell and Adams (1987)] and suggests a model selection process where a combining does not beat the best method chosen.

An overall MAPE of 13.2 % is found by choosing all methods whose MAPE is within the range of the MAPE of the best method plus the error bound that is found by taking into account the sampling variation (as measured by the standard error) of the MAPE's of the methods involved [see row (24) "best in confidence interval" in *Table 16.2*]. Another selection criterion used was to choose a method that prior experience with the results of the M-competition and their accuracy has indicated to be the most appropriate, called reference method, and to retain it unless evidence shows that another method produces forecasts which are statistically more accurate; that is, there is statistical evidence to reject the null hypotheses postulating that the reference method is not the best. Several other selection criteria were utilized, and their results can be seen in *Table 16.2*. The differences in forecasting accuracy caused by most of these criteria do not seem to be large. The important factor is the selection of a "best" method(s) among the four, run in parallel, based on the out-of-sample accuracy performance of these four methods for each of the $m$ forecasting horizons.

### 16.3.2 Within-method model selection

Given a particular method, some optimization criterion can be used to select an appropriate model. This means finding seasonal indices and optimal parameter values. This selection can be done in two ways. The first requires using the first $s$ data points and finding the model that best fits these data points. This "best" model can be subsequently used to make $m$ forecasts and compute various errors measured during the sliding simulation. Then, the first $s + 1$ data points can be used to find the model that best fits this augmented set and make $m$ new forecasts. This optimization process can continue, each time computing the optimal model until all but one of the data points have been used in the model-fitting process. Alternatively, the "best" model can be found by using *all* $n$ data points, and then this model, once found, can be utilized to make $m$ forecasts with $s$ data points, $s + 1$ data points, ..., and so on until all the data points but one have been utilized in the simulation.

Originally, the first approach was used. Then it was decided to compute the seasonal indices using all data points (finding seasonal indices requires at least 3-4 years of data which put a serious constraint on the starting period of the simulation) and still optimize the model used at each step of the simulation. The results were comparable. Later on, the parameters of the "best" model for all $n$ data points were found and used at each step of the simulation without re-optimizing the model's parameters at each step. To my surprise overall post-sample forecasting accuracy did not change. It was, therefore, decided to use the second approach, which involved considerably fewer computations. The results of *Table 16.2* are based on such an approach to within-methods model selection.

[The method of Brown's quadratic exponential smoothing was used instead of the Long AR model, to compute out-of-sample forecasts. At present, I am exploring ways of finding out-of-sample forecasts using the Long AR method.]

The best model for each of the four methods can be used to make $m$ forecasts at each step of the sliding simulation. Although the model selection procedure is the same as that of the prevalent approach, it allows us to know the *actual forecasting performance* of the optimally selected model for each of $m$ *forecasting horizons*. Such performance can be subsequently used to choose the best of the four methods, as described above, by comparing their out-of-sample accuracy at each of the $m$-forecasting horizons. This can be done by basing the model selection *not* on past data, but on the out-of-sample forecasting performance. This means using the first $s$ data points, making $m$ forecasts, and recording the errors for each of the $m$ horizons. Then, using the first $s+1$ data points to make $m$ forecasts and obtain their MSE's or MAPE's. This procedure would continue with $s+2$, $s+3, \ldots, n-1$ data points. If this simulation, starting with $s$ and ending at $n-1$ data points, is repeated using alternative models of the same method, the model that minimizes the one-period-ahead, out-of-sample forecasting errors can be selected, as can the model that minimizes two-periods-ahead, three-periods-ahead, ..., and $m$-period-ahead errors. Although the specific model is fitted to past data, the selection is done on out-of-sample forecasting performance for *each forecasting horizon*.

*Table 16.3* shows the results of the prevalent approach to model selection based on minimizing one-period-ahead MSE for the model fitting, and that of model selection based on out-of-sample MSE performance for each of the $m$-horizons. Although the results for a one-period-ahead forecasting horizon are similar between the two approaches, the improvement of the proposed out-of-sample, within-method model selection starts at forecasting horizon two and becomes larger as the length of the forecasting horizons increases. The improvement is considerable. The accuracy of single exponential smoothing with the new approach is slightly better than that of Parzen's ARARMA models, the best method of the M-competition. Furthermore, if yearly data (for which single exponential smoothing is inappropriate since it assumes no trend) are excluded, the performance of single exponential smoothing is much better than that of ARARMA models. The accuracy of ARARMA models, or any other method, will probably also improve if the model selection is based on out-of-sample data for *each* of the $m$ forecasting horizons. (Research for selecting other models than smoothing based on out-of-sample performance is being done at present.)

## 16.4 Discussion and Directions for Future Research

The approach to statistical forecasting proposed in this chapter makes theoretical sense. Equally important, when tested empirically with the M-competition data, it provides superior results in terms of improved forecasting accuracy. More academic research and usage of such an approach in applied settings is, therefore, required to verify its value.

*Table 16.3:* Within-method model selection using as criteria the model that minimizes the MSE for each forecasting horizon.

| Method | Fitting model | Forecasting horizons 1 | 2 | 3 | 4 | 6 | 8 | 12 | 18 | Average of forecasting horizons 1-4 | 1-6 | 1-8 | 1-12 | 1-18 |
|---|---|---|---|---|---|---|---|---|---|---|---|---|---|---|
| | Average MAPE: all data (111) | | | | | | | | | | | | | |
| (1) | 8.9 | 10.6 | 10.7 | 10.7 | 13.5 | 14.7 | 16.0 | 13.7 | 26.5 | 11.4 | 12.4 | 13.3 | 13.4 | 15.4 |
| (2) | 8.6 | 7.8 | 10.8 | 13.1 | 14.5 | 17.2 | 16.5 | 13.6 | 30.1 | 11.6 | 13.2 | 14.1 | 14.0 | 16.8 |
| (3) | n.a | 7.8 | 10.4 | 12.2 | 13.5 | 15.7 | 14.9 | 13.4 | 25.1 | 11.0 | 12.3 | 13.1 | 13.0 | 15.2 |
| (4) | 10.1 | 7.8 | 10.2 | 12.4 | 14.4 | 16.8 | 18.1 | 14.0 | 30.6 | 11.2 | 12.9 | 14.2 | 14.3 | 17.2 |
| (5) | n.a | 8.2 | 9.6 | 12.1 | 12.7 | 17.2 | 18.6 | 15.5 | 27.3 | 10.6 | 12.4 | 13.9 | 14.2 | 16.6 |
| (6) | 8.6 | 7.9 | 10.5 | 13.2 | 15.1 | 19.0 | 23.1 | 16.5 | 35.2 | 11.7 | 13.8 | 16.1 | 16.4 | 19.7 |
| (7) | n.a | 7.9 | 10.3 | 11.9 | 13.6 | 15.7 | 17.8 | 14.0 | 26.8 | 10.9 | 12.4 | 13.8 | 14.1 | 16.4 |
| (8) | 8.7 | 8.8 | 11.8 | 15.0 | 16.9 | 24.1 | 35.7 | 29.7 | 63.6 | 13.1 | 16.4 | 20.3 | 22.2 | 30.2 |
| (9) | n.a | 8.8 | 11.0 | 14.9 | 14.6 | 18.6 | 22.3 | 26.5 | 47.2 | 12.3 | 14.1 | 15.8 | 16.9 | 22.3 |
| | Average MAPE: yearly data (20) | | | | | | | | | | | | | |
| (1) | 9.6 | 7.6 | 7.7 | 12.8 | 16.0 | 18.0 | 0.0 | 0.0 | 0.0 | 11.0 | 13.8 | 13.8 | 13.8 | 13.8 |
| (2) | 11.4 | 6.2 | 9.1 | 16.3 | 21.0 | 25.4 | 0.0 | 0.0 | 0.0 | 13.1 | 16.9 | 16.9 | 16.9 | 16.9 |
| (3) | n.a | 6.1 | 8.0 | 15.7 | 20.9 | 24.5 | 0.0 | 0.0 | 0.0 | 12.7 | 16.4 | 16.4 | 16.4 | 16.4 |
| (4) | 15.1 | 6.9 | 9.6 | 15.2 | 20.3 | 20.9 | 0.0 | 0.0 | 0.0 | 13.0 | 16.0 | 16.0 | 16.0 | 16.0 |
| (5) | n.a | 6.6 | 7.1 | 11.9 | 18.7 | 24.4 | 0.0 | 0.0 | 0.0 | 11.1 | 15.8 | 15.8 | 15.8 | 15.8 |
| (6) | 12.9 | 5.6 | 7.2 | 11.9 | 16.2 | 16.5 | 0.0 | 0.0 | 0.0 | 10.2 | 12.7 | 12.7 | 12.7 | 12.7 |
| (7) | n.a | 5.7 | 6.7 | 10.9 | 14.1 | 15.9 | 0.0 | 0.0 | 0.0 | 9.4 | 11.8 | 11.8 | 11.8 | 11.8 |
| (8) | 11.1 | 7.0 | 8.6 | 11.8 | 16.0 | 17.4 | 0.0 | 0.0 | 0.0 | 10.9 | 13.6 | 13.6 | 13.6 | 13.6 |
| (9) | n.a | 7.3 | 6.8 | 11.3 | 12.4 | 15.1 | 0.0 | 0.0 | 0.0 | 9.5 | 11.4 | 11.4 | 11.4 | 11.4 |
| | Average MAPE: quarterly data (23) | | | | | | | | | | | | | |
| (1) | 7.7 | 6.8 | 7.6 | 12.0 | 16.5 | 20.4 | 21.0 | 0.0 | 0.0 | 10.7 | 14.1 | 16.7 | 16.7 | 16.7 |
| (2) | 7.7 | 9.0 | 12.0 | 14.4 | 20.5 | 21.9 | 22.6 | 0.0 | 0.0 | 14.0 | 16.5 | 18.5 | 18.5 | 18.5 |
| (3) | n.a | 9.1 | 10.7 | 11.4 | 15.6 | 17.4 | 18.1 | 0.0 | 0.0 | 11.7 | 13.6 | 15.4 | 15.4 | 15.4 |
| (4) | 9.6 | 8.8 | 8.6 | 11.9 | 19.7 | 24.8 | 26.6 | 0.0 | 0.0 | 12.2 | 16.0 | 19.3 | 19.3 | 19.3 |
| (5) | n.a | 9.5 | 8.6 | 12.7 | 13.7 | 21.9 | 22.0 | 0.0 | 0.0 | 11.1 | 13.9 | 16.6 | 16.6 | 16.6 |
| (6) | 7.2 | 9.2 | 10.4 | 17.1 | 25.1 | 32.2 | 39.2 | 0.0 | 0.0 | 15.4 | 20.7 | 25.9 | 25.9 | 25.9 |
| (7) | n.a | 7.4 | 10.6 | 14.4 | 20.3 | 21.5 | 21.2 | 0.0 | 0.0 | 13.5 | 15.9 | 17.9 | 17.9 | 17.9 |
| (8) | 7.9 | 11.1 | 12.5 | 21.1 | 32.0 | 46.0 | 66.6 | 0.0 | 0.0 | 19.2 | 27.0 | 35.6 | 35.6 | 35.6 |
| (9) | n.a | 10.0 | 12.1 | 22.4 | 25.2 | 29.5 | 34.0 | 0.0 | 0.0 | 17.4 | 21.2 | 24.6 | 24.6 | 24.6 |
| | Average MAPE: monthly data (68) | | | | | | | | | | | | | |
| (1) | 9.0 | 12.7 | 12.6 | 9.6 | 11.7 | 11.8 | 14.3 | 13.7 | 26.5 | 11.7 | 11.4 | 12.1 | 12.6 | 15.4 |
| (2) | 8.0 | 7.9 | 10.9 | 11.7 | 10.6 | 13.2 | 14.4 | 13.6 | 30.1 | 10.3 | 11.0 | 12.0 | 12.6 | 16.5 |
| (3) | n.a | 7.9 | 11.1 | 11.4 | 10.6 | 12.5 | 13.8 | 13.4 | 25.1 | 10.2 | 10.7 | 11.6 | 12.0 | 15.1 |
| (4) | 8.7 | 7.8 | 11.0 | 11.8 | 10.9 | 12.9 | 15.2 | 14.0 | 30.6 | 10.4 | 11.0 | 12.1 | 13.0 | 17.1 |
| (5) | n.a | 8.2 | 10.7 | 12.0 | 10.5 | 13.5 | 17.4 | 15.5 | 27.3 | 10.4 | 10.9 | 12.5 | 13.4 | 16.7 |
| (6) | 7.9 | 8.2 | 11.5 | 12.3 | 11.4 | 15.2 | 17.7 | 16.5 | 35.2 | 10.9 | 11.8 | 13.5 | 14.8 | 19.5 |
| (7) | n.a | 8.7 | 11.3 | 11.3 | 11.2 | 13.7 | 16.7 | 14.0 | 26.8 | 10.6 | 11.4 | 12.8 | 13.6 | 16.6 |
| (8) | 8.2 | 8.6 | 12.5 | 13.8 | 12.1 | 18.7 | 25.3 | 29.7 | 63.6 | 11.7 | 13.7 | 16.6 | 20.4 | 31.0 |
| (9) | n.a | 8.8 | 11.8 | 13.5 | 11.6 | 16.0 | 18.4 | 26.5 | 47.2 | 11.4 | 12.5 | 13.9 | 16.0 | 23.0 |

The methods are: (1) Parzen's ARARMA (M-competition); (2) single (optimal model fitting); (3) single (optimal out-of-sample); (4) dampen-trend (optimal model fitting); (5) dampen-trend (optimal out-of-sample); (6) Holt (optimal model fitting); (7) Holt (optimal out-of-sample); (8) quadratic (optimal model fitting); (9) quadratic (optimal out-of-sample).

In the last five years countless hours of CPU time have been used to come up with new ways of forecasting more accurately. Innumerable decision rules, combining procedures, error measures (MSE, MAPE, median, MAD, geometric mean, etc.), new methods, classification schemes (e.g., use method A for macro data, method B for micro data, method C for industry, or method A for monthly, B for quarterly, and C for yearly data), and method and model decision rules were attempted. As in similar work reported in the literature [Schnaars (1986)] any gains in post-sample accuracy were found to be marginal. Once the best method, however, among four runs in parallel, was selected based on out-of-sample accuracy measures, important gains in post-sample accuracy were observed (see *Table 16.2*). Furthermore, when the out-of-sample within-method model selection (see *Table 16.3*) was used important improvements in forecasting accuracy were also observed. At present, additional work is being done on model-method selection that combines the within- and among-methods rules using out-of-sample criteria for each of the $m$-forecasting horizons. The initial results seem encouraging although much more work is required before definite conclusions can be drawn. My hope is that replications and more research will provide additional insights to contribute to a better theoretical foundation for statistical forecasting and to even greater improvements in forecasting accuracy in organizations.

The idea of this sliding simulation provides additional possibilities beyond improved forecasting accuracy. First, realistic confidence intervals can be built for each of the $m$ forecasting horizons. Such intervals need not be symmetric since information is collected about underestimates as well as overestimates around the most likely out-of-sample forecasts. In addition, through an analysis of extreme errors it is possible to warn forecasting users about *unusual* errors and help them think of ways to be prepared to face similar errors when they occur in the future. Moreover, the analysis of such errors can be done for each of $m$-forecasting horizons since unexpected errors tend to become larger as the forecasting horizons become longer.

Additional improvements in forecasting accuracy are possible by the appropriate choice of the methods to be run in parallel. Moreover, the forecasts of advanced methods might prove to be superior to those of simple ones, if the best model among such methods is selected based on out-of-sample performance. In addition, other decision rules for selecting methods based on out-of-sample forecasting errors might further improve forecasting accuracy. These and similar issues need to be investigated through additional theoretical and empirical research. A rather encouraging conclusion of the empirical work carried out in this chapter is that combining methods seem to provide little or no gains in forecasting accuracy. This means that individual methods adequately capture underlying patterns in a way that their forecasts cannot be improved through combining.

## 16.5 Conclusions

In this chapter a new approach to statistical forecasting has been proposed and tested. Such an approach aims at eliminating the deficiencies of the prevalent approach to statistical forecasting. It is based on the principle that model selection must be done on actual, out-of-sample forecasting performance. Such selection can be made at two levels. First, the best model (within-method selection) of a single method can be chosen. Second, the best method among several, run in parallel, can be selected. Both the within-method and among-methods selection is done on out-of-sample comparisons. It is not assumed that there exist a unique method that can forecast best for all series and forecasting horizons. This means that a different model or method or both can be selected for each series and for each forecasting horizon, based on the actual out-of-sample forecasting performance of the model and method for this specific series and forecasting horizon.

The empirical testing of the proposed approach shows important improvements in forecasting accuracy. Such improvements extend to short-, medium-, and long-forecasting horizons; different types of data (yearly, quarterly, monthly); and other classifications. Finally, the improvements come both within-method, when the best model is chosen based on out-of-sample information (see *Table 16.3*), and among-methods, when the best method is chosen based on out-of-sample performance (see *Table 16.2*). Combining the within- *and* among-methods selection might further improve post-sample forecasting accuracy, and research currently aims at exploring and measuring the advantages of such combining.

## Acknowledgments

I would like to thank Scott Armstrong, Tim Davidson, Chris Chatfield, Maurice Desplas, Q. Duong, Robert Fildes, Ev Gardner and his Ph.D. students, Jan de Grooijer, Michèle Hibon, Ira Horowitz, Essam Mahmoud, Vic McGee, Steve Schnaars, Rob Weitz, and Bob Winkler for their valuable comments on earlier drafts of this chapter.

## References

Ahlers, D. and Lakonishok, J. (1983), A study of economists' consensus forecasts. *Management Science*, **29**, 1113-1125.

Andersen, A., and Carbone, R. (1983), One-step versus multiple-steps forecasts. Working Paper, University Laval.

Armstrong, J.S. (1986), Research on forecasting: A quarter-century review, 1960-1984. *Interfaces*, **16**, 89-109.

Box, G.E.P. and Jenkins, G.M. (1970). *Time Series Analysis: Forecasting and Control.* San Francisco: Holden-Day.

Brown, R.G. (1963), *Smoothing, Forecasting and Prediction of Disctrete Time Series.* Englewood Cliffs: Prentice-Hall.

Carbone, R., and Makridakis, S. (1986), Forecasting when pattern changes occur beyond the historical data. *Management Science*, **32**, 257-271.

Clemen, R.T. and Winkler, R.L. (1986), Combining economic forecasts. *Journal of Business and Economic Statistics*, **4**, 39-46.

Dawes, R.M. (1979), The robust beauty of improper linear models in decision making. *American Psychologist*, **34**, 571-582.

Dawes, R.M. (1986), Forecasting one's own preferences. *International Journal of Forecasting*, **2**, 5-14.

Fildes, R. (1986), Accuracy gains through individual univariate forecasting model selection. Working Paper No. 132, Manchester Business School.

Gardner, E.S., and Makridakis, S. (eds.) (1988), The future of forecasting. *International Journal of Forecasting*, **4** (forthcoming).

Gardner, E.S. and McKenzie, E. (1985), Forecasting trends in time series. *Management Science*, **31**, 1237-1246.

Goldberg, L.R. (1970), Man versus model of man: A rationale, plus some evidence for a method of improving on clinical inferences. *Psychological Bulletin*, **73**, 422-432.

Gupta, S. and Wilton, P.C. (1987), Combination of forecasts: An extension. *Management Science*, **33**, 356-373.

Hogarth, R.M. and Makridakis, S. (1981), Forecasting and planning: An evaluation. *Management Science*, **27**, 115-138.

Jenkins, G.M. (1982), Some practical aspects of forecasting in organizations. *Journal of Forecasting*, **1**, 3-23.

Johnston, J. (1972), *Econometric Methods, Second edition.* New York: McGraw-Hill.

Mahmoud, E. (1984), Accuracy in forecasting: A survey. *Journal of Forecasting*, **3**, 139-159.

Makridakis, S. (1981), Forecasting accuracy and the assumption of accuracy. *OMEGA*, **9**, 307-311.

Makridakis, S. (1986), The art and science of forecasting: An assessment and future directions. *International Journal of Forecasting*, **2**, 15-39.

Makridakis, S. (1987), Metaforecasting: Ways of improving forecasting accuracy and usefulness. Keynote address presented at the 7th International Symposium on Forecasting, Boston (INSEAD Working Paper Nr. 87/15).

Makridakis, S. and Wheelwright, S. (1977), Adaptive filering: An integrated moving average filter for autoregressive time series forecasting. *Operational Research Quarterly*, **28**, 877-888.

Makridakis, S. and Hibon, M. (1979), Accuracy of forecasting: An empirical investigation. *Journal of the Royal Statistical Society*, **A-142**, 97-125.

Makridakis, S. *et al.* (1982), The accuracy of extrapolation (time series) methods: Results of a forecasting competition. *Journal of Forecasting*, **1**, 111-153.

Makridakis, S. and Winkler, R. (1983), Averages of forecasts: Some empirical results. *Management Science*, **29**, 987-996.

Makridakis, S. and Winkler, R. (1988), Sampling distribution of post-sample forecasting errors. *Applied Statistics* (forthcoming).

Meese, R. and Geweke, J. (1984), A comparison of autoregressive univariate forecasting procedures for macroeconomic time series. *Journal of Business and Economic Statistics*, **2**, 191-200.

Parzen, E. (1982), ARARMA models for time series analyses and forecasting. *Journal of Forecasting*, **1**, 67-82.

Priestley, M.B. (1979), Discussion of the paper by Professor Makridakis and Dr. Hibon. *Journal of the Royal Statistical Society*, **A-142**, 127-129.

Russell, T.D. and Adams Jr., E.E. (1987), An empirical evaluation of alternative forecasting combinations. *Management Science*, **33**, 1267-1276.

Schnaars, S.P. (1986), An evaluation of rules of selecting on extrapolation model on yearly sales forecasts. *Interfaces*, **16**, 100-107.

Williams, W.H. and Goodman, M.L. (1971), A simple method for the construction of empirical confidence limits for economic forecasts. *Journal of the American Statistical Association*, **66**, 752-754.

Zarnowitz, V. (1984), The accuracy of individual and group forecasts from business outlook surveys. *Journal of Forecasting*, **3**, 11-26.

# Part IV

# Economic Modeling and the Use of Empirical Data

## CHAPTER 17

# The Relation between Economic Growth and Structural Change

*Wilhelm Krelle*

## Summary

The relation between economic growth and structural change is analyzed by a dynamic multisector world model, where production follows demand and demand changes with higher income. It shows that economic growth is inevitably connected with structural change. If structural change is restrained, the growth rate is reduced. The model covers not only structural change in the sense of changes in the proportion of production and demand of different commodities. Structural change in the statistical sense, i.e., change in the parameter values of the behavior equation, is not considered.

## 17.1 The Problem

There is no question that the long-term economic growth of an economy is connected with different kinds of structural change. What are the reasons for this interdependence? In this chapter we try to answer this question and to model the interrelationships in such a way that they can be tested econometrically. Some tests have already been performed and will be presented below.

Economic growth is measured by the growth rate of real GDP (or NMP in the case of CMEA countries) and its components, or by growth rates of GDP per capita. The concept of structural change is more difficult. It comprises many economic fields, e.g.:

1. Change in the sectoral composition of GDP, e.g., measured by the ratio of the value-added in specific industries to total GDP. This includes also the appearance of new

sectors of production.

2. Change in the commodity composition of demand. This also includes the emergence of new products.
3. Change in the size, composition, and employment of the labor force.
4. Change in trade relations, i.e., in the direction, size, and commodity composition of trade flows between different countries and in the ensuing asset and debt positions of countries.
5. Change in the relative size and international importance of different countries.
6. Change in the distribution of income and wealth of households and social groups.
7. Change in the general economic and political order, e.g., the degree of planning, state intervention, and regulation.
8. Change in the location of industries and households.

The structural changes may be smooth and continuous or abrupt and discrete, so that separate regimes may be identified with different behavioral characteristics in each regime. In this chapter we shall concentrate on points 1 and 2, give some hints with respect to 3, 4, 5, and omit 6, 7, and 8.

The statistical data and the econometric estimations presented in this chapter are mostly results of the IIASA-University of Bonn Research Project on Economic Growth and Structural Change, a common undertaking of the International Institute for Applied Systems Analysis (IIASA) in Laxenburg, Austria, and of the University of Bonn (Department of Economics). The research was carried out from 1985 to 1987 in Bonn under my supervision. The results are published in Krelle (1989).

## 17.2 Economic Growth and Commodity Composition of GDP and of Production Functions: A Look at the Statistics

*The growth rates of GDP* declined from the mid-1960s until the early 1980s in almost all countries and started to rise after that in most countries. *Figure 17.1* shows this declining phase in terms of the growth rates. *Figure 17.2* shows the observations in terms of absolute values of real GDP per capita in 1975 $ and some trend forecasts of the IIASA-University of Bonn Research Project [see Krelle (1989, p. 70)]. The forecasts are made under the assumption that the statistical sources for the CMEA countries are reliable and that there are no wars or internal troubles. Judging from *Figures 17.1* and *17.2* we may state that the long-term growth process has continued worldwide, though at a changing pace. This growth in aggregated terms was accompanied by important structural changes in the *production structure.* These changes are visible even on a highly aggregated sectoral

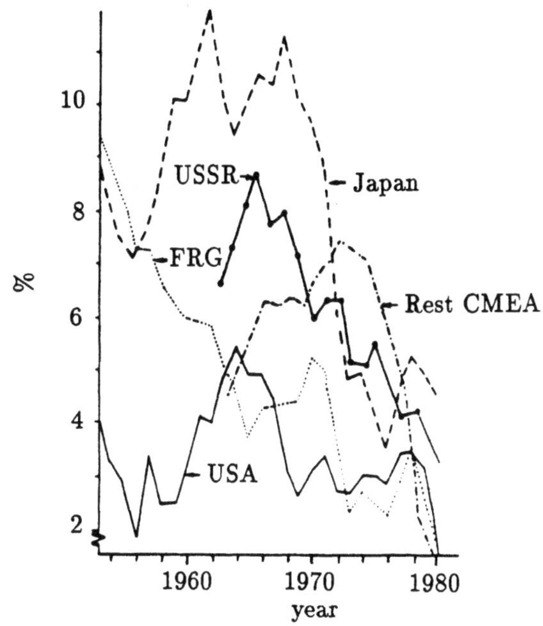


*Figure 17.1:* Growth rates (in percent) of GDP (for OECD countries) or NMP (for CMEA countries), five-year moving avarages. Source for OECD countries: *OECD, National Accounts, Volume I: Main Aggregates*, Paris, different years. Sources for CMEA countries: *UN Yearbook of National Accounts Statistics*, 1970-1982; *Statistical Yearbook of the CMEA (Statistics)*, 1970-1982; *Statistical Yearbook of the CMEA (Statisticheskij Ezhegodni Stran Chlenov SEV)*, 1971-1983; *United Nations Yearbook of International Trade Statistics*, 1977, 1982; National Statistical Yearbooks of USSR, Bulgaria, CSFR, GDR, Hungary, and Poland for various years; own calculations. The figure is taken from Krelle (1987, p. 380).

basis. *Figure 17.3* shows this graphically for the USA. For the most important OECD and CMEA countries actual figures and forecasts are given in *Tables 17.1* and *17.2*. These figures show that agriculture is declining in almost all countries. Manufacturing gives a mixed picture; it is declining in some OECD countries (UK and Canada), expanding in others (Japan, France, Italy), and more or less stable in the rest. Services are expanding almost everywhere with the exception of Japan. Since NMP does not contain services, the figures in *Table 17.2* are not directly comparable with the OECD figures of *Table 17.1*. With respect to the NMP concept, agriculture is also declining, but industry is expanding everywhere. The same is true for developing countries [see Krelle (1989, p. 677)].

These figures refer to real value-added and real GDP. They change if we apply nominal terms. In this case services are expanding and agriculture and manufacturing are declining in almost all OECD countries [see OECD (1989, pp. 120-122)].

If we go into more detail and consider a longer time range, much more structural change

*Table 17.1:* Sectoral composition of GDP, OECD countries (% of GDP).

| | | Agric. | Min. & Quar. | Manuf. | Public Services[a] | Constr. | Services |
|---|---|---|---|---|---|---|---|
| USA | 1962 | 4.3 | 3.0 | 23.9 | 2.2 | 6.6 | 59.9 |
| | 1984 | 3.0 | 2.5 | 24.0 | 2.7 | 4.9 | 63.1 |
| 1999 | medium scenario | 2.1 | 2.1 | 25.9 | 2.6 | 3.3 | 64.5 |
| | pessimistic scenario | 2.3 | 2.2 | 24.9 | 2.0 | 3.3 | 65.8 |
| FRG | 1962 | 4.0 | 3.0 | 36.3 | 1.8 | 8.4 | 46.6 |
| | 1984 | 2.7 | 0.9 | 36.4 | 2.9 | 5.9 | 52.2 |
| 1999 | medium scenario | 1.2 | 0.5 | 35.9 | 3.6 | 5.8 | 54.1 |
| | pessimistic scenario | 1.5 | 0.6 | 36.5 | 3.5 | 4.8 | 53.1 |
| Japan | 1962 | 11.9 | 1.1 | 22.9 | 1.9 | 7.9 | 54.3 |
| | 1984 | 4.8 | 0.7 | 30.4 | 2.2 | 9.7 | 52.3 |
| 1999 | medium scenario | 4.3 | 0.5 | 34.3 | 2.3 | 10.5 | 49.7 |
| | pessimistic scenario | 5.4 | 0.5 | 34.3 | 2.1 | 9.9 | 49.6 |
| France | 1962 | 8.9 | 2.1 | 26.8 | 1.3 | 7.7 | 53.2 |
| | 1984 | 4.5 | 0.7 | 28.7 | 2.1 | 5.5 | 58.4 |
| 1999 | medium scenario | 3.2 | 0.4 | 30.0 | 2.2 | 6.1 | 57.4 |
| | pessimistic scenario | 3.5 | 0.5 | 26.9 | 2.2 | 4.2 | 62.6 |
| UK | 1962 | 2.7 | 3.1 | 27.2 | 2.2 | 8.4 | 56.4 |
| | 1984 | 2.9 | 3.0 | 25.3 | 3.3 | 6.3 | 59.1 |
| 1999 | medium scenario | 2.9 | 3.3 | 25.4 | 3.7 | 5.4 | 55.2 |
| | pessimistic scenario | 3.2 | 7.9 | 22.6 | 3.5 | 4.6 | 54.3 |
| Italy | 1962 | 10.6 | 2.5 | 25.2 | 3.9 | 12.0 | 45.7 |
| | 1984 | 7.4 | 2.4 | 31.2 | 5.0 | 7.3 | 46.9 |
| 1999 | medium scenario | 6.1 | 2.2 | 31.1 | 5.9 | 6.6 | 47.0 |
| | pessimistic scenario | 6.8 | 2.4 | 34.5 | 4.8 | 5.9 | 46.6 |
| Canada | 1962 | 7.6 | 4.3 | 21.3 | 1.8 | 8.6 | 56.4 |
| | 1984 | 4.4 | 3.0 | 19.5 | 3.5 | 6.6 | 62.3 |
| 1999 | medium scenario | 3.8 | 3.1 | 18.6 | 4.2 | 6.5 | 64.8 |
| | pessimistic scenario | 4.2 | 3.0 | 19.2 | 3.3 | 6.1 | 63.7 |

[a] Electricity, water, and gas.

Source: Krelle (1989, p. 675).

*Table 17.2:* Sectoral composition of NMP, CMEA countries (% of NMP).

| | | Agric. | Industr. | Constr. | Trade | Traffic Comm. | Others |
|---|---|---|---|---|---|---|---|
| USSR | 1960 | 38.3 | 37.4 | 11.5 | 7.8 | 5.0 | – |
| | 1985 | 14.5 | 60.3 | 10.4 | 8.3 | 6.4 | – |
| 1999 | medium scenario | 6.9 | 67.7 | 10.3 | 8.3 | 6.8 | – |
| | pessimistic scenario | 7.9 | 68.3 | 9.3 | 7.9 | 6.6 | – |
| Bulgaria | 1960 | 57.5 | 24.3 | 6.3 | 3.1 | 3.9 | 5.0 |
| | 1985 | 19.9 | 48.4 | 9.1 | 10.7 | 8.4 | 3.5 |
| 1999 | medium scenario | 14.3 | 52.6 | 9.8 | 11.2 | 9.0 | 3.0 |
| | pessimistic scenario | 16.2 | 51.1 | 9.3 | 11.4 | 8.9 | 3.0 |
| CSFR | 1960 | 16.9 | 51.4 | 11.0 | 15.2 | 4.9 | 0.6 |
| | 1985 | 9.8 | 55.4 | 12.2 | 16.8 | 4.7 | 0.7 |
| 1999 | medium scenario | 8.6 | 55.5 | 13.0 | 17.6 | 4.6 | 0.7 |
| | pessimistic scenario | 9.0 | 56.7 | 13.0 | 15.6 | 4.7 | 1.0 |
| GDR | 1960 | 17.6 | 58.2 | 5.9 | 10.5 | 5.1 | 2.7 |
| | 1985 | 7.4 | 69.7 | 5.4 | 10.1 | 4.5 | 3.0 |
| 1999 | medium scenario | 5.2 | 71.0 | 5.9 | 10.1 | 4.3 | 3.4 |
| | pessimistic scenario | 5.3 | 71.3 | 5.7 | 10.0 | 4.3 | 3.3 |
| Hungary | 1960 | 35.5 | 36.7 | 10.5 | 12.3 | 4.8 | 0.4 |
| | 1985 | 17.9 | 52.2 | 10.3 | 13.0 | 5.7 | 1.0 |
| 1999 | medium scenario | 15.4 | 52.8 | 11.3 | 12.3 | 6.6 | 1.6 |
| | pessimistic scenario | 17.4 | 54.2 | 11.0 | 9.2 | 6.9 | 1.4 |
| Poland | 1960 | 38.2 | 31.9 | 13.2 | 13.1 | 3.0 | 0.7 |
| | 1985 | 18.2 | 48.9 | 12.3 | 15.0 | 4.4 | 1.3 |
| 1999 | medium scenario | 11.9 | 53.1 | 12.7 | 16.0 | 4.8 | 1.5 |
| | pessimistic scenario | 9.5 | 54.4 | 13.1 | 16.5 | 4.8 | 1.7 |
| Romania | 1960 | 53.2 | 22.9 | 8.7 | 8.2 | 4.0 | 3.0 |
| | 1985 | 22.0 | 54.3 | 7.8 | 6.0 | 7.7 | 2.2 |
| 1999 | medium scenario | 16.0 | 60.1 | 7.5 | 6.2 | 7.8 | 2.3 |
| | pessimistic scenario | 14.0 | 61.4 | 6.5 | 6.4 | 8.6 | 3.1 |

Source: Krelle (1989, p. 676).

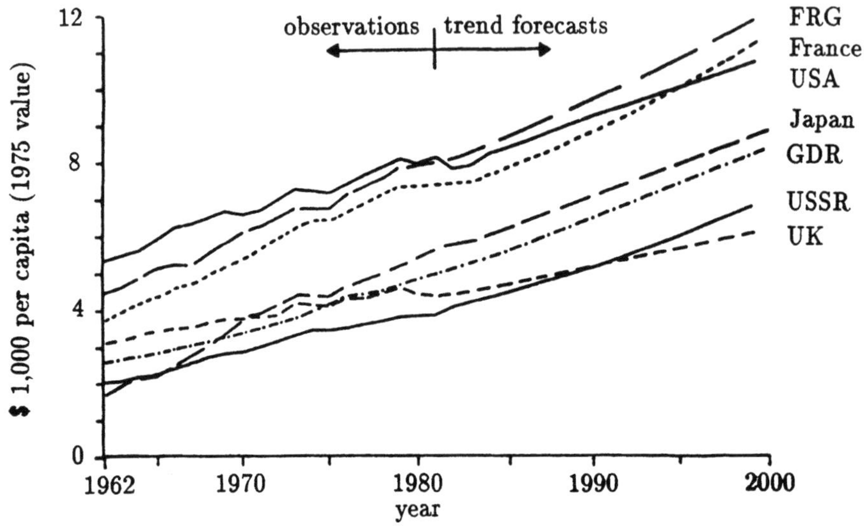


*Figure 17.2:* Real GDP per capita (OECD) or GDP "type 2" (CMEA) under the assumptions that the optimistic scenario will be realized for the USA, USSR, and UK; the pessimistic scenario for the FRG and Japan; and the medium scenario for all other contries. Source: Krelle (1989, p. 70).

becomes visible. New products and (connected with these) new production processes appear. *Figure 17.4* shows this for the sector of telecommunication, *Figures 17.5* and *17.6* for the iron and steel industry. We see that there are discontinuities in the structural composition of production: Starting from a certain date a new product and (or) a new production procedure appears and joins the series of older products and substitutes for them partly or (very rarely) totally. Thus structural breaks in the growth process may be partially or totally concealed by aggregation.

This type of structural change induced by technical progress is superimposed by changes due to the changing advantages of different locations, e.g., low-grade steel production is moving to the developing countries so that steel production in OECD countries is declining [see *Figure 17.7*].

It is clear from these figures that economic growth demands and is based on structural change. If structural change is hampered by social resistance, there will be no growth; the society will fall behind in international competition, and in 100 years will end up as an underdeveloped country.

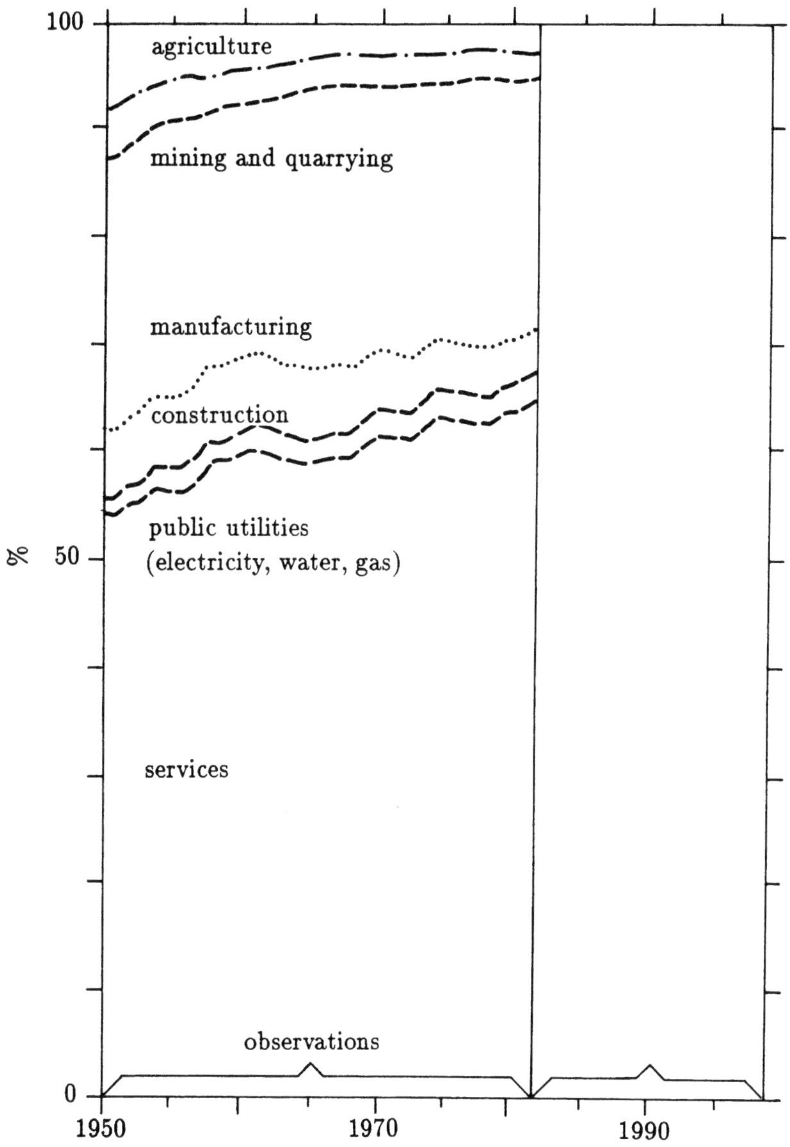


*Figure 17.3:* The structure of production, USA. Source: Krelle (1989, p. 11).

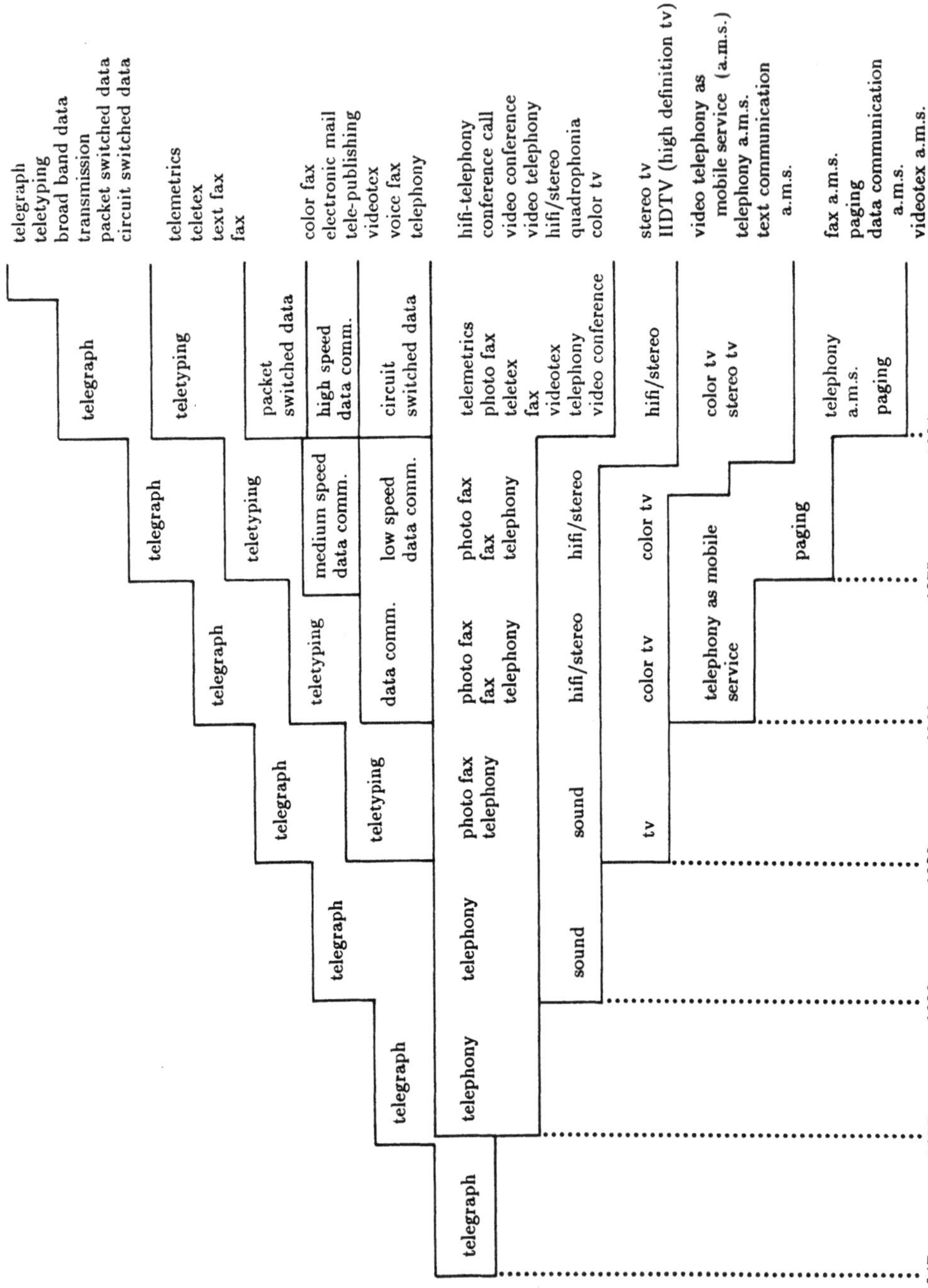


*Figure 17.4:* The development of telecommunication services from 1847 until the year 2000.

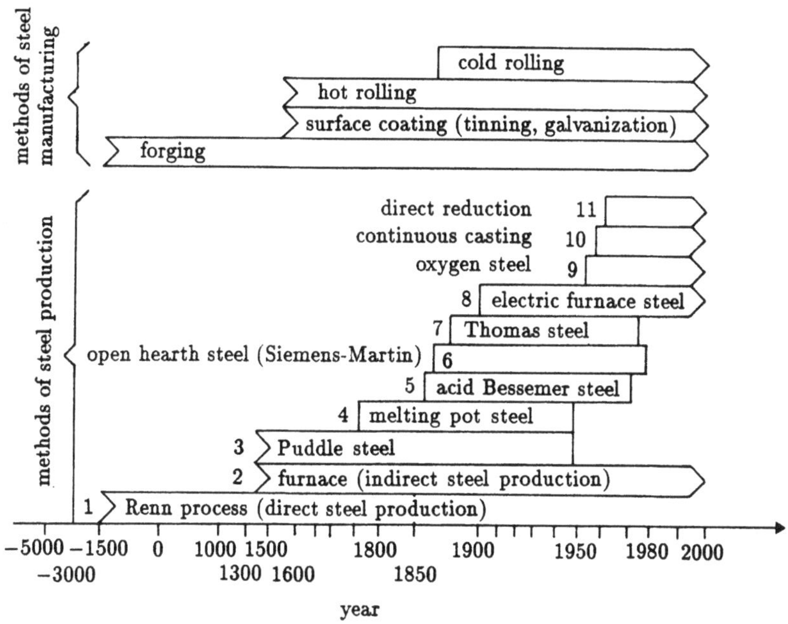


*Figure 17.5:* Innovations in steel production and manufacturing.

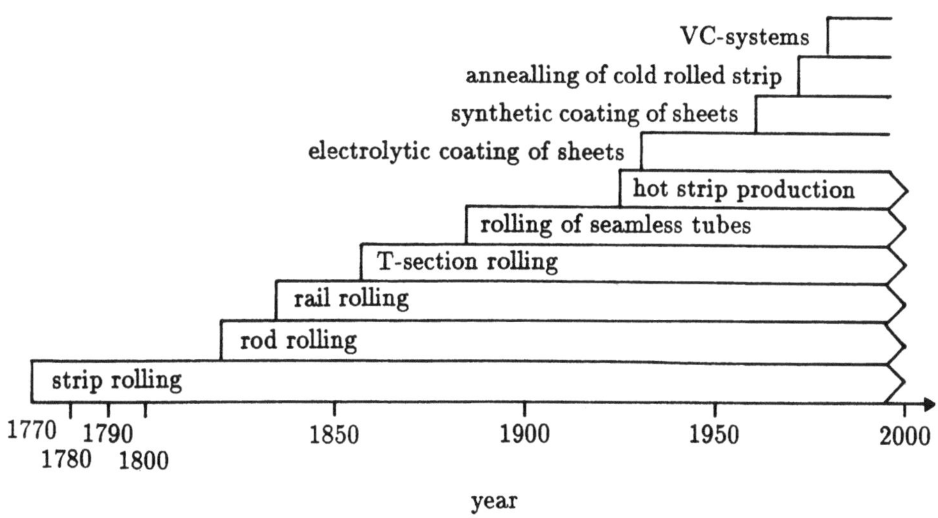


*Figure 17.6:* Rolled steel production; there are about 2500 different types of steel.

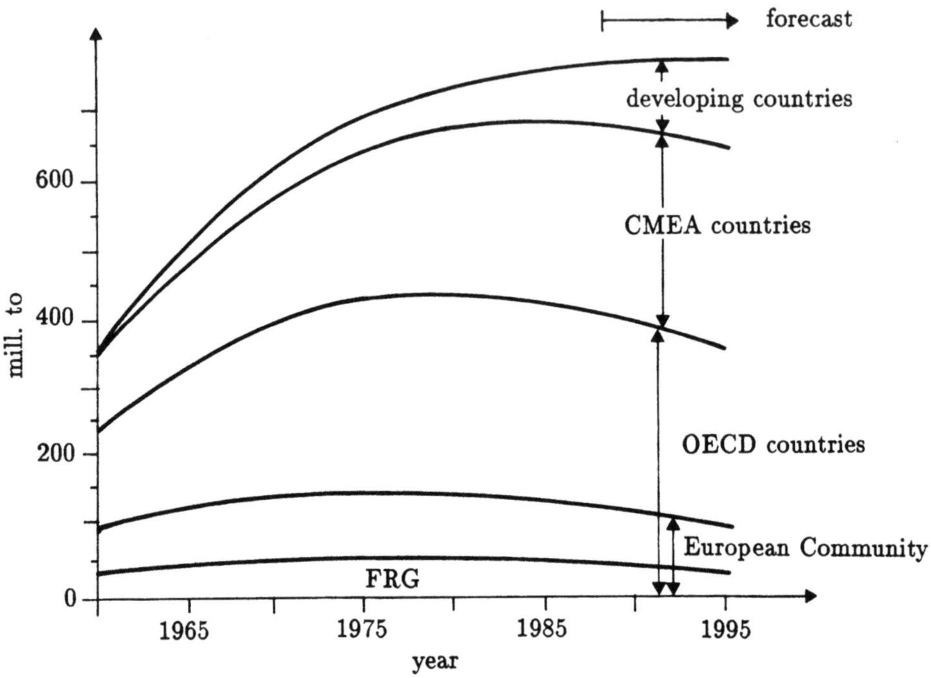


*Figure 17.7:* Trend in crude steel production. Sources: *Fünfzehn Fragen und Fakten zur Europäischen Stahlpolitik*, Wirtschaftsvereinigung Eisen- und Stahlindustrie, Düsseldorf, Dezember 1987; *Marktversorgung mit Stahl*, Wirtschaftsvereinigung Eisen- und Stahlindustrie, Düsseldorf, 01. 02. 1988.

## 17.3 A Theory of the Interrelation of Growth and Structural Change

### 17.3.1 The real part of the model

In this section I present a theory where growth and structural change (in the sense of points 1 and 2 listed in Section 17.1) are simultaneous and inseparable. For econometric estimations this theory has to be simplified. This will be done in Section 17.4. The theory assumes the form of a disequilibrium growth model for a closed economy, since the growth rates of the different sectors do not stay constant and are not equal. But it will be an equilibrium model in the sense of short-run income and employment theory, since we assume that demand equals supply in each period. The total model can only be solved numerically, but for parts of it analytical solutions are possible.

Let us suppose that for a certain number of periods (say, years) there are $n$ production sectors in the economy, each one producing one commodity (or category of commodities).

In the course of technical progress new commodities and thus new productive sectors emerge. We model this fact by always assuming that one new sector comes into being after the "state of technology" $\pi$ passes certain thresholds. Before that time the new product and the new production functions are unknown. Each production sector is described by a neoclassical production function with Harrod-neutral technical progress. In principle, each commodity may serve as a capital good as well as a consumption good. Thus, if there are $n$ sectors working in the period considered, say for $t \leq t_0$ and $\pi \leq \pi_0$, we assume for the production side that

$$Y_i = F_{i,n^*}(\pi L_i, K_{1i}, \ldots, K_{ni}, \delta_1 K_{n+1,i}, \delta_2 K_{n+2,i}, \ldots) \tag{17.1}$$

with $i = 1, \ldots, n$, in the current period, and $i = 1, \ldots, n,\ \delta_1(n+1), \delta_2(n+2), \ldots$, in future periods, where (for $\bar{\delta} > 0$ and $\pi$ being exogenous)

$$\delta_i = \begin{cases} 0 & \text{for } \pi \leq \pi_0 + i\bar{\delta} \\ 1 & \text{for } \pi > \pi_0 + i\bar{\delta} \end{cases} \quad i = 1, 2, \ldots, N, \ldots$$

and $n^*$ is the maximal number of commodities in the period considered [$n^* = n$ in equation (17.1) for the current period], $Y_i$ is the production (value-added) in sector $i$, $L_i$ is the labor employed in sector $i$, and $K_{ij}$ is the capital of type $j$ used in sector $i$.

The functions $F_{i,n^*}$ are assumed to be homogeneous of degree one in all nonzero variables and quasi-concave. We consider a period, where $\pi \leq \pi_0 + \bar{\delta}$, i.e., $n^* = n$. The productivity index $\pi$ is taken to rise in time: $\pi = \pi(t)$, $\pi' > 0$. We call a time span where all $\delta_i$ are constant a *regime* (i.e., there is a fixed number of commodities known and produced in the society). We may rewrite equation (17.1) as

$$y_i = \pi \cdot g_i(\kappa_{1i}, \ldots, \kappa_{ni}, \ldots), \tag{17.2}$$

where $y_i = \frac{Y_i}{L_i}$, $\kappa_{ji} = \frac{K_{ji}}{\pi L_i}$, $g_i(\cdot) = F_i(1, \kappa_{1i}, \ldots, \kappa_{ni}, \ldots)$. We start with some *basic definitions.* Production of commodity $i$ per capita is defined by

$$\bar{y}_i = \frac{Y_i}{L}, \quad \text{so that } \bar{y}_i = \pi\, \alpha_i\, g_i(\cdot), \tag{17.3}$$

where $\alpha_i = \frac{L_i}{L}$ is the proportion of labor employed in sector $i$. The actual price level $\bar{p}$ is defined as a weighted average of all market prices $p_i^*$:

$$\bar{p} = \sum_{j=1}^{\infty} x_j p_j^* \tag{17.4}$$

the weights $x_j$ obeying $\sum_j x_j = 1$ and $x_j = x_j(\delta_1, \delta_2, \ldots) \geq 0$, and

$$\begin{aligned}
j &= 1, \ldots, n \quad \text{if } \delta_1 = \delta_2 = \ldots = 0; \quad \text{then } x_{n+1} = x_{n+2} = \ldots = 0 \\
j &= 1, \ldots, n+1 \quad \text{if } \delta_1 = 1, \delta_2 = \delta_3 = \ldots = 0; \quad \text{then } x_{n+2} = x_{n+3} = \ldots = 0 \\
&\vdots \\
j &= 1, \ldots, n+m \quad \text{if } \delta_1 = \delta_2 = \ldots = \delta_m = 1, \quad \delta_{m+1} = \delta_{m+2} = \ldots = 0; \\
&\qquad \text{then } x_{n+m+1} = x_{n+m+2} = \ldots = 0\,.
\end{aligned}$$

In the following we use normalized prices $p_i$ such that the price level $p_Y$ of real domestic product is always unity:

$$p_i = \frac{p_i^*}{\sum_j x_j p_j^*} \,. \tag{17.5}$$

Therefore the following restrictions are valid for the normalized prices:

$$p_Y := \sum_i x_i p_i = 1 \,. \tag{17.6}$$

Let $w^*$ be the nominal wage rate (equal for all sectors). Then $w = \frac{w^*}{\bar{p}}$ is the real wage rate (to be used in the following); $r$ is the real-interest rate. Furthermore we define:

$$\begin{aligned} Y &= \textstyle\sum_i p_i Y_i = \text{ GDP} \\ K_{\cdot i} &= \textstyle\sum_j p_j K_{ji} = \text{ total capital used in sector } i \\ K_{j\cdot} &= \textstyle\sum_i K_{ji} = \text{ capital goods of type } j \text{ used in all sectors} \\ K &= \textstyle\sum_j \sum_i p_j K_{ji} = \text{ total capital in the economy} \end{aligned} \tag{17.7}$$

and

$$g_{ij}(\cdot) = \frac{\partial g_i(\kappa_{1i}, \ldots, \kappa_{ni}, \ldots)}{\partial \kappa_{ji}} \,. \tag{17.8}$$

We now return to the *supply side*. We assume that there are no pure profits in the economy, i.e., that the proceeds from selling each commodity cover the costs of production (including interest on capital):

$$w L_i + r \sum_j p_j K_{ji} = p_i Y_i \quad \text{or} \tag{17.9}$$

$$\frac{w}{\pi} + r \sum_j p_j \kappa_{ji} = p_i g_i(\cdot) \,. \tag{17.10}$$

Furthermore, we assume that the entrepreneurs employ labor and capital such that the production costs are minimized, given the factor prices and the price of the final product:

$$w L_i + r \sum_j p_j K_{ji} = \min_{L_i, K_{1i}, \ldots, K_{ni}} ! \qquad \text{s.t. } Y_i - \pi L_i \, g\left(\frac{K_{1i}}{\pi L_i}, \ldots, \frac{K_{ni}}{\pi L_i}, \ldots\right) = 0 \,.$$

This yields

$$\frac{w}{\pi} + \mu_i \sum_j g_{ij} \kappa_{ji} = \mu_i g_i \quad \text{and} \tag{17.11}$$

$$r p_j = \mu_i g_{ij} \,, \tag{17.12}$$

where $\mu_i$ is a Lagrangian multiplier. These are the marginal product rules. Substituting (17.12) into (17.11) and comparing it with (17.10) yields: $\mu_i = p_i$. Thus, considering (17.12), equation (17.11) is equivalent to (17.10). Therefore, we retain the equations (17.10) and (17.13) where

$$r p_j = p_i g_{ij} \,. \tag{17.13}$$

Since we are in a period where $\pi = \pi_0$, i.e., only $n$ commodities are known and produced, we have $n^2$ equations (17.13), $n$ equations (17.10), and one equation (17.6) to determine the $n^2$ unknowns $\kappa_{ji}$, the $n$ unknowns $p_i$, and the two unknowns $w$ and $r$. One equation is lacking. It will be provided by the demand side. The same applies for the absolute values of $L_i$ and $K_{ji}$. From the supply side we may solve for all but one of the above-mentioned unknowns as a function of the remaining one (say, $\kappa_{11}$), e.g.,

$$p_i = p_i(\kappa_{11}), \quad i = 1, \ldots, n. \tag{17.14}$$

In Krelle (1988, p. 461), this has been done for CES and Cobb-Douglas production functions and for arbitrary $n$. I reproduce the result for $n = 2$ and Cobb-Douglas production functions

$$Y_i = (\pi L_i)^{\beta_{oi}} K_{1i}^{\beta_{1i}} K_{2i}^{\beta_{2i}}, \quad i = 1, 2, \quad \beta_{ji} > 0, \quad \sum_{j=0}^{2} \beta_{ji} = 1,$$

where the remaining unknown is chosen to be $\kappa_{11}$. The results are

$$\begin{aligned}
\kappa_{1i} &= \frac{\beta_{01}\beta_{1i}}{\beta_{11}\beta_{0i}} \kappa_{11}, \quad \kappa_{2i} = \beta^{1/\beta} \left[\frac{\beta_{2i}}{\beta_{0i}}\right] \left[\frac{\beta_{01}}{\beta_{11}} \kappa_{11}\right]^{\beta_1/\beta_2} \\
1/p_1 &= x_1 + x_2 \beta^{-1/\beta_2} \left[\frac{\beta_{01}}{\beta_{11}} \kappa_{11}\right]^{1-(\beta_1/\beta_2)} \\
1/p_2 &= x_1 \beta^{1/\beta_2} \left[\frac{\beta_{01}}{\beta_{11}} \kappa_{11}\right]^{(\beta_1/\beta_2)-1} + x_2 \\
r &= \beta_{11} \kappa_{11}^{\beta_{11}-1} \kappa_{21}^{\beta_{21}} = \beta_{22} \kappa_{12}^{\beta_{12}} \kappa_{22}^{\beta_{22}-1} \quad \text{[from equation (17.13) for } i = j\text{]} \\
w &= \pi r \lambda, \quad \text{where } 1/\lambda = x_1 \left[\frac{\beta_{01}}{\beta_{11}} \kappa_{11}\right]^{-1} + x_2 \beta^{-1/\beta} \left[\frac{\beta_{01}}{\beta_{11}} \kappa_{11}\right]^{-\beta_1/\beta_2} \\
\beta &= \frac{\beta_{02}^{\beta_{02}} \beta_{12}^{\beta_{12}} \beta_{22}^{\beta_{22}}}{\beta_{01}^{\beta_{01}} \beta_{11}^{\beta_{11}} \beta_{21}^{\beta_{21}}}, \quad \beta_1 = 1 - \beta_{11} + \beta_{12} = \beta_{01} + \beta_{12} + \beta_{21} \\
\beta_2 &= 1 - \beta_{22} + \beta_{11} = \beta_{02} + \beta_{12} + \beta_{21};
\end{aligned}$$

$\lambda$ is the wage-interest ratio (or wage-profit ratio, as it is often called) where the wage rate refers to labor in efficiency units. It is interesting to note (but not surprising) that the prices $p_1$ and $p_2$ also depend on the weights $x_1$ and $x_2$ used to normalize the price level to unity. This is due to the fact that we are using real values. Their definitions depend on the weights. A proof of the existence and nonnegativity of the solution for CES and Cobb-Douglas production functions may be found in Krelle (1988, p. 463).

For CES and Cobb-Douglas production functions and for an arbitrary number $n$ of sectors the following relations may be proved:

1. $\frac{d\lambda}{dr} < 0$: the wage-interest ratio declines with rising interest rates.

2. $\frac{d\kappa_{ji}}{d\lambda} > 0$: all capital-labor ratios (labor in efficiency units) increase with a rising wage-interest ratio.

For a proof see Krelle (1988, p. 464).

We now turn to the *demand side*. Again we start with some *definitional equations*:

$$C = \sum_i C_i p_i = (1 - s)Y , \tag{17.15}$$

where $C$ is the real consumption, $s$ is the exogeneous savings ratio,

$$I = \sum_i I_i p_i = sY , \tag{17.16}$$

and $I$ is the real gross investment. Thus

$$Y = C + I , \quad Y_i = C_i + I_i . \tag{17.17}$$

From definitions (17.2) and (17.3) follows

$$K_{ij} = \pi L \alpha_j \kappa_{ij} , \quad L_i = \alpha_i L , \quad Y_i = \pi L \alpha_i g_i \tag{17.18}$$

and

$$\sum_{i=1}^{\infty} \alpha_i = 1 \tag{17.19}$$

$\alpha_i$ is the proportion of labor employed in sector $i$, where $\alpha_i \geq 0$ and

$$\begin{aligned} \alpha_{n+1} &= \alpha_{n+2} = \ldots = 0 , \text{ if } \delta_1 = 0 \\ \alpha_{n+1} &> 0 , \quad \alpha_{n+2} = \alpha_{n+3} = \ldots = 0 , \quad \text{if } \delta_1 = 1 \\ \alpha_{n+1} &> 0 , \quad \alpha_{n+2} > 0 , \quad \alpha_{n+3} = \alpha_{n+4} = \ldots = 0 , \text{ if } \delta_2 = 1, \text{ etc.} \end{aligned}$$

Capital accumulation is defined by

$$K_{ij,t} = K_{ij,t-1}(1 - d_{ij}) + I_{ij,t} , \tag{17.20}$$

where $d_{ij}$ is the rate of depreciation of $K_{ij}$. In (17.15)–(17.20) and in the following equations the indices $i$ and $j$ run from 1 to $n$ in the current period and to $\delta_1\,(n+1)$, $\delta_2\,(n+1), \ldots$ in later periods.

The *consumption function* is assumed to be dynamic:

$$C_i p_i = f_i(p_1, \ldots, p_n, C_{1,-1}, \ldots, C_{n,-1}, Y) , \tag{17.21}$$

where $f_i$ is inhomogeneous in $p$ and $Y$ [but homogeneous in $p_1^*, \ldots, p_n^*$; see (17.5)].

For economic estimations we may specify $f_i$ as a *dynamic linear expenditure system*:

$$C_i p_i = \bar{C}_i p_i + \gamma_i p_i C_{i,-1} + \beta_i \left[ (1 - s) \sum_j Y_j p_j - \sum_j (\bar{C}_j p_j + \gamma_j p_j C_{j,-1}) \right] , \tag{17.22}$$

where $Y_j = \pi L \alpha_j g_j$, $\bar{C}_i \geq 0$, $\gamma_i \geq 0$, $\beta_i \geq 0$ are given constants,

$$\sum_{i=1}^{\infty} \beta_i = 1\,, \tag{17.23}$$

and

$$\begin{aligned}
\bar{C}_{n+1} &= \bar{C}_{n+2} = \ldots = \gamma_{n+1} = \gamma_{n+2} = \ldots = \beta_{n+1} = \beta_{n+2} = \ldots = 0\,, \text{ if } \delta_1 = 0 \\
\bar{C}_{n+1} &> 0,\ \gamma_{n+1} > 0,\ \beta_{n+1} > 0 \text{ and} \\
\bar{C}_{n+2} &= \bar{C}_{n+3} = \ldots = \gamma_{n+2} = \gamma_{n+3} = \ldots = \beta_{n+2} = \beta_{n+3} = \ldots = 0, \text{ if } \delta_1 = 1 \\
\bar{C}_{n+1} &> 0,\ \bar{C}_{n+2} > 0,\ \gamma_{n+1} > 0,\ \gamma_{n+2} > 0,\ \beta_{n+1} > 0,\ \beta_{n+2} > 0 \text{ and} \\
\bar{C}_{n+3} &= \bar{C}_{n+4} = \ldots = \gamma_{n+3} = \gamma_{n+4} = \ldots = \beta_{n+3} = \beta_{n+4} = \ldots = 0, \text{ if } \delta_2 = 1\,,
\end{aligned}$$

etc. The $\bar{C}_i$, $\gamma_i$, and $\beta_i$ depend on $\delta_1, \delta_2, \ldots$. Equation (17.22) may be written as

$$\begin{aligned}
\zeta_i p_i = \bar{\zeta}_i p_i + \gamma_i p_i \zeta_{i,-1} \frac{1}{\widehat{\pi L} + 1} \\
+ \beta_i \left[ (1-s) \sum\nolimits_j \alpha_j g_j p_j - \sum\nolimits_j \left( \bar{\zeta}_j p_j + \gamma_j p_j \zeta_{j,-1} \frac{1}{\widehat{\pi L} + 1} \right) \right],
\end{aligned} \tag{17.24}$$

where $\zeta_i = \frac{C_i}{\pi L}$, $\bar{\zeta}_i = \frac{\bar{C}_i}{\pi L}$ (exogenous), $\widehat{\pi L} = \frac{\pi L - \pi_{-1} L_{-1}}{\pi_{-1} L_{-1}}$ the growth rate of $\pi L$ (exogenous).

The *investment function* is constructed under the assumption that in each period capital goods are produced that are necessary to provide the cost-minimizing capital equipment in all sectors, given the existing capital stock and its depreciation:

$$I_i p_i = p_i \sum\nolimits_j \left[ K_{ij} - K_{ij,-1}(1 - d_{ij}) \right], \tag{17.25}$$

which, considering equation (17.18), may be rewritten as

$$I_i p_i = p_i \pi L \sum\nolimits_j \alpha_j \kappa_{ij} - p_i \sum\nolimits_j K_{ij,-1}(1 - d_{ij})\,. \tag{17.26}$$

Because of equation (17.16) we require

$$\sum\nolimits_i I_i p_i = sY = s \pi L \sum\nolimits_j \alpha_j g_j p_j\,. \tag{17.27}$$

Thus, in the case of the dynamic expenditure system, total demand for commodity $i$ is

$$\begin{aligned}
Y_i p_i &= C_i p_i + I_i p_i = \pi L \alpha_i p_i g_i \\
&= \bar{C}_i p_i + \gamma_i p_i C_{i,-1} + \beta_i \left[ (1-s) \pi L \sum\nolimits_j \alpha_j p_j g_j - \sum\nolimits_j (\bar{C}_j p_j + \gamma_j p_j C_{j,-1}) \right] \\
&\quad + p_i \pi L \sum\nolimits_j \alpha_j \kappa_{ij} - p_i \sum\nolimits_j K_{ij,-1}(1 - d_{ij})
\end{aligned} \tag{17.28}$$

or

$$\alpha_i p_i g_i = \bar{\zeta}_i p_i + \frac{1}{\widehat{\pi L}+1}\gamma_i p_i \zeta_{i,-1} \\ + \beta_i \left[(1-s)\sum_j \alpha_j p_j g_j - \sum_j (\bar{\zeta}_j p_j + \frac{1}{\widehat{\pi L}+1}\gamma_j p_j \zeta_{j,-1})\right] \\ + p_i \sum_j \alpha_j \kappa_{ij} - p_i \frac{1}{\widehat{\pi L}+1}\sum_j \kappa_{ij,-1}\alpha_{j,-1}(1-d_{ij}) \tag{17.29}$$

under the constraint of (17.27), which may be rewritten as

$$\sum_i \left[p_i \pi L \sum_j \alpha_j \kappa_{ij} - p_i \sum_j K_{ij,-1}(1-d_{ij})\right] = s\pi L \sum_j \alpha_j p_j g_j \tag{17.30}$$

or

$$\sum_i \left[p_i \sum_j \alpha_j \kappa_{ij} - \frac{1}{\widehat{\pi L}+1} p_i \sum_j \kappa_{ij,-1}\, \alpha_{j,-1}(1-d_{ij})\right] = s \sum_j \alpha_j p_j g_j \, . \tag{17.31}$$

In addition we have from (17.19)

$$\sum_i \alpha_i = 1 \tag{17.32}$$

where $s$, $\pi$, and $L$ are exogenous. In the current period where we have $n$ sectors the equations (17.28), (17.30), and (17.32) or (17.29), (17.31), and (17.32) consist of $n+2$ equations, one equation more than the $n+1$ unknowns $\alpha_1, \ldots, \alpha_n$ and $\kappa_{11}$. To obtain the absolute values of $Y_i, L_i$ and $K_{ij}$, the definitional equations (17.18) must be used in addition.

We can get rid of the superfluous equation by Walras's law. One equation in the system (17.28) depends linearly on the others, taking into account equations (17.30) and (17.24). Thus one equation (say, the $n$-th equation) may be dropped. This completes the model. It can be solved given the initial conditions (which are the solutions of the last period) and given the exogenous variables.

We illustrate this solution for $n = 2$ and for the dynamic expenditure system. From the supply side we get $p_1$, $p_2$, $\kappa_{12}$, $\kappa_{21}$, $\kappa_{22}$, $w$, and $r$ as functions of $\kappa_{11}$, see above. Similarly, we get $g_1$ and $g_2$ as functions of $\kappa_{11}$. Because of equation (17.32), $\alpha_2 = 1 - \alpha_1$. Thus the only equation left in the system (17.28) for $i = 1$ may be written as

$$\Phi_1(\kappa_{11}, \alpha_1, \pi L) = 0 \, ,$$

given the initial conditions. Similarly, (17.30) now becomes

$$\Phi_2(\kappa_{11}, \alpha_1, \pi L) = 0 \, .$$

These equations determine $\kappa_{11}$ and $\alpha_1$, given the initial conditions and the exogenous variables $\pi$ and $L$. Thus the total system is solved for each period. Unfortunately, only a numerical solution is possible.

The *general solution* of the total system is provided by the following algorithm:

1. For a given period, where $n$ commodities are produced, solve the system (17.6), (17.10), (17.13) (where $i,j = 1, \ldots, n$) for $p_1, \ldots, p_n, \kappa_{12}, \ldots, \kappa_{nn}$, $w$, $r$ as a function of $\kappa_{11}$: $p_i(\kappa_{11})$, $\kappa_{ij}(\kappa_{11})$, $w(\kappa_{11})$, $r(\kappa_{11})$. From $\kappa_{12}(\kappa_{11}), \ldots, \kappa_{nn}(\kappa_{11})$ determine $g_i(\cdot) = g_i(\kappa_{11})$ from the definition of $g_i$ in equation (17.2), $i = 1, \ldots, n$. Of course, all solutions depend also on the weights $x_1, \ldots, x_n$, see (17.6), but we do not indicate this explicitly.
2. Introduce these functions into equations (17.28) and (17.30).
3. Solve the system (17.28) (where $i = 1, \ldots, n-1$), (17.30), and (17.32) for $\alpha_1, \ldots, \alpha_n$ and $\kappa_{11}$, given the initial conditions $C_{1,-1}, \ldots, C_{n,-1}$, $K_{11,-1}, \ldots, K_{nn,-1}$ and the exogenous variables $s, \pi, L$.
4. Determine $p_1, \ldots, p_n, \kappa_{11}, \ldots, \kappa_{nn}, g_i$ from $p_i = p_i(\kappa_{11})$, $\kappa_{ij} = \kappa_{ij}(\kappa_{11})$, $g_i = g_i(\kappa_{11})$, $i, j = 1, \ldots, n$.
5. Determine $C_1, \ldots, C_n$ from (17.22), where $Y_i = \pi L \alpha_j g_j$, and $K_{11}, \ldots, K_{nn}$ from (17.18). This provides the initial conditions for the next period.
6. Start the solution for the next period at 1.

Instead of steps 2, 3, and 5 the following equivalent algorithm may be used:

2a. Introduce these functions into equations (17.29) and (17.31).

3a. Solve the system (17.29) (where $i = 1, \ldots, n-1$), (17.31), and (17.32) for $\alpha_1, \ldots, \alpha_n$, and $\kappa_{11}$, given the initial conditions $\zeta_{1,-1}, \ldots, \zeta_{n,-1}, \kappa_{11,-1}, \ldots, \kappa_{nn,-1}, \alpha_{1,-1}, \ldots,$ $\alpha_{n,-1}$ and the exogenous variables $s, \widehat{\pi L}$.

5a. Determine $\zeta_1, \ldots, \zeta_n$ from equations (17.24), $i = 1, \ldots, n$. Thus the initial conditions $\zeta_1, \ldots, \zeta_n, \kappa_{11}, \ldots, \kappa_{nn}, \alpha_1, \ldots, \alpha_n$ are provided for the next period.

From the definitions $C_i = \pi L \zeta_i$, $Y_i = \pi L \alpha_i g_i$, $K_{ij} = \pi L \alpha_i \kappa_{ij}$ we obtain the absolute figures.

### 17.3.2 Asymptotic behavior within each regime

The *supply system*, (17.6), (17.10), and (17.13), is a static one. It determines the wage and interest rates, the prices, and the capital-labor ratios as functions of one of these variables, say, $\kappa_{11}$. Thus this system yields

$$w = w(\kappa_{11}), r = r(\kappa_{11}), p_i = p_i(\kappa_{11}), \kappa_{ij} = \kappa_{ij}(\kappa_{11}),\ i, j = 1, \ldots, n,\ \kappa_{ij} \neq \kappa_{11}.$$

The *demand system*, (17.24), (17.29), (17.31), and (17.32), is a dynamic one. It determines the consumption demand for each commodity, the distribution of labor over the production sectors, and the capital-labor ratio $\kappa_{11}$ as a function of time:

$$\zeta_i = \zeta_i(t), \quad \alpha_i = \alpha_i(t), \quad \kappa_{11} = \kappa_{11}(t).$$

Assume that each regime exists for such a long time that the asymptotic behavior of the system becomes interesting. The dynamic properties follow from the demand system. If it converges, then it converges to the solution of system (17.24), (17.29), (17.31), and (17.32) where $\zeta_i = \zeta_{i,-1}$, $\alpha_i = \alpha_{i,-1}$, $\kappa_{11} = \kappa_{11,-1}$. In this case (17.24) becomes a linear system in $\zeta_i$, given $\kappa_{11}$, the solution of which is $\zeta_i^*(\kappa_{11})$. Equation (17.29) becomes a linear system in $\alpha_1, \ldots, \alpha_{n-1}$, given $\zeta_i^*$ and $\kappa_{11}$, the solution of which is $\alpha_i^*(\kappa_{11})$, $i = 1, \ldots, n-1$. Equation (17.32) provides $\alpha_n^*$ as a function of $\alpha_1^*, \ldots, \alpha_{n-1}^*$. Finally, equation (17.31) is a nonlinear implicit function of $\kappa_{11}$ alone, which has to be solved for $\kappa_{11}$ to yield $\kappa_{11}^*$. Thus we get the convergence values $\zeta_i^*, \alpha_i^*, \kappa_{11}^*$.

But will the system converge? This cannot be proved in general, though there are some indications that this is true. To test the stability one has to estimate the parameters and to solve the system numerically. In this case two general tests are available, which are easy to apply. For both tests the system (17.24), (17.29), and (17.31) [after the substitution of $\alpha_n = 1 - \sum_{j=1}^{n-1} \alpha_j$, see (17.32)] has to be linearized at the convergence point $\zeta_i^*, \alpha_i^*, \kappa_{11}^*$, $i = 1, \ldots, n$. This yields the system

$$A\xi_t = B\xi_{t-1} + c\,, \quad \text{or, if } |A| \neq 0\text{:} \quad \xi_t = D\xi_{t-1} + d\,, \tag{17.33}$$

where $\xi' = (\zeta_1, \ldots, \zeta_n, \alpha_1, \ldots, \alpha_{n-1}, \kappa_{11})$, $A$ and $B$ are $(2n \times 2n)$ matrices, and $c' = (c_1, \ldots, c_{2n})$ is a $2n$-vector; $D = A^{-1}B$, $d = A^{-1}c$.

This system has the static solution $\xi^* = (I - D)^{-1}d$, which reproduces the convergence values of the dynamic demand system (17.24), (17.29), (17.31), and (17.32). The general solution of (17.33) is

$$\xi_t = a_1\rho_1^t + \ldots + a_n\rho_n^t + \xi^*\,,$$

where $a_i$ are vectors that also depend on the initial conditions and $\rho_i$ are the characteristic roots of $D$, i.e., the solutions of $|\rho I - D| = 0$. The system converges to $\xi^*$ if $|\rho_i| < 1$ for all $i$.

There are systems for which this may easily be checked. Assume that $E = I - D$ has the property $e_{ii} > 0$, $e_{ij} \leq 0$ for $i \neq j$ and that it has a *dominant main diagonal*, i.e., there are positive real numbers $\hat{d}_1, \ldots, \hat{d}_n$ such that for all $j = 1, \ldots, n$

$$\hat{d}_j e_{jj} > \sum_{i \neq j} -\hat{d}_i e_{ij} \geq 0\,.$$

Then all $|\rho_i| < 1$, and the system converges.

A second test consists of the *Brauer-Solow conditions* (which are sufficient, but not necessary). If $D \geq 0$ and indecomposable and if all $r_i = \sum_{j=1}^{n} d_{ij} \leq 1$ and $r_i < 1$ for at least one $i$ (i.e., if all row sums or equivalently column sums are smaller than or equal to 1), then all $|\rho_i| < 1$ and the system converges. For details, see Krelle (1988, pp. 750 and 760), or Takayama (1974, p. 381).

Whether these conditions hold cannot be said without numerical specifications. But there are some indications that the system may converge. First, we observe that the production structure changes only slowly and that this change is mostly due to the introduction of new products and new production methods. If our system reflects this feature appropriately, we may expect it to converge if there is no technical progress in the sense stated above. Second, the two main parts of our dynamic system treated separately converge under mild conditions, so that we may also expect the total system to converge under these (or similar) conditions.

The dynamic system consists of the partial system (17.24), (17.29), and (17.31), after equation (17.32) has been eliminated by introducing $\alpha_n = 1-\sum_{j=1}^{n-1}\alpha_j$ into the other equations. Consider first the system (17.24); assume that $(1-s)y := \frac{C}{\pi L} = (1-s)\sum_j \alpha_j p_j g_j$ stays constant during the period in question, as well as during the period of the growth rate $\widehat{\pi L}$. From equation (17.15) we get

$$\sum_{i=1}^{n} p_i\zeta_i = y(1-s) \quad \text{or} \quad \zeta_n = \frac{1}{p_n}\left[(1-s)y - \sum_{j=1}^{n-1}\zeta_j p_j\right].$$

This will be introduced into system (17.24) so that we only need the first $n-1$ equations to determine $\zeta_1,\ldots,\zeta_n$ as functions of $\zeta_{1,-1},\ldots,\zeta_{n-1,-1}$. If all the other variables $\alpha_i, g_i, p_i$ have already converged so that they may be considered as constant, the system (17.24) becomes a linear difference equation of the first order:

$$\zeta_t = D\zeta_{t-1} + d\,,$$

with

$$D = \begin{bmatrix} a_{11} - \beta_1\Delta a_1 & -\beta_1\frac{p_2}{p_1}\Delta a_2 & \cdots & -\beta_1\frac{p_{n-1}}{p_1}\Delta a_{n-1} \\ -\beta_2\frac{p_1}{p_2}\Delta a_1 & a_{12} - \beta_2\Delta a_2 & \cdots & -\beta_2\frac{p_{n-1}}{p_2}\Delta a_{n-1} \\ \vdots & \vdots & \vdots & \vdots \\ -\beta_{n-1}\frac{p_1}{p_{n-1}}\Delta a_1 & -\beta_{n-1}\frac{p_2}{p_{n-1}}\Delta a_2 & \cdots & a_{1,n-1} - \beta_{n-1}\Delta a_{n-1} \end{bmatrix}$$

$$d = \begin{bmatrix} a_{01} - \frac{\beta_1}{p_1}\sum_{j=1}^{n} a_{0j}p_j + \frac{\beta_1}{p_1}(1-a_{1n})y(1-s) \\ a_{02} - \frac{\beta_2}{p_2}\sum_{j=1}^{n} a_{0j}p_j + \frac{\beta_2}{p_2}(1-a_{1n})y(1-s) \\ \vdots \\ a_{0,n-1} - \frac{\beta_{n-1}}{p_{n-1}}\sum_{j=1}^{n} a_{oj}p_j + \frac{\beta_{n-1}}{p_{n-1}}(1-a_{1n})y(1-s) \end{bmatrix} > 0, \quad \zeta_t = \begin{bmatrix} \zeta_{1t} \\ \zeta_{2t} \\ \vdots \\ \zeta_{n-1,t} \end{bmatrix},$$

where $\Delta a_i := a_{1i} - a_{1n}$, and $a_{0i} := \bar{\zeta}_i$, $0 \le a_{0i} < 1$, $a_{1i} := \frac{\gamma_i}{\widehat{\pi L}+1}$. Now we regroup the sectors such that $a_{1n} \ge a_{1j}$, $j = 1,\ldots,n-1$. In this case $D \ge 0$. If the $a_{1i}$ are not too different, i.e., if

$$a_{1i} + \beta_i\sum_{j=1}^{n-1}(a_{1n} - a_{1j})\frac{p_j}{p_i} < 1 \quad \text{for all } i = 1,\ldots,n-1\,, \tag{17.34}$$

the Brauer-Solow conditions are fulfilled and $I - D$ has a dominant main diagonal. Thus the system converges. Since the $a_{1j}$ are essentially the $\gamma_i$, the condition (17.34) means that the remanence effect of the former consumption should not be too different between commodities.

Similarly, we consider the equations (17.29) as a system of difference equations for $\alpha_1, \ldots, \alpha_{n-1}$, where $\alpha_n = 1 - \sum_{j=1}^{n-1} \alpha_j$; see (17.32). This has to be introduced into the $n-1$ equations in (17.29). All other variables $\zeta_i$, $p_i$, $g_i$, $\kappa_{ij}$ are assumed to be constant at their convergence values. We also keep $y = \sum_j \alpha_j p_j g_j$ and $\kappa_{i.} = \sum_j \alpha_j \kappa_{ij}$ constant. Thus (17.29) may be written as

$$\alpha_t = D^* \alpha_{t-1} + d^* ,$$

where

$$D^* = \begin{bmatrix} \frac{1}{g_1(\widehat{\pi L}+1)}(\kappa_{1n} - \kappa_{11}) & \cdots & \frac{1}{g_1(\widehat{\pi L}+1)}(\kappa_{1n} - \kappa_{1,n-1}) \\ & \vdots & \\ \frac{1}{g_{n-1}(\widehat{\pi L}+1)}(\kappa_{n-1,n} - \kappa_{n-1,1}) & \cdots & \frac{1}{g_{n-1}(\widehat{\pi L}+1)}(\kappa_{n-1,n} - \kappa_{n-1,n-1}) \end{bmatrix}$$

$$0 < d^* = \begin{bmatrix} b_{01} + \frac{\beta_1}{p_1 g_1}(1-s)y - \beta_1 \sum_{j=1}^n b_{0j} + \frac{\kappa_{1.}}{g_1} - \frac{\kappa_{1n}}{g_1(\widehat{\pi L}+1)} \\ \vdots \\ b_{0,n-1} + \frac{\beta_{n-1}}{p_{n-1} g_{n-1}}(1-s)y - \beta_{n-1} \sum_{j=1}^n b_{0j} + \frac{\kappa_{n-1,.}}{g_{n-1}} - \frac{\kappa_{n-1,n}}{g_{n-1}(\widehat{\pi L}+1)} \end{bmatrix}$$

$$\alpha_t = \begin{bmatrix} \alpha_{1t} \\ \vdots \\ \alpha_{n-1,t} \end{bmatrix} \quad \text{and} \quad b_{0i} = \frac{\bar{\zeta}_i}{g_i} + \frac{\gamma_i}{g_i(\widehat{\pi L}+1)} \zeta_i .$$

If a sector $n$ exists, which is more capital intensive than all the others, the sectors may be regrouped such that

$$\kappa_{in} \geq \kappa_{ij} \quad \text{for all } i,j = 1, \ldots, n-1 .$$

Then we have $D^* \geq 0$. If the differences in the capital intensities are not too large, i.e., if

$$\frac{1}{g_i(\widehat{\pi L}+1)} \cdot \sum_{j=1}^{n-1} (\kappa_{in} - \kappa_{ij}) < 1 \quad \text{for } i = 1, \ldots, n-1 ,$$

then the system converges.

Finally, consider equation (17.31) as a linear difference equation in $\kappa_{11}$, whereas the other variables $p_i, \alpha_i, \kappa_{ij}$ $(i,j = 1, \ldots, n;\ \kappa_{ij} \neq \kappa_{11})$ are assumed to stay at their convergence levels. This yields

$$\kappa_{11} = a \kappa_{11,-1} + b ,$$

where

$$a = \frac{1-d_{11}}{\widehat{\pi L}+1} < 1$$

$$b = sy - \sum_{i=1}^{n}\left[p_i \sum\nolimits_{j\in J} \alpha_j \kappa_{1j} - p_i \sum\nolimits_{j\in J} \kappa_{1j}\alpha_j \frac{1-d_{ij}}{\widehat{\pi L}+1}\right] > 0$$

$$J = \begin{cases} 1,\ldots,n & \text{if} \quad i \neq 1 \\ 2,\ldots,n & \text{if} \quad i = 1 . \end{cases}$$

The solution is

$$\kappa_{11,t} = \kappa_{11,0}\, a^t + \frac{b}{1-a},$$

where $b/(1-a)$ is the convergence value.

Thus we have shown that there is reason to believe that the real system will converge in each regime, if the regime lasts long enough and if $y = Y/(\pi L)$ stays constant during this period.

### 17.3.3 Introduction of money into the model

The model presented above does not contain a monetary sector and therefore does not explain the absolute prices $p_i^*$ and the general price level $\bar{p}$ [see equation (17.4)]. But these values are needed to explain exports and imports and the exchange rates, when one proceeds from the model of a closed economy to a world model with several open economies. In this section I give some hints as to the theoretical background of the approach and present some empirical results.

Tobin (1965) was the first to introduce money into an aggregated growth model. Hahn (1969), Fischer (1972), Stein (1971), and others followed. Ramanathan (1982) reviews this development. Krelle (1988) analyzed different models with and without money illusion. Most of these models suffer from instability because they do not limit the velocity of money from above. There are no analytically tractable models that include money and consider the growth process of disaggregated economies connected by foreign trade.

Here we extend the approach of Krelle (1988, p. 416), in the following way. Money supply $M$ in a country (I took $M2$) was assumed to be exogenous and forecast separately under the assumption that the supply behavior of the central banks would continue in the future as it was observed in the past. Money demand was estimated in the form of a function for the velocity, $v$, of money. It is defined as

$$v = \frac{pY}{M}$$

and estimated by

$$\tilde{v}_t = a_1 \tilde{v}_{t-1} + a_2(\tilde{k}_t - \tilde{k}_{t-3}) + a_3(\tilde{\pi}_t - \tilde{\pi}_{t-3}) + a_4 \frac{A_t^n}{Y_t^n} + a_5 ,$$

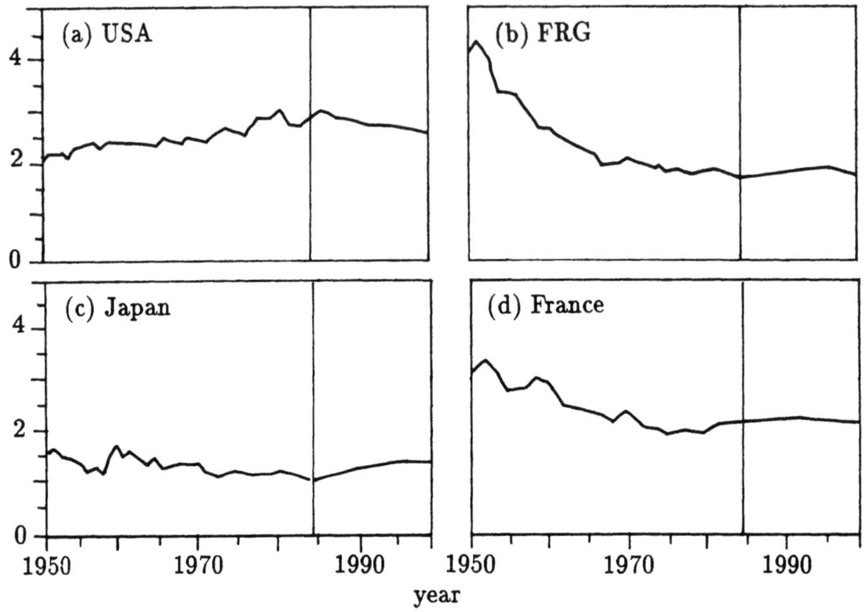


*Figure 17.8:* Velocity of money ($M2$). Source: Ross (1989, p. 116).

where $\tilde{x}$ means the logarithm of $x$, $k = K/Y$ is the capital coefficient, $\pi = \frac{Y}{L}$ is the labor productivity, $A^n$ is the accumulated foreign trade surplus, and $Y^n$ is the GDP at current prices. The estimates showed that $0 < a_1 < 1$, $a_2 \leq 0$, $a_4 \leq 0$. For details see Ross (1989, p. 95). The actual development and forecasts for the velocity of money are shown in *Figure 17.8.*

## 17.4 A Simplified Model of the Interrelation of Growth and Structural Change

### 17.4.1 The upside-down approach

The real part of the model presented in Section 17.3 is much too complicated to be implemented in practice and estimated econometrically. If we want to test the theory on actual data, the model has to be simplified such that the necessary statistical data are available for all countries considered and that such a project can be carried out with the available means in the available time. On the other hand, for many important countries foreign trade plays a decisive role so that it cannot be disregarded. Thus the model of Section 17.3 has to be extended to cover also exports and imports and capital flows. These corrections are the topic of this section. We start with the simplification of the real part of the model; see equations (17.1) to (17.32).

The endogenous unknowns of this model are $\kappa_{ij}$, $\alpha_i$, $p_i$, $w$, and $r$, $i, j = 1, \ldots, n$. They are determined as functions of the exogenous variables $\pi$, $L$, $s$; the parameters of the production, consumption, and investment functions; and the lagged variables. Let there be $N$ such exogenous variables. We change their notation to $z_1, \ldots, z_N$. Thus the solution of the model gives us

$$\kappa_{ij} = \kappa_{ij}(z),\ \alpha_i = \alpha_i(z),\ p_i = p_i(z),\ w = w(z),\ r = r(z),\ z := (z_1 \ldots, z_N)\,.$$

From (17.3) we get

$$L_i = \alpha_i L\,,$$

and therefore from (17.2)

$$K_{ji} = \kappa_{ji}\alpha_i \pi L\,.$$

Assuming that the production function $F_i$ in equation (17.1) is linear homogeneous we may rewrite equation (17.1) for a specific period as

$$\frac{Y_i}{Y} = F_i \left[\alpha_i(z)\,\frac{\pi L}{Y}, \frac{K_{1i}}{L}\,\frac{L}{Y}, \ldots, \frac{K_{ni}}{L}\,\frac{L}{Y}, \ldots\right]. \tag{17.35}$$

Since only figures for the aggregate capital $K$ in the economy are available, we have to aggregate the different capital goods to a compound capital $K_{.i}$ in sector $i$, where

$$K_{.i} = \sum_j p_j(z) K_{ji} = \sum_j p_j(z)\, \kappa_{ji}(z)\, \alpha_i(z)\, \pi L\,,$$

and relate it to the total capital $K$ in the society by

$$\frac{K_{.i}}{K} = \frac{\sum_j p_j(z)\, \kappa_{ji}(z)\, \alpha_i(z)}{\sum_i \sum_j p_j(z)\, \kappa_{ji}(z)\, \alpha_i(z)} =: \varphi_i(z)\,.$$

Now we approximate equation (17.35) as

$$\frac{Y_i}{Y} = G_i \left[\alpha_i(z)\, \pi\, \frac{L}{Y}\,,\ \varphi_i(z)\,\frac{K}{L}\,\frac{L}{Y}, \ldots\right] + \epsilon_i\,, \tag{17.36}$$

where $\epsilon_i$ is the error term. This may be formulated as

$$\frac{Y_i}{Y} = \tilde{G}_i \left(\frac{L}{Y}, \frac{K}{L}, \pi, z_1, \ldots, z_N, \ldots\right) + \epsilon_i\,, \tag{17.37}$$

where, e.g., $z_1 = s, z_2 = \ldots$ (see above).

If $\tilde{G}_i$ is of the Cobb-Douglas type, the equation explaining the production structure would be

$$\frac{Y_i}{Y} = \left(\frac{L}{Y}\right)^{\alpha_{0i}} \pi^{\alpha_{1i}} \left(\frac{K}{L}\right)^{\alpha_{2i}} s^{\alpha_{3i}}\, z_2^{\alpha_{4i}} \ldots + \nu_i\,, \tag{17.38}$$

when $\sum_i \frac{Y_i}{Y} = 1$ and $\nu_i$ is an error term.

Thus equation (17.37) or (17.38) gives us the production structure, if total GDP $Y$, total capital $K$, and total employment $L$ are given. We estimate these totals first by solving the system (17.1) to (17.32) for the aggregate economy; the index $i$ describing the sector has to be dropped everywhere in equation (17.1) to (17.32). Thus we have instead of (17.1)

$$Y = F(\pi L, K, \ldots) \tag{17.39}$$

and instead of (17.2)

$$Y = \pi g(\kappa, \ldots). \tag{17.40}$$

Equations (17.4), (17.5), (17.6), and (17.7) are not applicable; instead of (17.8)

$$g' = \frac{\partial g}{\partial \kappa}, \tag{17.41}$$

and so on.

This provides the "upside-down" approach. We first estimate the totals and afterward introduce the structure by applying equation (17.37) or (17.38).

### 17.4.2 The introduction of foreign trade

The model of Section 17.3 as well as the simplified model of the foregoing subsection refers to a closed economy. If we want to test the theory we have to consider the interrelationships of different economies by foreign trade and capital flows. This means that the aggregate production function (17.39) must include foreign factor imports:

$$Y = F(\pi L, K, IM_R, \ldots), \tag{17.42}$$

where $IM_R$ means imported secondary inputs (e.g., raw materials).

The GDP identity (17.17) for the closed economy has to be substituted by

$$Y = C + I + EX - IM, \tag{17.43}$$

where $EX$ are the total real exports, and $IM$ are the total real imports. There is a relation $\mu(z)$ such that

$$IM_R = \mu(z)\, IM.$$

Exports and imports also influence the sectoral composition of GDP. The vector $z = (z_1, \ldots, z_N)$ must be reinterpreted to include also foreign prices and exchange rates which codertermine exports and imports. Thus equation (17.36) becomes

$$\frac{Y_i}{Y} = G_i \left[ \alpha_i(z)\, \pi \frac{L}{Y}, \varphi_i(z) \frac{K}{L} \frac{L}{Y}, \mu(z) \frac{IM}{Y}, \ldots \right] + \epsilon_i^*, \tag{17.44}$$

which may be reformulated as

$$\frac{Y_i}{Y} = \tilde{G}_i\left(\frac{L}{Y}, \frac{K}{L}, \cdot\frac{IM}{Y}, \pi, s, z_2, \ldots, z_N, \ldots\right) + \epsilon_i^\star . \tag{17.45}$$

The exogenous variables $\pi, s, z_2, \ldots z_N$ codetermine the export ratio $EX/Y$:

$$\frac{EX}{Y} = R(\pi, s, z_2, \ldots, z_N) .$$

Therefore we change equation (17.45) to

$$\frac{Y_i}{Y} = \check{G}\left(\frac{L}{Y}, \frac{K}{L}, \frac{IM}{Y}, \frac{EX}{Y}, \pi, s, z_2, \ldots, z_N, \ldots\right) + \epsilon_i^{**} . \tag{17.46}$$

We assume $\check{G}$ to be of Cobb-Douglas type. Since $\frac{L}{Y}$ and $\pi$ are correlated, we omit $\frac{L}{Y}$. Thus our structural equation to be estimated finally becomes

$$\frac{Y_i}{Y} = \pi^{\alpha_{1i}}\left(\frac{K}{L}\right)^{\alpha_{2i}}\left(\frac{EX}{Y}\right)^{\alpha_{3i}}\left(\frac{IM}{Y}\right)^{\alpha_{4i}} s^{\alpha_{5i}}\, a_{6i} + y_i . \tag{17.47}$$

Ross (1989) carried out the estimations.

Donges *et al.* (1989) used a simplified approach that explains the % ratio $A_i = (Y_i/Y)\,100$ as a function of per capita GDP $\frac{Y}{P}$, where $P$ is the population. They estimated the parameters of the equation

$$\log\frac{A_i}{100 - A_i} = a_0 + a_1\frac{Y}{P} + a_2 \log\frac{Y}{P} + a_3 . \tag{17.48}$$

The structural changes forecast by equation (17.47) are presented for some OECD countries in *Table 17.1* and for some CMEA countries in *Table 17.2*. The figures for 1962 and 1984 are the actual figures; those for 1999 are forecasts under different assumptions. For details, see Krelle (1989, p. 42) and Ross (1989, p. 97).

The forecasts of Donges *et al.* (1989) by equation (17.48) are reproduced in *Table 17.3*. There are some differences in the definition of sectors and of future scenarios between Ross and Donges *et al.*, but there are also substantial differences in the results. Donges *et al.* predict a decline of the manufacturing sector in West Germany from about 33% in 1985 to about 25% in the year 2000. Ross predicts a constancy of this figure at about 36%. It remains to be seen which approach is more appropriate. I have more confidence in the economic approach of equation (17.47).

We used the upside-down approach also in foreign trade: First we estimated total real exports and imports, and afterward we disaggregated the totals to get their commodity compositions. The usual foreign trade theory is static and therefore not suitable for analyzing economic growth and structural change; c.f., e.g., Ohlin (1967), Takayama (1972), Schittko (1976). There are some dynamic approaches in Gabisch (1976) and Ramanathan

*Table 17.3:* West Germany: Sectoral and branches shares (%) of GDP and employment.

| | Gross Domestic Product | | | | | Employment | | | | |
|---|---|---|---|---|---|---|---|---|---|---|
| | Actual | | Predicted for 2000[a] | | | Actual | | Predicted for 2000[a] | | |
| Sectors/branches | 1970 | 1985 | A | B | C | 1970 | 1985 | A | B | C |
| Primary sector[b] | 4.6 | 2.8 | 2.5 | 2.3 | 2.3 | 9.7 | 6.3 | 4.3 | 3.7 | 3.6 |
| Secondary sector[b] | 50.4 | 41.1 | 33.0 | 30.2 | 28.8 | 47.7 | 40.1 | 32.1 | 29.8 | 28.5 |
| Tertiary sector[b] | 45.0 | 56.1 | 64.5 | 67.5 | 68.9 | 42.6 | 53.6 | 63.6 | 68.5 | 67.9 |
| Agric., forestry, fishing | 3.4 | 1.7 | 1.5 | 1.4 | 1.4 | 8.5 | 5.4 | 3.5 | 3.0 | 2.9 |
| Mining and quarrying | 1.3 | 1.1 | 1.0 | 0.9 | 0.9 | 1.2 | 0.9 | 0.8 | 0.7 | 0.7 |
| Manufacturing: | 40.2 | 33.2 | 27.5 | 25.2 | 24.0 | 38.1 | 32.0 | 27.2 | 25.1 | 23.9 |
| Intermediate goods | 11.8 | 8.8 | 6.2 | 5.5 | 5.0 | 8.4 | 6.5 | 4.7 | 4.2 | 3.8 |
| Investment goods | 15.0 | 15.7 | 14.8 | 14.0 | 13.7 | 16.3 | 15.9 | 15.1 | 14.3 | 14.0 |
| Consumer goods | 12.5 | 8.7 | 6.5 | 5.7 | 5.3 | 13.3 | 9.6 | 7.4 | 6.6 | 6.1 |
| Electricity, gas, water | 2.2 | 2.8 | 2.5 | 2.3 | 2.2 | 0.9 | 1.1 | 0.9 | 0.9 | 0.9 |
| Construction | 8.0 | 5.1 | 3.0 | 2.7 | 2.6 | 8.7 | 7.0 | 4.0 | 3.8 | 3.7 |
| Wholesale, retail trade, restaurants, and hotels | 11.4 | 10.9 | 9.7 | 9.4 | 9.2 | 15.1 | 16.2 | 15.9 | 15.7 | 15.5 |
| Transport, storage, and communications: | 5.9 | 5.9 | 6.6 | 7.4 | 7.7 | 5.3 | 5.6 | 6.4 | 7.1 | 7.4 |
| Transport and storage | 4.1 | 3.4 | 3.1 | 2.9 | 2.7 | 3.6 | 3.6 | 3.3 | 3.0 | 2.8 |
| Communications | 1.8 | 2.5 | 3.5 | 4.6 | 5.0 | 1.7 | 2.0 | 3.1 | 4.1 | 4.6 |
| Finance, insur., real estate, and business services: | 10.7 | 15.9 | 21.6 | 22.4 | 23.4 | 4.3 | 6.2 | 12.2 | 13.2 | 14.1 |
| Financial institutions | 2.6 | 4.5 | 5.2 | 5.5 | 5.7 | 1.5 | 2.3 | 3.0 | 3.3 | 3.5 |
| Insurance | 0.7 | 1.1 | 1.4 | 1.4 | 1.5 | 0.7 | 0.8 | 1.2 | 1.2 | 1.3 |
| Real estate, busin. serv. | 7.4 | 10.3 | 15.0 | 15.5 | 16.2 | 2.1 | 3.1 | 8.0 | 8.7 | 9.3 |
| Personal services, other private producers | 7.2 | 11.8 | 15.2 | 17.1 | 17.7 | 6.7 | 9.6 | 13.3 | 14.8 | 15.4 |
| Government services | 9.7 | 11.6 | 11.4 | 11.2 | 10.9 | 11.2 | 16.0 | 15.8 | 15.7 | 15.5 |

[a] The predicted percentages of the year 2000 refer to three different assumptions on the expected annual growth rate of per capita income: $A = 2\%$; $B = 3\%$; $C = 4\%$. For method of projection, see text.
[b] For definition see *Table 16.1* in Donges *et al.* (1989, p. 387).
Source: Donges *et al.* (1989, p. 396).

(1982). A dynamic theory of international trade may be found in Krelle (1988, p. 502). But all these theories consider only two commodities and two countries of approximately equal size or two commodities and one relatively small country vis-à-vis a large world market. There is one exception. In a recent paper Grossman and Helpman (1989) deal with a two-country three-sector dynamic model of a very specific kind. One sector is the *R&D* sector, which produces non-tradable blueprints as well as disembodied knowledge, which is disseminated immediately and without cost all over the world. The second sector produces intermediate goods, the third consumption goods. There are no capital goods, and thus there is no capital accumulation. The production functions for the two countries are equal. Consumers all over the world have equal homothetic preferences. Money is not included in the model. Equilibrium growth paths are analyzed. The advantage of this model is that technical progress is endogenous and that the model can be solved analytically. The disadvantages are evident: The model is too removed from reality to be tested econometrically.

Fortunately, there is a whole literature on import and export functions [see, e.g., Welsch (1989, pp. 257-258)]. In the IIASA-University of Bonn Research Project we used a dynamic linear expenditure system to explain total imports of a country:

$$IM = a_0 IM_{-1} + a_1 \frac{C^n}{p_{IM}} + a_2 \frac{I^n}{p_{IM}} + a_3 \frac{EX^n}{p_{IM}} - a_4 \frac{p_y}{p_{IM}} Y_{-1} ,$$

where $C^n$ is the nominal consumption (in current prices), $I^n$ is the nominal investment, $EX^n$ are the nominal exports, $Y$ is the real GDP, $p_y$ is the price level of GDP, and $p_{IM}$ is the price level of imports. Exports are explained as imports of the rest of the world.

The commodity breakdown of exports and imports is not identical with that of production, owing to statistical problems. We also used a dynamic linear expenditure system to explain and forecast the structural composition of exports and imports. We only present the estimation equation for exports:

$$EX_j^n = a\, p_j\, EX_{j,-1} + b\, EX^n - \sum_{k \neq j} c_k\, p_k\, EX_{k,-1} , \tag{17.49}$$

where $EX^n$ are the total expenditures on commodity exports, $EX_j^n$ the same for commodity $j$, $EX_j$ the volume of exports of type $j$, and $p_j$ its export price. The commodity breakdown is SITC 0+1, SITC 2+4, SITC 3, SITC 5+...+9. For imports a similiar equation has been used. The results for OECD countries may be found in Welsch (1989a, 1989b). They are quite satisfactory but not reproduced here, owing to lack of space.

## 17.5 The Driving Forces of Economic Growth and Structural Change

The driving forces of economic growth and structural change are the rate of technical progress $\dot{\pi}/\pi = \hat{\pi}$ (which measures the inventiveness and imaginative power of a society and its willingness to put new ideas into practice), the rate of saving $s$ (which measures the sacrifice in current consumption that a society is willing to make for the advantage of future generations), and the rate of growth $\dot{L}/L = \hat{L}$ of the labor force which is related to population growth. All three forces are deeply embedded in the fabric of a society and cannot be explained by economic variables alone. They show characteristic trends and long-term fluctuations that generate Kondratieff-type cycles in the economic sphere. *Figure 17.9* shows the rate of technical progress for the USA, the FRG, Japan, and the USSR from the 1950s to the beginning of the 1980s. A long wave is visible with an upswing till the end of the 1960s and a downswing after that. The savings ratio shows a similar long wave superimposed by shorter waves originating from the business cycle, see *Figure 17.10*. Labor input follows more complicated rules. *Table 17.4* shows some observations and forecasts. Labor input ($L$, in billions of working hours) follows from

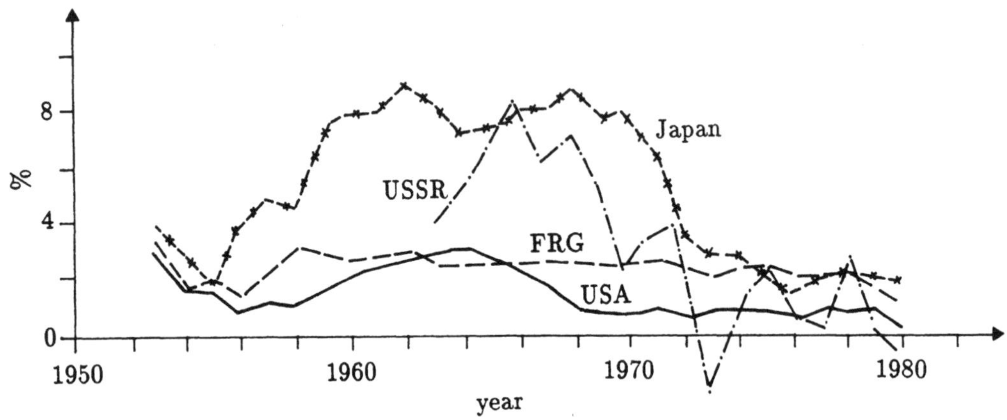


*Figure 17.9:* Rate of technical progress. Source: Krelle (1989, p. 26).

the size of the population ($p$, in million persons), the labor participation rate $\lambda$, the employment rate $\epsilon$, and the hours of work $h$ per week:

$$L = p\,\lambda\,\epsilon\,h\,52/1000\,.$$

Each determinant follows its own rules and has to be estimated separately.

We cannot deal with the complicated problems of population and employment growth. Starting with Malthus (1798) quite a few models of population growth have been developed, mostly in connection with growth models [see, e.g., Hagen (1959), Beckmann (1965), Niehans (1963), Krelle (1988, p. 196)]. But we shall briefly indicate a possible explanation of the long-term cycles of the rate of technical progress and the savings rate, a special research project on this under way at the University of Bonn.

The rate of technical progress depends basically on the generation of new knowledge and on the transfer of this knowledge into practice in the form of new commodities and new production processes. Knowledge production is a stochastic process that depends on capital and labor allocated to the research and development sectors of the economy. *Figure 17.11* shows estimates of the rate of growth $w_F$ of knowledge, of the transfer coefficient $\delta$, and of the rate $w_{\sim}$ of technical progress. There are no cycles to speak of in the process of knowledge creation, but there are long-term and short-term cycles in the economic application of this knowledge. It needs entrepreneurship to put theoretical knowledge into practice, i.e., energy, optimism, willingness to undertake risks. But it also needs enough freedom in the society and discipline and diligence in the labor force. These are psychologically and sociologically determined variables that are not directly measurable. We group them into two aggregate latent variables: organizational efficiency and degree of activity of a population. There are many indicators for these that allow one to estimate them by LISREL, PLS, or other methods. This latent-variable research

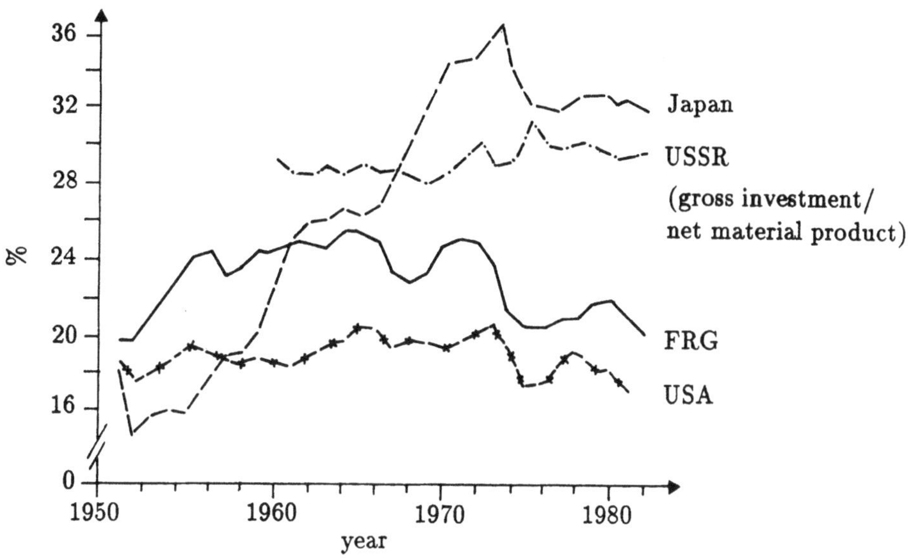


*Figure 17.10:* Development of the savings rate since 1950. Source: Data base of the IIASA-University of Bonn Research Project; see Ross (1989, p. 101).

is now in progress. First results concerning the "degree of activity" are already available [see Krelle (1989, p. 31)].

The relation between the rate of growth $\hat{\pi}$ of technical progress, the degree of activity $\alpha$, and the organizational efficiency $\omega$ may be modeled as follows: Let $\bar{\alpha}, \bar{\omega}, \bar{\hat{\pi}}$ be the "normal" or average values of $\alpha, \omega, \hat{\pi}$. Retracting forces in the society retard economic activity and organizational efficiency if they are "too large", and improve them if they are considered "too low". Similarly, "too high" organizational efficiency and "too high" technical progress tend to reduce economic activity and organizational efficiency since society resists too much change, whereas a higher degree of economic activity may improve organizational efficiency. The rate of technical progress is positively related to $\alpha$ and $\omega$. In the linear case this yields the system

$$\begin{aligned}
\dot{\alpha} &= -a_{11}(\alpha - \bar{\alpha}) - a_{12}(\omega - \bar{\omega}) - a_{13}(\hat{\pi} - \bar{\hat{\pi}}) \\
\dot{\omega} &= a_{21}(\alpha - \bar{\alpha}) - a_{22}(\omega - \bar{\omega}) - a_{23}(\hat{\pi} - \bar{\hat{\pi}}) \\
\hat{\pi} &= b_1\alpha + b_2\omega\,,
\end{aligned}$$

where all $a_{ij}$ and $b_i > 0$. After substitution of the third equation for the first two equations, the system may be rewritten as

$$\begin{aligned}
\dot{x} &= -c_{11}x - c_{12}y + c_{11}\bar{x} + c_{12}\bar{y} \\
\dot{y} &= c_{21}x - c_{22}y - c_{21}\bar{x} + c_{22}\bar{y}\,,
\end{aligned}$$

*Table 17.4:* Observations and forecasts of labor input (OECD countries).

| Country | | $w_{POP}$ Observed means 61–84 | 71–84 | 76–84 | Forec. 1985–2000 | $w_{LPR}$ Observed means 61–84 | 71–84 | 76–84 | Forec. 1985–2000 |
|---|---|---|---|---|---|---|---|---|---|
| 01 | USA | 1.1 | 1.0 | 1.0 | 1.0 | 0.9 | 1.2 | 1.1 | 0.8 |
| 02 | FRG | 0.4 | 0.06 | −0.1 | −0.2 | −0.3 | 0.1 | 0.4 | 0.4 |
| 03 | Japan | 1.0 | 1.0 | 0.8 | 0.6 | 0.1 | 0.0 | 0.4 | 0.1 |
| 04 | France | 0.8 | 0.6 | 0.4 | 0.3 | 0.1 | 0.2 | 0.4 | 0.3 |
| 05 | UK | 0.3 | 0.1 | 0.1 | −0.1 | 0.1 | 0.4 | 0.4 | 0.3 |
| 06 | Italy | 0.6 | 0.4 | 0.2 | 0.3 | −0.3 | 0.4 | 0.8 | 0.6 |
| 07 | NL | 1.0 | 0.7 | 0.6 | 0.4 | 0.4 | 0.7 | 1.3 | 0.6 |
| 08 | B/L | 0.3 | 0.2 | 0.1 | 0.1 | 0.3 | 0.6 | 0.6 | 0.4 |
| 09 | Canada | 1.4 | 1.2 | 1.1 | 1.4 | 1.4 | 1.6 | 1.3 | 0.9 |
| 10 | Rest of OECD[a] | 1.4 | 1.3 | 1.2 | 1.2 | −0.1 | 0.0 | 0.0 | 0.0 |

| | $w_{ER}$ Observed means 61–84 | 71–84 | 76–84 | Forec. 1985–2000 | $w_{WH}$ Observed means 61–84 | 71–84 | 76–84 | Forec. 1985–2000 | $w_L^b$ Observed means 61–84 | 71–84 | 76–84 | Forec. 1985–2000 |
|---|---|---|---|---|---|---|---|---|---|---|---|---|
| 01 | −0.1 | −0.2 | −0.2 | −0.1 | −0.4 | −0.4 | −0.2 | −0.3 | 1.6 | 1.7 | 2.0 | 1.4 |
| 02 | −0.3 | −0.6 | −0.5 | −0.2 | −0.9 | −0.8 | −0.4 | −0.4 | −1.1 | −1.2 | −0.7 | −0.4 |
| 03 | −0.04 | −0.1 | −0.1 | −0.1 | 0.0 | −0.3 | 0.5 | −0.2 | 0.6 | 0.6 | 1.6 | 0.4 |
| 04 | −0.4 | −0.6 | −0.7 | −0.5 | −0.7 | −1.0 | −1.0 | −0.8 | −0.25 | −0.8 | −0.9 | −0.7 |
| 05 | −0.4 | −0.7 | −1.0 | −0.5 | −0.4 | −0.3 | −0.03 | −0.2 | −0.4 | −0.5 | −0.5 | −0.5 |
| 06 | −0.2 | −0.4 | −0.6 | −0.4 | −1.0 | −1.0 | 0.3 | −0.2 | −1.0 | −0.6 | 0.8 | 0.3 |
| 07 | −0.6 | −1.0 | −1.1 | −0.5 | −0.8 | −0.6 | −0.2 | −0.3 | −0.04 | −0.2 | 0.6 | 0.2 |
| 08 | −0.5 | −0.9 | −1.2 | −0.7 | −0.9 | −1.2 | −0.3 | −0.5 | −0.7 | −1.3 | −0.8 | −0.7 |
| 09 | −0.2 | −0.4 | −0.5 | −0.4 | −0.2 | −0.2 | −0.02 | −0.2 | 2.4 | 2.2 | 1.9 | 1.7 |
| 10 | −0.2 | −0.4 | −0.6 | −0.4 | −0.7 | −0.7 | −0.1 | −0.5 | 0.3 | 0.1 | 0.4 | 0.3 |

[a] The means for country 10 are calculated only up to 1982 (instead of 1984).
[b] By definition:

$$w_L = w_{POP} + w_{LPR} + w_{ER} + w_{WH} ,$$

where $w_L$ is the growth rate of employed labor (in working hours), $w_{POP}$ is the growth rate of population, $w_{LPR}$ is the growth rate of the labor participation rate, $w_{ER}$ is the growth rate of employment rate, and $w_{WH}$ is the growth rate of average working hours. The precision is up to the rounding errors.
Source: Krelle (1989, p. 74).

where $x := \alpha$, $y := \omega$, $c_{11}, c_{12}, c_{22} > 0$, $c_{21} = a_{21} - a_{23}b_1 < 0$. If the characteristic roots of $C = \begin{pmatrix} c_{11} & c_{12} \\ -c_{21} & c_{22} \end{pmatrix}$ are conjugate complex (which happens if $\frac{a^2}{4} - b < 0$, $a = c_{11} + c_{22} > 0$, $b = c_{11}c_{22} + c_{12}c_{21}$), we get as a solution

$$\begin{aligned} x(t) &= e^{\gamma t}(A_{11} \cos \phi t + A_{12} \sin \phi t) + \bar{x} \\ y(t) &= e^{\gamma t}(A_{21} \cos \phi t + A_{22} \sin \phi t) + \bar{y} . \end{aligned}$$

Here $\gamma = -a/2$, $\phi = \arcsin \delta/r = \arccos \gamma/r$, $\delta = \sqrt{b - a^2/4}$, $r = \sqrt{\gamma^2 + \delta^2}$. This theory could explain long-term cycles in the driving forces of economic growth that in turn yield long-term fluctuations in the structure of production and foreign trade.

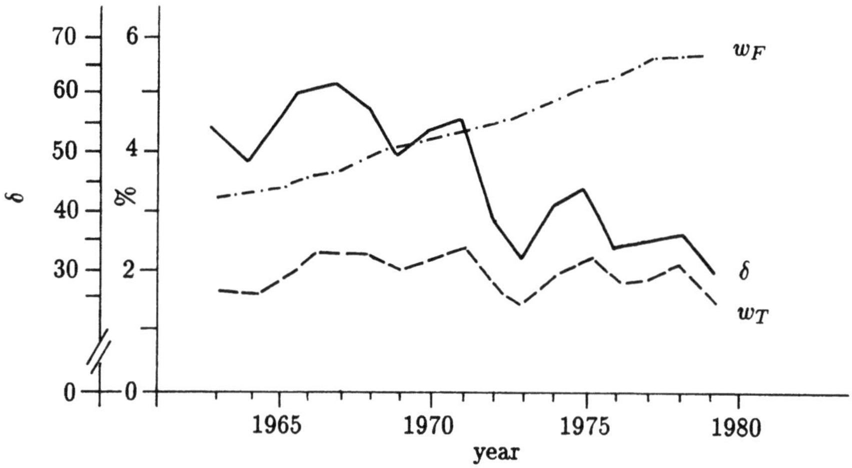


*Figure 17.11:* Rates of growth of technical progress, $w_T$, and knowledge, $w_F$, and their ratio $\delta$, the "transfer coefficient" for West Germany. Source: Krelle (1987, p. 390).

## 17.6 Empirical Results

In this section results with regard to the interdependence of economic growth and structural change are presented. They are derived from the econometric world model outlined above. Econometric models exist for almost all countries of the world. All of them include foreign trade, many of them also a monetary sector. But they do not take into account the repercussions of their own economic activities on other economies. They are built under the assumption of "a small country in a very large world". Only world models allow for the general interdependence that actually exists. The best-known and currently used world models are those of Project Link (directed by Lawrence Klein), the Wharton-Econometric Model, and the EPA model of the Economic Planning Agency of Japan. But there are others as well. I only mention the Globus Model developed by K.W. Deutsch and published recently by Bremer (1987), the world model of Mesarovic and Pestel (1974), and the UN model constructed by Leontief *et al.* (1977).

The latest model of this kind is the IIASA-University of Bonn Research Project published in Krelle (1989). It comprises models for nine OECD countries and an aggregated "country" called other developed market economies, seven CMEA countries and the rest of the European CMEA countries, India, six groups of developing countries, and the rest of the world. The approach is basically that of Section 17.3 and 17.4. We estimated forecasts for three scenarios: An optimistic one with relatively high rates of technical progress and high savings ratios, a medium one, and a pessimistic one with low figures for these two driving forces. *Figure 17.2* shows the growth paths of some countries under plausible assumptions for the scenarios. Of course, possible wars, revolutions, and other internal troubles are not taken into account. The accompanying structural change of production is indicated in *Tables 17.1* and *17.2* for some OECD and CMEA countries. More detailed

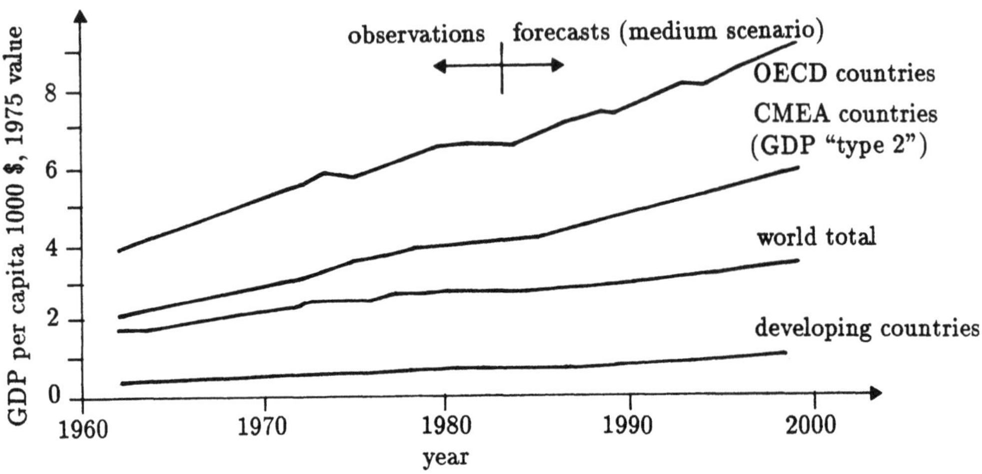


*Figure 17.12:* Development of the standard of living.

results follow from similar estimates of Donges *et al.* (1989, p. 385), and are reproduced in *Table 17.3*. Though the percentage changes seem to be small, the changes in absolute values are large. Many persons have to change their profession, lose their job, look for another one, and move to other places. This is not easy to accomplish and becomes socially more difficult the higher the standard of living is.

*Figure 17.12* shows structural changes in the average standard of living between OECD, CMEA, and developing countries. The developed market economies are still running faster than the CMEA countries. The fate of the developing countries is deplorable: Their average progress is very slow, though their growth rates of GDP are higher than those of the OECD countries. The excessive population growth prevents them from keeping pace with the OECD countries. Thus the international distribution of income will become more unequal in the future—with all the unpleasant political and social consequences. *Figure 17.13* shows the current trade balances of these three groups of countries in the past and forecasts up to the year 2000. According to these forecasts the debts of the developing countries will continue to rise—also an unpleasant outlook.

## 17.7 Conclusion

Economic growth (in the sense of a rise in GDP per capita) is inevitably connected to structural change. If the structural change is restrained, the growth rate is reduced. We demonstrated the relation between growth and structural change by a dynamic multisector world model, where production follows demand and demand changes with higher income. But this model could only cover structural change in the sense of changes in the proportions of production and of demand for different commodities, not the emergence of new products

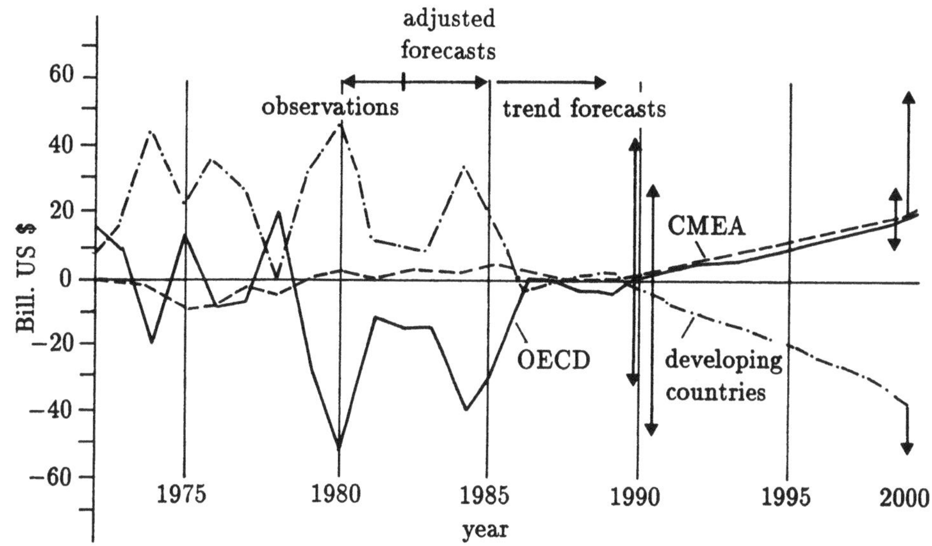


*Figure 17.13:* Current trade balance (including services), world regions.

or new production methods. That means structural change in the statistical sense, i.e., a sudden or slow change in the parameter values in the behavior equations is not considered. Slow changes may be treated by time dependent structural parameters, using a Kalman filter approach. This has been done by Kirchen (1988) at Bonn University. But structural breaks in the sense of sudden parameter changes can till now only be discovered *ex post.* The theory of Section 17.3 may give some hints as to the forecasts of such events—of course, only in probability.

## References

Beckmann, M. (1965), Wirtschaftliches Wachstum bei abnehmendem Skalenertrag, pp. 99-112 in W. Krelle (ed.), Theorien des einzel- und gesamtwirtschaftlichen Wachstums. Berlin: Duncker & Humblot.

Bremer, St.A. (ed.) (1987), *The Globus-Model: Computer Simulation of Worldwide Political and Economic Developments.* Frankfurt and Boulder, Colorado: Campus/Westview.

Donges, J.B., Klodt, H., and Schmidt, K.-D. (1989), Prospective structural changes in the West German economy, chapter 16 in W. Krelle (ed.), *The Future of the World Economy.* Berlin: Springer-Verlag.

Fischer, S. (1972), Keynes-Wicksell and neoclassical models of money and growth. *American Economic Review*, **62**, 880-890.

Gabisch, G. (1976), *Außenhandel und Wirtschaftswachstum.* Tübingen: J.C.B. Mohr.

Grossman, G.M. and Helpman, E. (1989), Comparative advantage and long-run growth. NBER Working Paper 2809, Cambridge, National Bureau of Economic Research.

Hagen, E.E. (1959), Population and economic growth. *American Economic Review*, **49**, 310-327.

Hahn, F.H. (1969), On money and growth. *Journal of Money, Credit and Banking*, **1**, 172-187.

Kirchen, A. (1988), *Schätzung zeitveränderlicher Strukturparameter in ökonometrischen Prognosemodellen.* Frankfurt/Main: Athenäum.

Krelle, W. (1987), Long-term fluctuations of technical progress and growth. *Journal of Institutional and Theoretical Economics*, **143**, 379-401.

Krelle, W. (1988), *Theorie des wirtschaftlichen Wachstums, Second edition.* Berlin: Springer-Verlag.

Krelle, W. (ed.) (1989), *The Future of the World Economy.* Berlin: Springer-Verlag.

Leontief, W., Carter, A., and Petri, P. (1977), *The Future of the World Economy. A United Nations Study.* Oxford: Oxford University Press.

Malthus, T.R. (1798), *An Essay on the Principles of Population.* London: Murray.

Mesarovic, M.D. and Pestel, E. (1974), *Mankind at the Turning Point.* New York: Dutton.

Niehans, J. (1963), Economic growth with two endogenous factors. *Quarterly Journal of Economics*, **77**, 349-371.

OECD (1989), *Economics in Transition: Structural Adjustment in OECD Countries.* Paris: OECD.

Ohlin, B. (1967), *Interregional and International Trade, Second edition.* Cambridge: Harvard University Press.

Ramanathan, R. (1982), *Introduction to the Theory of Economic Growth.* Berlin: Springer-Verlag.

Ross, H. (1989), Economic growth and structural change of OECD countries, chapter 3 in W. Krelle (ed.), *The Future of the World Economy.* Berlin: Springer-Verlag.

Schittko, U. (1976), *Lehrbuch der Außenwirtschaftstheorie.* Stuttgart: Gustav Fischer Verlag.

Stein, J. (1971), *Money and Capacity Growth.* New York: Columbia University Press.

Takayama, A. (1972), *International Trade.* New York: Rinehart and Winston.

Takayama, A. (1974), *Mathematical Economics.* Hinsdale: Dryden.

Tobin, J. (1965), Money and economic growth. *Econometrica*, **33**, 671-684.

Welsch, H. (1989a), Habit persistance and the structure of international commodity trade. *Empirical Economics*, **14**, 21-42.

Welsch, H. (1989b), Modeling exchange rates and foreign trade of developed market economies, chapter 8 in W. Krelle (ed.), *The Future of the World Economy.* Berlin: Springer-Verlag.

CHAPTER 18

# Structural Change in Swedish and Finnish Monthly Industrial Output Series

*Timo Teräsvirta*


## Summary

This chapter considers the hypothesis of no structural change in Swedish and Finnish industrial output series. The alternative hypothesis to a linear regression model is a rather flexible parametric form of structural change called smooth transition regression. Its parameters may thus be estimated if the null hypothesis is rejected. The null hypothesis may be tested without actually estimating the alternative using a simple $F$-test. The test procedure is derived in this chapter. The null hypothesis is rejected for both Swedish and Finnish autoregressions. There may be other reasons for rejection than structural change; however, a smooth transition model corresponding to the alternative is successfully estimated by nonlinear least squares for both countries. The results indicate that the structural change has been similar in both countries but has occurred in Sweden about a decade earlier than in Finland. The two time series are also found not to be cointegrated.


## 18.1 Introduction

In a recent paper, Luukkonen and Teräsvirta (1990) investigated the asymmetry of business cycles in a number of OECD economies. The business cycle indicators used in their paper were the quarterly unemployment rate and the industrial production, and they applied univariate time series techniques to analyze their data. Like some other authors who used different techniques [see, e.g., Brock and Sayers (1988)] they found nonlinearities in some of the analyzed series. However, the Swedish and Finnish unemployment and industrial output series clearly passed the linearity tests, so that no evidence in favor of asymmetry of business cycles in these two neighboring countries was found. The observation period comprised the years 1960-1986.

In this chapter we shall reconsider the issue for Sweden and Finland using monthly time series of industrial production starting from 1960(i) and extending to 1989(iv). Granger and T.-H. Lee (1989) recently argued that even if a discrete process is nonlinear, its temporal aggregates may sometimes seem linear. Thus, even if the quarterly industrial output series for the two countries look linear, the monthly processes may display nonlinear behavior. In fact, this is exactly what happens here. When linearity of the two processes is tested against smooth transition autoregression it is rejected. [However, in practice one may sometimes reject linearity with monthly but not with quarterly data simply because the number of monthly observations happens to be larger than that of quarterly ones unless the significance level of the test is an appropriate function of the number of observations.] The auxiliary regression used in these Lagrange multiplier-type tests suggests, however, that in both cases the nonlinearity may be related to changes in seasonality rather than to cyclical asymmetry. This leads to testing for structural change with a new technique with the advantage that a rejection of the constant parameter hypothesis at the same time suggests a time-varying parameter alternative. In this chapter the constancy is in fact rejected and the alternative model for the monthly industrial output is estimated for both countries.

After analyzing the two time series separately the question of their possible cointegration is raised. Although the industrial structures of the two countries are rather similar, the graphs of the series indicate that the two output series do not have a common trend. For the years preceding the first oil crisis the series seem to follow a rather similar pattern, but the hypothesis of no cointegration between them cannot be rejected with pre-1974 data either.

The plan of the chapter is the following. In Section 18.2 the smooth transition regression model is considered. Section 18.3 discusses the results of testing linearity against smooth transition autoregression. Section 18.4 contains results of the analysis of possible structural change in the framework of smooth transition regression models. Section 18.5 is devoted to the cointegration analysis, and Section 18.6 concludes the chapter.

## 18.2 Smooth Transition Regression Model

Consider the nonlinear model

$$y_t = \beta' x_t + (\theta' x_t) F(\tilde{z}_t) + u_t \ , \quad t = 1, \ldots, T \ , \tag{18.1}$$

where $\beta = (\beta_1, \ldots, \beta_p)'$ and $\theta = (\theta_1, \ldots, \theta_p)'$ are $p \times 1$ parameter vectors, and $x_t = (x_{t1}, \ldots, x_{tp})'$, $x_{t1} \equiv 1$, say, is a $p \times 1$ observation vector. Furthermore, $u_t \sim \text{nid}(0, \sigma^2)$, $Eu_t x_t = 0$, $Eu_t \tilde{z}_t = 0$, for all $t$; and $F$ is a bounded, monotonically increasing function of $\tilde{z}_t$.

A parametric functional form for $F$ may be assumed to estimate (18.1); for instance, Chan and Tong (1986) proposed the cumulative distribution function of the standard normal distribution. Luukkonen *et al.* (1988) suggested the cumulative distribution function of the logistic distribution, i.e.,

$$F(\tilde{z}_t) = (1 + \exp\{-\tilde{z}_t\})^{-1}, \quad -\infty < \tilde{z}_t < \infty, \tag{18.2}$$

where $\tilde{z}_t = \gamma(z_t - c)$, $\gamma > 0$; $\gamma$ and $c$ are parameters. If $\gamma \to \infty$, (18.2) becomes a Heaviside function

$$F(\tilde{z}_t) = \begin{cases} 0 & \text{if } z_t \leq c \\ 1 & \text{if } z_t > c\,. \end{cases} \tag{18.3}$$

In that case (18.1) with (18.2) becomes a switching regression model in which switching is deterministic and based on the variable $z_t$. For a discussion of switching regression models, see Goldfeld and Quandt (1973).

Testing linearity of (18.1) constitutes an interesting and relevant statistical problem. When $F$ is a Heaviside function and $c$ unknown, the likelihood function of the model does not meet the conventional regularity conditions. The likelihood ratio test statistic thus does not have the usual $\chi^2$ distribution under the null hypothesis. This is one reason why the smooth transition regression is often preferable to the switching regression: Testing linearity is easier if the alternative is smooth transition regression than if it is switching regression. In testing linearity of (18.1) the functional form of $F$ need not be fully specified. The following assumptions [see Luukkonen *et al.* (1988)] suffice:

- Function $F(z)$ is odd, monotonically increasing, possessing a nonzero derivative of order $(2s+1)$ in an open interval $(-a, a)$ for $a > 0$, $s \geq 0$.
- $F(0) = 0$ and $[d^h F(z)/dz^h]_{z=0} \neq 0$ for $h$ odd and $1 \leq h \leq k = 2s+1$.

The condition $F(0) = 0$ is convenient in deriving the testing procedure, but it is not restrictive. For instance, in connection with (18.2) it only means replacing $F(z)$ by $F(z) - F(0)$.

A difficulty inherent in (18.1) is that the model is not identified under the linearity hypothesis. If $\theta = 0$ in (18.1), $\gamma$ and $c$ in the argument of $F$ can take any value. Conversely, the null hypothesis may be formulated as $H_0$: $\gamma = 0$, which leaves $\theta$ and $c$ unidentified. This problem may be circumvented by approximating $F$ at $z = 0$ by a Taylor series expansion. Assuming $s = (k-1)/2$, one may apply the $k$-th order approximation

$$T_k(z) = g_1 z + g_3 z^3 + \ldots + g_k z^k,$$

where $g_j = F^{(j)}(0)/j!$, $j = 1, 3, \ldots, k$. Furthermore, as $z = \gamma(z_t - c)$ we may write

$$T_k(z) = \alpha_0 + \alpha' s_t, \tag{18.4}$$

where $\alpha = (\alpha_1, \alpha_2, \ldots, \alpha_k)'$ and $s_t = (z_t, z_t^2, \ldots, z_t^k)'$. Substituting (18.4) for $F$ in (18.1) yields

$$y_t = \phi' x_t + (\theta' x_t)(\alpha' s_t) + \eta_t = \phi' x_t + \sum_{i=1}^{k} \sum_{j=1}^{p} \psi_{ij} z_t^i x_{tj} + \eta_t \,, \tag{18.5}$$

where $\phi = \beta + \alpha_0 \theta$.

The null hypothesis of no structural change is

$$H_0\colon \psi_{ij} = 0, \quad i = 1, \ldots, k, \quad j = 1, \ldots, p\,. \tag{18.6}$$

Luukkonen *et al.* (1988) applied (18.5) in the case where

$$x_t = (1, y_{t-1}, \ldots, y_{t-p})' \quad \text{and} \quad z_t = \sum_{j=1}^{p} a_j y_{t-j}\,. \tag{18.7}$$

In (18.7) $a_j$, $j = 1, \ldots, p$, are unknown parameters with $a_j = 0$ if $j \neq d$ and $a_d = 1$ if $d \leq p$. This leads to combining terms and reparameterizing (18.5) and amounts to testing linearity against a (univariate) smooth transition autoregressive (STAR) model with the delay parameter unknown [see Luukkonen *et al.* (1988) for details]. The STAR model was the nonlinear alternative to autoregression in the business-cycle investigation of Luukkonen and Teräsvirta (1990).

If $z_t$ and $x_t$ are fixed, testing (18.6) against (18.5) involves testing a linear hypothesis in a linear regression model (18.5). If we assume $\eta_t \sim \text{nid}(0, \sigma_\eta^2)$ in (18.5), then under (18.6)

$$\text{SC}_k = \frac{(\text{SSE}_R - \text{SSE})/kp}{\text{SSE}/[T - (k+1)p]}\,, \tag{18.8}$$

where $\text{SSE}_R$ is the residual sum of squares from the restricted and SSE that from the unrestricted model, follows an $F[kp, T - (k+1)p]$ distribution. If $z_t$ is a linear function of time, for instance, the time index itself, then (18.8) is a test of linearity against structural change. The alternative assumes that except for $\sigma^2$, the parameters of the model change monotonically over time. This way of defining the alternative and of using a Taylor series approximation of $F$ circumvents the complications arising in testing for structural change with a switching regression as the alternative [see Krämer and Sonnberger (1986, p. 49) for discussion about this problem]. If the first-order Taylor series approximation is applied [$k = 1$ in (18.4)], the test statistic (18.8), $\text{SC}_1$, is the same as that of Farley *et al.* (1975) with one exception: These authors exclude a change in the intercept. A further discussion of the test for structural change based on smooth transition regression and its properties is deferred to a forthcoming paper.

## 18.3 Analyzing Output Series: Nonlinearity

In this section I shall begin the study of the monthly Swedish and Finnish industrial output series. The focus will be on testing linearity against STAR; the STAR models are capable of generating asymmetric realizations. The time series are the monthly volume of industrial production in Sweden and Finland. They are published in the *OECD Main Economic Indicators*, comprise the months 1960(i)–1989(iv), and are seasonally unadjusted but adjusted for the number of working days in the month. I shall consider logarithmed series and denote them $y_t$.

Both series are trending, and the first question then is how to difference them to achieve stationarity. The autocorrelation functions of both series display prominent peaks at lags $12, 24, \ldots$, and they die out very slowly. A heuristic argument [Box and Jenkins (1970)] would be that seasonal (12-month) differencing would stationarize the logarithmic series. This may be checked more formally by defining the moving sum $Y_t = \sum_{j=0}^{11} y_{t-j}$ and testing if it is integrated of order one [$I(1)$] against the hypothesis that it is trend-stationary. Kunst (1989) recently adopted this approach. Because $\nabla Y_t = Y_t - Y_{t-1} = \nabla_{12} y_t = y_t - y_{t-12}$, $Y_t$ being $I(1)$ means that $\nabla_{12} y_t$ is stationary. This may be done by considering the following regression model

$$\nabla Y_t = \mu + \alpha t + \beta_0 Y_{t-1} + \sum_j \beta_j \nabla Y_{t-j} + u_t \,, \quad u_t \sim \text{nid}(0, \sigma^2) \tag{18.9}$$

and testing $H_0$: $\beta_0 = 0$. The Dickey-Fuller test does not reject the null hypothesis for either country at the significance level 0.05, and $\hat{\alpha}$ does not appear significant; see *Table 18.1.* Thus I shall work with the seasonally differenced series $\nabla_{12} y_t$.

Consider now the possible cyclical asymmetry of these series, which is done by testing linearity of $\nabla_{12} y_t$ against STAR. Luukkonen and Teräsvirta (1990) explained why a STAR model is a feasible alternative in investigating cyclical asymmetry. First, an adequate linear autoregressive representation is estimated for both countries. For Sweden (setting $x_t = \nabla_{12} y_t$),

$$\begin{aligned} x_t = \; & \underset{(0.0024)}{0.0059} + \underset{(0.050)}{0.35}\, x_{t-1} + \underset{(0.055)}{0.18}\, x_{t-3} + \underset{(0.056)}{0.15}\, x_{t-4} \\ & + \underset{(0.056)}{0.17}\, x_{t-6} + \underset{(0.056)}{0.17}\, x_{t-7} - \underset{(0.047)}{0.24}\, x_{t-12} + \hat{u}_t \,, \end{aligned} \tag{18.10}$$

$$R^2 = 0.547 \,, \quad s = 0.036 \,, \quad z(12) = 5.65\ (0.93),$$

where the figures in parentheses are estimated standard deviations of the parameter estimates, and $z(12)$ is the Ljung-Box test ($p$-value in parentheses) against autocorrelation of order 12 at most. For Finland,

$$x_t = \underset{(0.0032)}{0.013} + \underset{(0.049)}{0.54}\, x_{t-1} + \underset{(0.050)}{0.25}\, x_{t-3} - \underset{(0.042)}{0.065}\, x_{t-12} + \hat{u}_t \tag{18.11}$$

$$R^2 = 0.490 \,, \quad s = 0.033 \,, \quad z(12) = 19.4\ (0.07).$$

*Table 18.1:* Results of testing the degree of integration and the cointegration of Swedish and Finnish monthly logarithmic industrial output series, 1962(i)-1989(iv).

| Country | Lags in model (18.9) | $\tau_\mu$ | $\tau_\tau$ | "$t$-value" of time trend |
|---|---|---|---|---|
| Sweden | 1,3,4,6,7,12 | −2.62 | −2.47 | 1.35 |
| Finland | 1,3,12 | −1.86 | −1.06 | 0.75 |
| Critical values ($\alpha = 0.05$) | | −2.87 | −3.43 | |

$\tau_\mu$ and $\tau_\tau$ are the Dickey-Fuller test statistics without and with a time trend, respectively [see Dickey and Fuller (1979)]. The lag structure of the models was simplified to include only significant lags before carrying out the tests. The critical values are response surface estimates of MacKinnon (1990).

| Cointegration Sweden/Finland | Lags in ADF autoregression | $t$ | Critical value ($\alpha = 0.05$) |
|---|---|---|---|
| | 1 to 7 | −1.88 | −3.36 |

The test is an augmented Dickey-Fuller test. The critical value is a response surface estimate of MacKinnon (1990).

Next I assume that lags in (18.10) and (18.11), respectively, constitute the linear part of (18.1) and test the null hypothesis of linearity using the first-order test $S_1$ (based on the first-order Taylor series approximation) and an augmented first-order test $S_A$ [see Luukkonen *et al.* (1988)]. The results are, in *Table 18.2.* In the case of Finland, linearity is rejected, extremely strongly when the test statistic is $S_A$. As to the Swedish series, the null is rejected at the 0.10 but not at the 0.05 level. Yet the number of regressors in the auxiliary regression for $S_A$ (28) is much larger than in the previous case (13), so that the evidence for nonlinearity is in part swamped by other, insignificant, regressors. A closer look at the auxiliary regressions reveals that the power of the test against the null hypothesis originates in "seasonal" regressors $x_{t-12}^2$ and $x_{t-12}^3$; see *Table 18.2.* This may indicate a misspecification of seasonality in a constant parameter linear autoregressive model rather than cyclical asymmetry. For quarterly series, i.e., four-quarter differences, Luukkonen and Teräsvirta (1990) did not reject linearity when the alternative was STAR.

## 18.4 Analyzing Output Series: Structural Change

The results of the previous section seem to suggest the possibility that the rejections of the linearity hypothesis are due to structural change. The viability of this idea can be investigated by testing linearity using (18.9) and (18.10) as a startingpoint against smooth transition regression where the transition variable is the time index or a linear transformation thereof. This means considering (18.1) with $x_t = (1, y_{t-1}, \ldots, y_{t-p})'$, $z_t$ being a linear function of the time index. The hypothesis to be tested is $H_0$: $\theta = 0$ in

*Table 18.2:* Results of testing linearity against STAR using Swedish and Finnish monthly logarithmic industrial output series, 1963(i)-1989(iv).

| | Sweden | | Finland | |
|---|---|---|---|---|
| Statistic | $S_1$ | $S_A$ | $S_1$ | $S_A$ |
| $p$-value | 0.248 | 0.081 | 0.294 | 0.00015 |
| Degrees of freedom | 21 | 27 | 6 | 9 |

Under $H_0$, both $S_1$ and $S_A$ asymptotically follow a $\chi^2$ distribution with $h(h+1)/2$ and $h(h+1)/2+h$ degrees of freedom, respectively, $h$ being the number of lags in the autoregressive part of the model. $S_1$ is identical to the linearity test of Tsay (1986). The following (nonlinear) regressors turned out to be significant (in parentheses: $t$-values of coefficient estimates) in the auxiliary regression for

$S_1$ (Sweden): $y_{t-12}^2$ (3.03)
$S_A$ (Sweden): $y_{t-12}^2$ (3.37), $y_{t-1}^3$ (−3.13), $y_{t-12}^3$ (1.99)
$S_1$ (Finland): $y_{t-1}y_{t-3}$ (2.17)
$S_A$ (Finland): $y_{t-12}^2$ (−3.40), $y_{t-12}^3$ (4.78)

*Table 18.3:* Results of testing linearity against smooth transition regression using Swedish and Finnish monthly logarithmic industrial output series, 1963(i)-1989(iv).

| | Sweden | | Finland | |
|---|---|---|---|---|
| Statistic | $SC_1$ | $SC_3$ | $SC_1$ | $SC_3$ |
| $p$-value | 0.263 | 0.608 | 0.423 | 0.012 |
| Degrees of freedom | 7 | 21 | 4 | 12 |

The following (nonlinear) regressors turned out to be significant (in parentheses: $t$-values of coefficient estimates) in the regression to compute

$SC_1$ (Sweden): $z_t^2 y_{t-12}$ (−2.37)
$SC_3$ (Sweden): $z_t^2$ (1.99); closest: $z_t^2 y_{t-12}$ (−1.75)
$SC_1$ (Finland): none
$SC_3$ (Finland): $z_t^2$ (3.39), $z_t y_{t-1}$ (−2.39), $z_t^2 y_{t-12}$ (−3.38); closest: $z_t^3 y_{t-12}$ (−1.89)

(18.1). This can be transformed into the problem of testing (18.6) in (18.5). Because $x_t$ contains lags of $y_t$, the null distribution of (18.8) is not an $F$ distribution, but the $F$ distribution is still a reasonable approximation. For the test statistics $SC_1$ and $SC_3$, see (18.8), the results are in *Table 18.3.*

For the Swedish industrial production series, the test statistics do not reject linearity. However, the individual coefficient estimates suggest that the seasonality might be changing over time. For Finland, using $SC_3$ leads to a rejection at the 0.05 level. Because even the Swedish data bear some evidence of seasonal structural change, I shall fit the smooth transition regression model (18.1) to both the Swedish and the Finnish data assuming that $F$ equals (18.2), where $z_t = 0.001(t-170)$ [$t = 1$ at 1960(i)] and $\theta = (0,\ldots,0,\theta_{12})'$. The estimated model for the Swedish series becomes

$$x_t = \underset{(0.0026)}{0.0050} + \underset{(0.050)}{0.34}\, x_{t-1} + \underset{(0.054)}{0.17}\, x_{t-3} + \underset{(0.055)}{0.12}\, x_{t-4} \qquad (18.12)$$

$$+ \underset{(0.055)}{0.17}\, x_{t-6} + \underset{(0.055)}{0.18}\, x_{t-7} - \underset{(0.097)}{0.043}\, x_{t-12}$$
$$- \underset{(0.11)}{0.27} (1 + \exp\{-\underset{(26.5)}{30.9} (z_t + \underset{(0.037)}{0.040})\})^{-1} x_{t-12} + \hat{u}_t$$

$R^2 = 0.497$ , $s = 0.033$.

[The conditional least squares estimation of smooth transition regression models appears very sensitive to the initial values. It seems that reasonably good startingvalues can often be obtained by conditioning on $\gamma$, estimating the model at various values of this parameter, and selecting the initial values from the model with the smallest sum of squared residuals.] The coefficient estimates for the linear part of (18.12) are similar to those in (18.10). The only exception is the estimate for the seasonality parameter $\beta_{12}$. Indeed, the seasonality seems to have changed during the observation period; note that $\hat{c} = -0.040$ corresponds to $t$ = 1970(x), and the standard deviation of $\hat{c}$, 0.037, is equivalent to 37 months. This estimate has to be interpreted together with $\hat{\gamma} = 30.9$ (which has a large standard deviation, 26.5) and the correlation between $\hat{c}$ and $\hat{\gamma}$ (which equals $-0.90$). Thus a small increase in $\hat{c}$ with a simultaneous decrease in $\hat{\gamma}$ may have little effect on the residual sum of squares. This indicates, together with the large standard deviations, that the structural change may have been smoother than the high value $\hat{\gamma}$ suggests. However, $\theta_{12}$ is estimated fairly accurately, so that an overall interpretation is available. In the 1970s, there was a smooth structural change in the seasonality of the monthly Swedish industrial production series. After being constant in the 1960s (no seasonality parameter $\beta_{12}$ was needed to complement seasonal differencing) in an autoregressive model, the seasonality became variable in the 1970s (the estimated seasonality parameter $\hat{\beta}_{12} + \hat{\theta}_{12} \approx -0.31 \pm 0.11$).

Estimating model (18.1) using the Finnish data yields the equation

$$x_t = \underset{(0.0034)}{0.015} + \underset{(0.049)}{0.53}\, x_{t-1} + \underset{(0.051)}{0.24}\, x_{t-3} - \underset{(0.044)}{0.043}\, x_{t-12} \tag{18.13}$$
$$- \underset{(0.12)}{0.23} (1 + \exp\{-\underset{(106)}{83.8} (z_t - \underset{(0.024)}{0.085})\})^{-1} x_{t-12} + \hat{u}_t$$

$R^2 = 0.499$ , $s = 0.033$.

This smooth transition regression model resembles in many respects its Swedish counterpart (18.12). Again, it seems that the seasonality was constant at first and became variable later. Here, $\hat{c} = 0.085$, which corresponds to $t$ = 1981(iii), and its standard deviation is equivalent to 24 months. Thus the change in seasonality indicated by the model occurred later in Finland than it did in Sweden, but the pattern is similar: $\hat{\beta}_{12}+\hat{\theta}_{12} \approx -0.27 \pm 0.11$. The estimates of $\hat{c}$ and $\hat{\gamma}$ in (18.13) are again correlated ($-0.79$), and the high standard deviation of $\hat{\gamma}$ indicates that the joint estimation of $c$ and $\gamma$ involves substantial uncertainty.

It is of course by no means clear that the sole reason for rejecting the null hypothesis when testing against structural change actually is structural change. Nevertheless, it is interesting to see that a smooth transition regression model can be successfully fitted to both data sets and that the results for the two output series are both interpretable and remarkably similar.

## 18.5 Cointegration Between the Swedish and Finnish Series

The analysis of the two industrial production time series has been univariate. In this section I shall briefly consider both processes together and see if they are cointegrated. Because seasonal variation is not of interest here, I shall consider the 12-month moving sums $Y_t^S$ (Sweden) and $Y_t^{SF}$ (Finland). Both economies are small and open, and are dependent on foreign trade. The countries are neighbors, and their main industries compete in several markets. This might speak in favor of some form of cointegration. On the other hand, a look at the output series is enough to see that they are not cointegrated. The effects of the first oil crisis and the consequent recession on Swedish manufacturing were more severe and lasted longer than those on Finnish industries. The formal cointegration test, see *Table 18.1*, confirms this assertion; the null hypothesis of no cointegration cannot be rejected when the observation period is 1962(i)-1989(iv). I also checked if the two series had been cointegrated before 1974, the oil crisis disrupting this harmony. However, the hypothesis of no cointegration cannot be rejected for the observation period 1962(i)–1973(xii) either.

The cointegration is a linear concept, and the above results thus relate to a search of a linear, constant parameter, long-term relationship between $Y_t^S$ and $Y_t^{SF}$. If one allows for time-variation (structural change) in the parameters, the concept of cointegration becomes much more flexible. Granger and H.S. Lee (1990) discuss time-varying cointegration in Chapter 10. The considerations in Section 18.2 could constitute a basis for an alternative approach to time-varying parameter cointegration. Nevertheless, the matter will not be pursued further here.

## 18.6 Conclusions

The results of the chapter show that the differenced monthly logarithmic time series of Swedish and Finnish industrial output since 1960 are nonlinear, although the corresponding quarterly series appear linear. It seems that the auxiliary regressions in many linearity tests contain information about causes of possible nonlinearity. In this chapter, that information hints at structural change in seasonal parameters of univariate autoregressive models. Indeed, a parametric model for the change can be estimated for both countries. The estimation results suggest that the seasonality in both Swedish and Finnish industrial production series has changed from constant to variable during the period of observation and that the change occurs a decade or so earlier in Sweden than in Finland.

## Acknowledgment

I wish to thank Wilhelm Krelle for his comments on an earlier version of the chapter and Pekka Pere for able research assistance. The responsibility for any errors is mine. Financial support from the Yrjö Jahnsson Foundation and the Academy of Finland is gratefully acknowledged.

## References

Box, G.E.P. and Jenkins, G.M. (1970), *Time Series Analysis, Forecasting and Control.* San Francisco: Holden-Day.

Brock, W.A. and Sayers, C.L. (1988), Is the business cycle characterized by deterministic chaos? *Journal of Monetary Economics*, **22**, 71-90.

Chan, K.S. and Tong, H. (1986), On estimating thresholds in autoregressive models. *Journal of Time Series Analysis*, **7**, 179-190.

Dickey, D.A. and Fuller, W.A. (1979), Distribution of the estimators for autoregressive time series with a unit root. *Journal of the American Statistical Association*, **74**, 427-431.

Farley, J.V., Hinich, M., and McGuire, T.W. (1975), Some comparisons of tests with a shift in the slopes of a multivariate linear time series model. *Journal of Econometrics*, **3**, 297-318.

Goldfeld, S.M. and Quandt, R.E. (1973), The estimation of structural shifts by switching regressions. *Annals of Economic and Social Measurement*, **2**, 475-485.

Granger, C.W.J. and Lee, H.S. (1990), An introduction to time varying parameter cointegration, in P. Hackl and A.H. Westlund (eds.), *Economic Structural Change. Analysis and Forecasting.* Berlin: Springer-Verlag.

Granger, C.W.J. and Lee, T.-H. (1989), The effect of aggregation on nonlinearity. Economics Discussion Paper No. 89-43, University of California, San Diego.

Krämer, W. and Sonnberger, H. (1986), *The Linear Regression Model Under Test.* Heidelberg: Physica-Verlag.

Kunst, R. (1989), Cointegration in macroeconomic systems: Seasonality and explosive roots. Research Memorandum No. 255, Institute for Advanced Studies, Vienna.

Luukkonen, R. and Teräsvirta, T. (1990), Testing linearity of economic time series against cyclical asymmetry. Discussion Paper No. 262 (revised version), Research Institute of the Finnish Economy.

Luukkonen, R., Saikkonen, P., and Teräsvirta, T. (1988), Testing linearity against smooth transition autoregressive models. *Biometrika*, **75**, 491-499.

MacKinnon, J.G. (1990), Critical values of cointegration tests. Economics Discussion Paper No. 90-4, University of California, San Diego.

Tsay, R.S. (1986), Nonlinearity tests for time series. *Biometrika*, **73**, 461-466.

CHAPTER 19

# On the Estimation of Time-varying Parameters in Transfer Function Models

*Lennart Claesson and Anders H. Westlund*

## Summary

Transfer function models are here used to identify empirically economic dynamic systems. An iterative recursive instrumental variables estimator for transfer function estimation, originally suggested by Young [see, e.g., Young (1984)], is here applied in estimating structurally varying transfer function models describing leading indicator relations for business-cycle forecasting. A series of numerical examples clearly indicate the good capability of the estimator to discover parameter changes. At least for single-input models the estimator converges well enough toward the correct and time-varying parameter values within a few iterations, as soon as the signal-to-noise ratio is not too small.

## 19.1 Introduction

The identification and estimation of dynamic structure represent one very significant element of economic structural analysis, besides being generally crucial to economic forecasting. For example, some basic features of business-cycle theory presume dynamic models, and business-cycle forecasting often focuses on so-called leading indicators and their dynamic relations with business-cycle reference series.

The identification of dynamic structures is essentially an empirical task. Economic theory seldom gives more than minor support to the specification efforts. At best, theory provides the analysis with a number of vague restrictions, such as whether leads or lags are to be expected. Obviously there is a need for efficient and general empirical procedures to assist in generating and testing hypotheses with respect to the dynamic elements of economic structures.

In econometric modeling, dynamics is often introduced by various types of finite distributed lags. Infinite lag structures are sometimes used to increase the generality. ARMAX and transfer function models are well-known examples, which represent very general dynamic structures. Such models, taken from time series analysis following the lines of Box and Jenkins, will be applied in this chapter to help analyze dynamic structural relations. For an excellent discussion on the relative importance of theory and empirical data to the modeling of dynamics, and for a survey of various models to be used in this connection, see Hendry *et al.* (1984).

Economic structures, in general, and dynamics, in particular, are traditionally assumed to be time invariant. There are, however, empirical as well as theoretical rationales to relax that assumption. Numerous contributions to methodological research, e.g., statistical test and estimation procedures to identify and determine parameter variability, have been published over the last decade [see Hackl and Westlund (1989)]. Furthermore, econometric and time series literature now also includes an increasing number of important empirical studies with the focus on structural variability problems. For example, the above-mentioned leading indicator dynamics often tends to vary between different time periods, in particular between the upswing and downswing business-cycle periods. This means that parameters related to a certain lag structure might be significant over one period and nonsignificant over another, i.e., some kind of parameter variability is present.

To render the analysis of parameter variability feasible, some limiting assumption is required. A lack of relevant theoretical *a priori* restrictions often means a tacit assumption that the parameters vary according to Markovian models. Kalman filtering is then often suggested as an approach to estimate structurally varying econometric models and time series models [see, e.g., Athans (1974) and Ledolter (1981)]. However, time series models with moving average terms include nonlinearities which make recursive estimation procedures such as Kalman filtering complex. Such nonlinearities also characterize ARMAX and transfer function models. One possible way to reduce complexity involves linearizing by instrumental variables and the use of an iterative recursive instrumental variables procedure [see, e.g., Young (1984)].

The purpose of this chapter is to introduce such a recursive procedure to econometric analysis of dynamics through transfer function models. In particular it will be used to analyze possible variabilities in the dynamic structures. As the need for recursive estimation of structurally varying transfer function models has been noted in relation, e.g., to some attempts to forecast business-cycle processes, the chapter starts with some comments on the theory of business-cycle dynamics in Section 19.2. Some comments are also given on the leading indicator approaches, which have gained general attention among business-cycle forecasters. Some empirical indications of varying dynamics among such leading indicator relations are pointed out. Section 19.3 introduces the transfer function model, and briefly discusses some related general identification and estimation problems. A recursive instrumental variable estimation procedure for transfer function estimation is then presented. In particular, parameter variability is assumed and necessary modifications with respect to the recursive estimation procedure are emphasized. The empirical application of transfer function models and the recursive estimation procedure is in general very difficult. Partly, this is due to the small data sets typically available, and partly to the often badly specified models (low signal-to-noise ratios). Such complications are generally accentuated in the case of varying parameter structures, e.g., when the dynamics are not time invariant. Section 19.4 provides, by a Monte Carlo study, some numerical illustrations to the earlier-given theoretical statements. Section 19.5 then returns to business-cycle forecasting by structurally varying transfer function models. A number of dynamic leading indicator relations are recursively estimated, and the results discussed. Section 19.6, finally, summarizes the study and gives some suggestions for further research.

## 19.2 Dynamic Models for Business-Cycle Forecasting

A majority of the commonly accepted theories of the business-cycle mechanisms entail dynamic models. This is the case both in "external" theories identifying the origins of the business cycle outside the economic system and in "internal" theories that try to explain business cycles through mechanisms within the economic process itself (for a review of business-cycle theories, see, e.g., Zarnowitz, 1985).

The "internal" or basically endogenous theories often focus on the multiplier-accelerator effects, i.e., on the internal dynamics of the economic system (involving lagged reactions among such components as consumption and investment levels). On the assumption that consumption is lagged behind income and that induced investment is lagged behind change in output, such a multiplier-accelerator mechanism is expected to produce a fluctuating output model. The theories thus postulate that as a result of such dynamics industrial economies will be characterized by recurrent fluctuations with a certain regularity.

Various exogenous and endogenous shocks are also expected to propagate cycles. Such influences, however, generally cannot by themselves generate fluctuations without some internal dynamics of the economic system. The monetarist interpretation that connects the cycle with contractions and expansions in the money supply is based on the assumption of lengthy (and possibly varying) lags of output behind the monetary changes. Laidler (1975) exemplifies these ideas by discussing a dynamic model where the lagged and current exogenous monetary growth and current inflation cause output changes, which in turn give impulses to inflation, etc.

Equilibrium theories also require some dynamics propagation mechanisms to explain the output and employment movements occurring over the business cycles. For example, it has been observed that rapid adjustments of production and employment rates are very costly. In general, order and construction lags are rather long with the consequence that investment often adds first to demand and later to capacity. Thus, the effects on supply from increased capacity are delayed.

Irrespective of the theoretical perspective used to explain the business-cycle mechanism, the elements of dynamics are always crucial. It is also obvious that the mechanism is complex, and that the set of factors and relations that constitute the business-cycle system will vary significantly over time.

Business-cycle theories touch a wide set of economic time series, e.g., production, employment, income, investment, various financial series. Several of these show recurrent cyclical fluctations that more or less consistently will lead an appropriate aggregate that represents the business-cycle series. So far this is consistent with different business-cycle theories, but the way in which leading indicators in practice are used for business-cycle forecasting is often just rudimentarily related to economic theory. Still, dynamics is the basic element of the model. Some of the used indicators will measure activities that actually are taking place, and that will lead the aggregate economic business-cycle series. Examples are housing starts and output from industries that first reflect economic recovery. Other leading indicators are taken from business surveys aimed at indicating future business tendencies.

One of the most important criteria according to which leading indicators are selected concerns degree of stability in the dynamic structure. As business-cycle forecasts based on leading indicators usually concentrate on turning points, the interest in the dynamics stability of the indicator will also focus on turning points. In principle, however, this stability requirement holds for the business-cycle process as a whole.

During the late 1970s, OECD created a system of leading indicators to be used for business-cycle forecasting in its member countries. An examination of the leading indicator system in the case of Sweden shows that the stability assumption hardly is satisfied. A simple measure of instability over a time period is the mean deviation from the observed median lead time between each leading indicator and the business-cycle reference series (here "industrial production"). *Figure 19.1* illustrates the lead instability as measured by this criterion for 12 leading indicators based on Swedish data since 1960. Obviously, the lead stability hypothesis does not receive much support.

One possible and partial explanation to the observed lead instability might be related to differences between the two categories of turning points (i.e., from upswings to downswings, and from downswings to upswings). *Figure 19.2* shows that this might be the case. Obviously, most of the indicators lead the reference series by shorter time periods at the points where upswings change into downswing phases. Consequently these two kinds of turning points should be predicted on the basis of different assumptions concerning the dynamic structure. This problem needs further theoretical as well as empirical analysis.

Below, varying dynamics will be discussed with reference to the general dynamic structure, and not be confined to turning point leads. In the following sections, transfer function models and recursive estimation procedures are applied to analyze possible dynamics.

## 19.3 Transfer Function Models and Recursive Estimation

### 19.3.1 The transfer function model

Modeling dynamics generally means the matching of theoretically postulated lag reactions to empirically observed auto- and cross-correlation structures among the relevant time

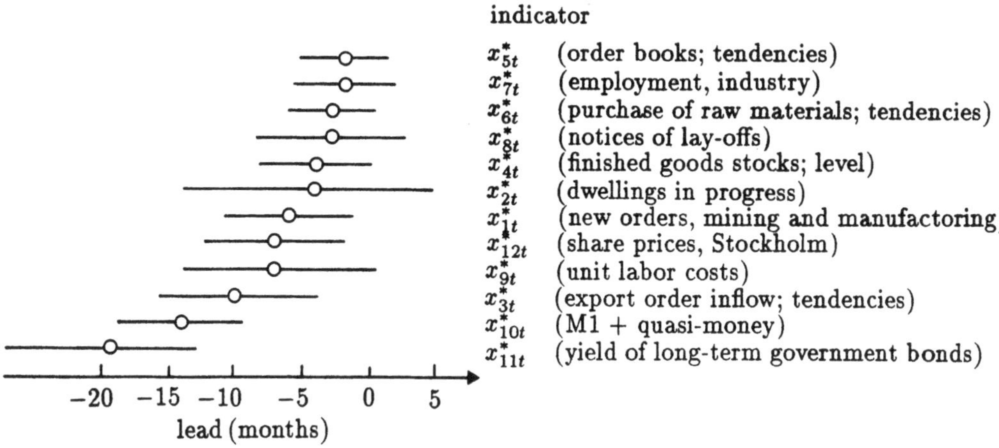


*Figure 19.1:* Median lead times for Swedish leading indicators at all business-cycle turning points 1960-1988 (intervals show mean deviations from median lead).

series data. Economic theory, however, provides little prior information about such lag reactions. As the time series to be applied often are very short, and the way they are linked together is characterized by a high degree of complexity, the paucity of dynamic theory causes problems in the modeling process. Completely unrestricted estimation is probably valueless, and will at least imply an imprecise and inefficient estimation of the underlying dynamic structure. Thus, although we focus on a very general class of dynamic models, it should be emphasized that all relevant theoretical information must be utilized in order to strive for parsimonious specification.

A general class of models used in describing dynamic economic structures is the autoregressive (AR) moving average (MA) model with explanatory (X) variables (i.e., the ARMAX models) with the following form:

$$f_0(B)y_t = \sum_{j=1}^{M} f_j(B)x_{j,t-\tau_j} + f_{M+1}(B)e_t\,, \tag{19.1}$$

where $y_t$ denotes the output at time $t$, $x_{jt}$ denotes the $j$-th input at time $t$, $\tau_j$ is an integer-valued pure delay of the $j$-th input, $e_t$ denotes a sequence of independent random variables $[e_t \sim IN(0,\sigma^2)]$, $B$ denotes the backward shift operator, i.e., $B^s x_t = x_{t-s}$, and $f_i(B) = \sum_{j=0}^{\ell_i} f_{ij}B^j$ $(i = 0,1,\ldots,M+1)$, where $f_{00} = f_{M+1,0} = 1$.

The transfer function model is given by re-expressing (19.1) as

$$y_t = \sum_{j=1}^{M} \frac{b_j(B)}{a_j(B)} x_{j,t-\tau_j} + \frac{d(B)}{c(B)} e_t\,, \tag{19.2}$$

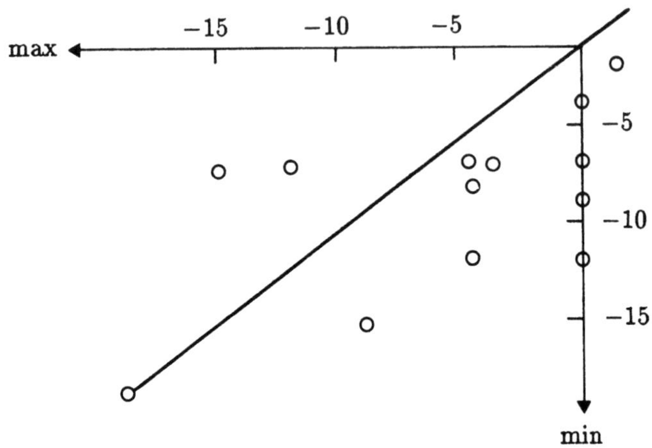


*Figure 19.2:* Median leads at turning points (max. and min., respectively) for Swedish leading indicators.

where all common factors have been cancelled in the polynomial ratios, and where

$$
\begin{aligned}
a_j(B) &= 1 + \sum_{i=1}^{n_j} a_{ji} B^i \\
b_j(B) &= \sum_{i=0}^{m_j} b_{ji} B^i \\
c(B) &= 1 + \sum_{i=1}^{p} c_i B^i \\
d(B) &= 1 + \sum_{i=1}^{q} d_i B^i
\end{aligned}
$$

Identification entails determining the integer set $\Omega = \{\tau_j, n_j, m_j, p, q; 1 \leq j \leq M\}$. Traditionally, so-called impulse response functions $\{\nu_{jk}; k \geq 0\}$, given by the infinite polynomial expansion of each of the $M$ single-input single-output systems (represented by $b_j(B)B^{\tau_j}/a_j(B) = \sum_{k=0}^{\infty} \nu_{jk} B^k$), are determined. Comparing each impulse response function with a catalogue of such functions, determined from known transfer function structures, yields information to help identify the model structure. Identification through evaluating observed impulse response functions is in practice often very complicated. The only parameter that is relatively easy to determine is the pure delay integer $\tau_j$. These identification complications are always accentuated when allowing for time-varying parameters. An alternative identification procedure that might be more efficient for such models is given by Young *et al.* (1984). The following discussion presupposes a well-identified model structure, i.e., our interest will focus on estimation problems. Given $\Omega$, maximum likelihood estimates of the parameters in (19.2) can be computed in a straightforward manner.

It is a well-known fact that ordinary least squares estimation of (19.2) will yield estimates that are asymptotically biased and inconsistent. To simplify, let us consider the single-input special case of (19.2), i.e.,

$$y_t = \frac{b(B)}{a(B)} x_t + \frac{d(B)}{c(B)}, e_t \tag{19.3}$$

where $a(B) = 1 + \sum_{i=1}^{n} a_i B^i$, $b(B) = \sum_{i=1}^{m} b_i B^i$, $c(B) = 1 + \sum_{i=1}^{p} c_i B^i$, and $d(B) = 1 + \sum_{i=1}^{q} d_i B^i$, and where to simplify notations it is assumed that $\tau_j = 0$. It is easily found that by multiplying both sides of (19.3) by $a(B)$

$$y_t = \mathbf{u}_t' \mathbf{a} + \eta_t , \tag{19.4}$$

where

$$\begin{aligned} \mathbf{u}_t' &= (-y_{t-1}, \ldots, -y_{t-n}, x_t, \ldots, x_{t-m}) \\ \mathbf{a} &= (a_1, \ldots, a_n, b_0, \ldots, b_m)' \\ \eta_t &= \Delta_t + \sum_{i=1}^{n} a_i \Delta_{t-i} , \end{aligned}$$

where $\Delta_t = \frac{d(B)}{c(B)} e_t$. But the vector $\mathbf{u}_t$ is contaminated by the noise

$$\mathbf{u}_t = \mathbf{Z}_t + \boldsymbol{\Delta}_t ; \tag{19.5}$$

here

$$\begin{aligned} \mathbf{Z}_t &= (-Z_{t-1}, \ldots, -Z_{t-n}, x_t, \ldots, x_{t-m})' \\ \boldsymbol{\Delta}_t &= (\Delta_{t-1}, \ldots, -\Delta_{t-n}, 0, \ldots, 0)' , \end{aligned}$$

where the unobservable $Z_t$ variables in $\mathbf{Z}_t$ constitute the deterministic output from (19.3), i.e.,

$$Z_t = \frac{b(B)}{a(B)} x_t .$$

Thus, if $\hat{\mathbf{a}}_t$ is the OLS estimate based on $t$-observations of $\mathbf{a}$ in (19.4), we find that

$$\begin{aligned} \operatorname{plim}(\hat{\mathbf{a}}_t - \mathbf{a}) &= \left[ \frac{1}{t} \sum_{i=1}^{t} (\mathbf{Z}_i + \boldsymbol{\Delta}_i)(\mathbf{Z}_i + \boldsymbol{\Delta}_i)' \right]^{-1} \frac{1}{t} \sum_{i=1}^{t} (\mathbf{Z}_i + \boldsymbol{\Delta}_i) \eta_i \\ &= \left[ R_{\mathbf{Z},\mathbf{Z}} + R_{\boldsymbol{\Delta},\boldsymbol{\Delta}} \right]^{-1} R_{\boldsymbol{\Delta},\eta} \neq 0 , \end{aligned}$$

where $R_{x,y} = \frac{1}{t} \sum_{i=1}^{t} x_i y_i$.

This shows that the OLS estimates for $a_t$ are asymptotically biased and inconsistent. A remedy for achieving consistency and, possibly, asymptotic efficiency is the introduction of instrumental variables.

### 19.3.2 Recursive estimation

One way of deriving recursive estimators of the parameters in (19.2) is to determine the estimates that minimize $\sum_{t=1}^{T} e_t^2$, where

$$e_t = \frac{c(B)}{d(B)}\left[y_t - \sum_{j=1}^{M} \frac{b_j(B)}{a_j(B)} x_{jt}\right] . \tag{19.6}$$

[To simplify notations without loosing generality, $\tau_j = 0$ is assumed in the following.] It is now possible to reformulate (19.6) as

$$e_t = \frac{c(B)}{a_\ell(B)d(B)}(\delta_{\ell t} - \boldsymbol{\delta}_{\ell t}'\mathbf{a}_\ell) , \tag{19.7}$$

where $\mathbf{a}_\ell = (a_{\ell 1}, \ldots, a_{\ell n_\ell}, b_{\ell 0}, \ldots, b_{\ell m_\ell})'$ [c.f. (19.4)],

$$\delta_{\ell t} = y_t - \sum_{j \neq \ell} \frac{b_j(B)}{a_j(B)} x_{jt} ,$$

and $\boldsymbol{\delta}_{\ell t} = (-\delta_{\ell,t-1}, \ldots, -\delta_{\ell,t-n_\ell}, x_{\ell t}, \ldots, x_{\ell,t-m_\ell})'$.

Prefiltering $\delta_{\ell t}$ and $x_{\ell t}$ by $\frac{c(B)}{a_\ell(B)d(B)}$ means creating

$$\delta_{\ell t}^* = \frac{c(B)}{a_\ell(B)d(B)} \delta_{\ell t}$$

and

$$x_{\ell t}^* = \frac{c(B)}{a_\ell(B)d(B)} x_{\ell t} .$$

As a function of $\delta_{\ell t}^*$ and $x_{\ell t}^*$, (19.7) is then rewritten as

$$e_t = \delta_{\ell t}^* - \boldsymbol{\delta}_{\ell t}^{*\prime}\mathbf{a}_\ell . \tag{19.8}$$

But since the $\delta_{\ell t}^*$ variables in (19.8) are contaminated by noise [c.f. (19.5)], we have here an errors-in-variables type of situation. One approach to obtain consistent estimates of the parameters in such models is to use

$$\hat{z}_{\ell t} = \frac{\hat{b}_\ell(B)}{\hat{a}_\ell(B)} x_{\ell t} \tag{19.9}$$

as instrumental variables for $\delta_{\ell t}$ [see Jakeman *et al.* (1980)]. The normal equations obtained when minimizing $\sum e_t^2$ subject to (19.8) may be written in recursive form as

$$\hat{\mathbf{a}}_{\ell t} = \hat{\mathbf{a}}_{\ell,t-1} - \hat{V}_{\ell,t-1}\hat{\mathbf{z}}_{\ell t}^* \left(\hat{\sigma}^2 + \hat{\mathbf{z}}_{\ell t}^{*\prime}\hat{V}_{\ell,t-1}\hat{\mathbf{z}}_{\ell t}^*\right)^{-1} \left(\hat{\boldsymbol{\delta}}_{\ell t}^{*\prime}\hat{\mathbf{a}}_{\ell,t-1} - \hat{\delta}_{\ell t}^*\right) , \tag{19.10}$$

where $\hat{\mathbf{z}}^*_{\ell t} = (-\hat{z}^*_{\ell,t-1}, \ldots, -\hat{z}^*_{\ell,t-n_\ell}, x^*_{\ell t}, \ldots, x^*_{\ell,t-m_\ell})'$ and

$$\hat{z}^*_{\ell t} = \frac{c(B)}{a_\ell(B)d(B)}\hat{z}_{\ell t}$$

and where $\hat{V}_{\ell t}$ denotes the estimated covariance matrix for $\hat{\mathbf{a}}_{\ell t}$, which in turn may be recursively updated by

$$\hat{V}_{\ell t} = \hat{V}_{\ell,t-1} - \hat{V}_{\ell,t-1}\hat{\mathbf{z}}^*_{\ell t}\left[\hat{\sigma}^2 + \hat{\mathbf{z}}^{*\prime}_{\ell t}\hat{V}_{\ell,t-1}\hat{\mathbf{z}}^*_{\ell t}\right]^{-1}\hat{\mathbf{z}}^{*\prime}_{\ell t}\hat{V}_{\ell,t-1}\,. \tag{19.11}$$

The corresponding "nonsymmetric" algorithms are obtained, if $\hat{\mathbf{z}}^{*\prime}_{\ell t}$ is replaced by $\hat{\boldsymbol{\delta}}^{*\prime}_{\ell t}$. In applications characterized by paucity of data and low signal-to-noise ratios (typical for economic processes), such algorithms often yield better robustness and convergence characteristics.

Prefiltering by $c(B)/[a_\ell(B)d(B)]$, which requires estimation of the residual process, proves $\hat{\mathbf{a}}_{\ell t}$ to be asymptotically efficient [see Jakeman *et al.* (1980)]. Although $e_t$ is colored noise, it is often possible to assume $c(B) = d(B) = 1$ and thus avoid the fully refined, but also rather complex, procedure. The estimation is still consistent and yields estimates that for most practical purposes are tolerably efficient statistically. The recursive formulas (19.10) and (19.11) are still applicable, but the prefilters are reduced to $1/a_\ell(B)$.

To apply this estimation procedure we obviously need estimates of the parameters in the polynomials to perform the prefiltering operations. These estimates should also be utilized to generate suitable instrumental variables from (19.9). In practice this can be achieved in an iterative-recursive manner, that is, by doing several iterations through the whole time series.

On some arbitrary $k$-th iteration, the recursive estimates $\hat{\mathbf{a}}_{\ell t}$ and $\hat{V}_{\ell t}$ are given by (19.10) and (19.11), respectively. The estimates $\hat{\sigma}^2$, $\hat{\mathbf{z}}^*_{\ell t}$ and $\hat{\boldsymbol{\delta}}^*_{\ell t}$ are based on the estimates of $\mathbf{a}_{\ell t}$ from the final recursion in the previous $(k-1)$-th iteration. The estimates of the noise-model polynomials $c(B)$ and $d(B)$ are obtained from recursive formulas similar to (19.10) and (19.11), using the residuals generated by the final estimates of $\mathbf{a}_{\ell t}$. Thus, the deterministic and noise models are estimated separately, but they are coordinated in each iteration. At the final recursion on iteration $k$, new estimates are obtained and the process is repeated until convergence is reached.

This iterative-recursive estimation procedure calls for some kind of initialization before the first iteration. Initial estimates of $\delta_{\ell t}$, $t = 1, \ldots, T$, could be obtained by estimating each input-output relationship separately. The parameter estimators $\hat{a}_{\ell t}$ are initialized at $t = 0$ with zero elements, and the initial $\hat{V}_{\ell t}$ matrices are chosen diagonally with all elements set to some large number. This choice of $\hat{V}_{\ell 0}$ can be interpreted as if we put little confidence in the initial estimates $\hat{\mathbf{a}}_{\ell 0}$. Since neither instrumental variables nor prefilters can be generated before the first iteration, $\hat{\mathbf{z}}^*_{\ell t}$ (and $\hat{\mathbf{z}}^{*\prime}_{\ell t}$) are replaced by $\hat{\boldsymbol{\delta}}^*_{\ell t}$ (and $\hat{\boldsymbol{\delta}}^{*\prime}_{\ell t}$) in (19.10) and (19.11) during the first iteration. The resulting estimates are biased but prove to be of sufficient quality to get the process started.

In the subsequent iterations $\hat{a}_{\ell 0}$ is set equal to the final estimates from the previous iteration. The matrices $\hat{V}_{\ell 0}$ are set to the values obtained in some specified recursion in the preceding iteration. This setting of $\hat{V}_{\ell 0}$ is heuristic and the appropriate recursion is typically given by $T/I$, where $T$ is the number of observations and $I$ the number of iterations.

A final iteration $(I+1)$ is recommended in which $\hat{V}_{\ell 0}$ again is set with large values in the diagonal. This is done to get a satisfactory estimate of the error-covariance matrix. Even if the nonsymmetric version of (19.10) and (19.11) has been used in the "regular" iterations, it seems preferable to utilize the symmetric form in this last iteration.

The need for time-variable parameters in (19.2) is indicated in Section 19.2. There are several ways in which such time-variable parameter modifications may be considered. A general form of stochastic models for parameter variations, adapted to the requirements by simple recursive estimation algorithms, is given by the following Markovian assumption:

$$\mathbf{a}_{jt} = \Phi_j \mathbf{a}_{jt-1} + \Gamma_j \varepsilon_{jt} \, , \tag{19.12}$$

where $\mathbf{a}_{jt} = (a_{j1t}, \ldots, a_{jn_jt}; b_{j0t}, \ldots, a_{jm_jt})'$, and where $\Phi_j$ and $\Gamma_j$ are assumed known, and $E[\varepsilon_{jt}] = 0$, $E[\varepsilon_{jt}\varepsilon'_{jt'}] = Q_j\delta_{tt'}$, and $E[\varepsilon_{jt}e_{t'}] = 0$ for all $t$, $t'$ and finally $\delta_{tt'}$ denotes the Kronecker delta.

Thus, it is now possible to proceed in line with Kalman filtering [see Kalman (1960)] by adding the following prediction relations to the previously given recursive updating formulas (19.10) and (19.11):

$$\hat{\mathbf{a}}_{\ell,t|t-1} = \Phi_\ell \hat{\mathbf{a}}_{\ell,t-1} \tag{19.13}$$

and

$$\hat{V}_{\ell,t|t-1} = \Phi_\ell \hat{V}_{\ell,t-1} \Phi'_\ell + \Gamma_\ell Q \Gamma'_\ell \, . \tag{19.14}$$

Furthermore, $\hat{\mathbf{a}}_{\ell,t-1}$ and $\hat{V}_{\ell,t-1}$ are replaced by $\hat{\mathbf{a}}_{\ell,t|t-1}$ and $\hat{V}_{\ell,t|t-1}$, respectively, in (19.10) and (19.11) (where $\hat{\mathbf{a}}_{\ell,t|t-1}$ and $\hat{V}_{\ell t|t-1}$ denote estimates for parameters at time $t$ based on data up to $t-1$).

## 19.4 A Monte Carlo Study

### 19.4.1 The design of the study

The recursive estimation procedure (19.10) and (19.11) has some attractive asymptotic qualities. The instrumental variable approach makes it consistent and sometimes asymptotically efficient. To judge its capacity to contribute to the analysis of dynamics in economic systems, it is also important to verify its small-sample properties and, perhaps above all, to study how the procedure may identify parameter variabilities.

As theoretical finite sample evaluation will be mathematically intractable, we must rely on empirical observations or on numerical Monte Carlo studies. The Monte Carlo simulations involve transfer function models with constant as well as time-varying parameters. The following models are considered:

Model I $\quad y_t = \frac{b_{10}}{1+a_{11}B} x_{1t} + \varepsilon_t.$

Model II $\quad y_t = \frac{b_{20}+b_{21}B}{1+a_{21}B+a_{22}B^2} x_{2t} + \varepsilon_t.$

Model III $\quad y_t = \frac{b_{10}}{1+a_{11}B} x_{1t} + \frac{b_{20}+b_{21}B}{1+a_{21}B+a_{22}B^2} x_{2t} + \varepsilon_t.$

In the case of constant parameters, true parameter values are as follows:

$$\begin{array}{llll} a_{11} = -.6 & & b_{10} = 2 & \\ a_{21} = -1.4 & a_{22} = .5 & b_{20} = -.1 & b_{21} = .2 \ . \end{array}$$

In the case of time-varying parameters the following assumptions are made:

| Model | True parameters |
|---|---|
| I | $a_{11} = \begin{cases} -.6 & t = 1,\dots,90,181,\dots,T \\ -.4 & t = 91,\dots,180 \end{cases}$ |
| | $b_{10}$ constant, as above |
| II | $a_{21} = \begin{cases} -1.4 & t = 1,\dots,180 \\ -1.2 & t = 181,\dots,T \end{cases}$ |
| | $b_{21} = \begin{cases} .2 & t = 1,\dots,180 \\ .35 & t = 181,\dots,T \end{cases}$ |
| | $a_{22}$, $b_{20}$ constant, as above |

The time series length $T$ equals 300 and, in cases of constant parameters and the one-input system, 100. In all experiments the smaller samples consist of the first 100 observations of the corresponding larger samples.

For all simulation experiments presented in this chapter, 500 replicates were performed. The same stationary input series $x_{1t}$ and $x_{2t}$ were used throughout all replicates. The noise series were generated as pseudorandom observations from a normal distribution with variance appropriately chosen to give the selected signal-to-noise ratios $S/N$. Two levels of this ratio selected for this study: a high level ($S/N \simeq 10$) and a (very) low level ($S/N \simeq .5$). To demonstrate the explicit effect of increased random disturbances, the noise series for the experiments with low signal-to-noise ratios were generated as $\sqrt{20}$ times the noise series used in the experiments with high signal-to-noise ratios.

In the estimation of model III the initialization before the first iteration, i.e., determining the starting values of $\delta_{\ell t}$, was based on a first estimation run through the data. Experimentation showed an improvement in convergence if the deviations from their respective means were used in place of the original variables in this first estimation run. This modification was not necessary in estimating models I and II since $\delta_{\ell t} = y_t$ in single input models.

### 19.4.2 The case of constant parameters

In this section the proposed estimation technique is applied to the models with constant parameters. Several results from the simulations are displayed in *Tables 19.1(a-c)*. The values given in the tables are the means and the standard deviations (in parentheses) of the 500 estimates of each coefficient. Some of the results are based on fewer replicates due to convergence problems (c.f. notes below the tables).

The results are in accordance with what was expected, with the exception of the estimates of the autoregressive parameters in model II, *Table 19.1(b)*, with $T = 100$ and low signal-to-noise ratio. The mean values differ markedly from the true parameter values, and the standard deviations are higher than expected in comparison with the other results. This indicates that the number of iterations are too few, at least for some replicates, to obtain fully converged estimates for these two parameters.

Model complexity, sample size, and signal-to-noise ratio all have their influence on the performance of the recursive estimation procedure. In some of the experiments, where less favorable combinations of these factors were tried, we had some problems with numerical instability. This showed itself in the form of exploding IV series. Since the autoregressive part of model II (or the second input of model III) are "close to" being nonstationary, the resulting estimates could with a little bad luck generate a nonstationary transfer function to be used in constructing the IV series for the next iteration. This could in turn result in even poorer estimates and eventually in numerical overflow. The risk of this of course depends on the quality of the estimates in previous iterations. Thus, the risk is higher for short and noisy series but decreases with each iteration since the quality of the estimates should be improved by the iterative process.

The problems encountered could surely, at least to some extent, be solved by a more sophisticated approach than the quite straightforward application of the estimation procedure described in Section 19.3, which was used here. Applying some kind of convergence criteria instead of a fixed number of iterations should, for example, solve the earlier-mentioned problem of too few iterations.

### 19.4.3 The case of varying parameters

The simulation results for the models with parameter shifts are presented in *Figures 19.3–19.8*. All results are based on 500 replicates, and the mean and standard deviation were calculated at each recursive step. The four series plotted in each figure are the mean (solid line), the mean ± one standard deviation (dotted lines), and the true coefficient (dashed line).

*Figures 19.3* and *19.4* show the recursive estimates for model I with high signal-to-noise ratio. In *Figure 19.3*, estimation was carried out under the usual assumption of constant parameters. As expected, the parameter shifts in coefficient $a_{11}$ are not reflected in the recursive estimates. *Figure 19.4* displays the same estimates when a random walk model for both coefficients are incorporated to allow for possible parameter changes. The shifts

*Table 19.1:* Simulation results for (*a*) model I, (*b*) model II, and (*c*) model III.

| (*a*) | $T = 100$ | | $T = 300$ | |
|---|---|---|---|---|
| | High $S/N$ | Low $S/N$ | High $S/N$ | Low $S/N$ |
| $a_{11} = -0.6$ | −0.599 | −0.594 | −0.600 | −0.598 |
| | (0.015) | (0.067) | (0.009) | (0.039) |
| $b_{10} = 2.0$ | 2.004 | 2.031 | 2.001 | 2.010 |
| | (0.074) | (0.333) | (0.043) | (0.193) |
| (*b*) | $T = 100$ | | $T = 300$ | |
| | High $S/N$ | Low $S/N$[a] | High $S/N$ | Low $S/N$ |
| $a_{21} = -1.4$ | −1.397 | −1.330 | −1.400 | −1.391 |
| | (0.032) | (0.224) | (0.018) | (0.086) |
| $a_{22} = 0.5$ | 0.498 | 0.436 | 0.500 | 0.492 |
| | (0.029) | (0.208) | (0.016) | (0.078) |
| $b_{20} = -0.1$ | −0.101 | −0.104 | −0.100 | −0.102 |
| | (0.011) | (0.050) | (0.006) | (0.028) |
| $b_{21} = 0.2$ | 0.201 | 0.210 | 0.200 | 0.203 |
| | (0.013) | (0.057) | (0.008) | (0.033) |
| (*c*) | $T = 300$ | | | |
| | High $S/N$ | Low $S/N$[b] | | |
| $a_{11} = -0.6$ | −0.603 | −0.605 | | |
| | (0.012) | (0.050) | | |
| $b_{10} = 2.0$ | 2.009 | 2.020 | | |
| | (0.045) | (0.196) | | |
| $a_{21} = -1.4$ | −1.400 | −1.389 | | |
| | (0.018) | (0.091) | | |
| $a_{22} = 0.5$ | 0.500 | 0.491 | | |
| | (0.017) | (0.084) | | |
| $b_{20} = -0.1$ | −0.101 | −0.104 | | |
| | (0.007) | (0.029) | | |
| $b_{21} = 0.2$ | 0.201 | 0.204 | | |
| | (0.008) | (0.033) | | |

[a] 13 replicates were omitted.
[b] 18 replicates were omitted.

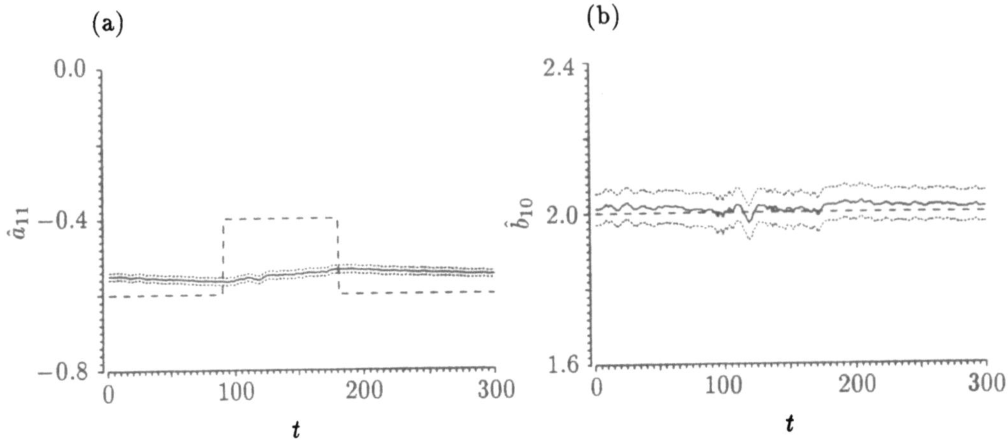


*Figure 19.3:* Recursive estimates of model I assuming constant parameters; high signal-to-noise ratio.

in $a_{11}$ are rather rapidly detected by the estimation procedure. Not surprisingly, the shifts cause a temporary turbulence in the recursive estimates of the constant parameter $b_{10}$ as well.

*Figures 19.5* and *19.6* show the corresponding results for model I with low signal-to-noise ratio. Consequently, the standard deviations are higher, but this is the only obvious difference.

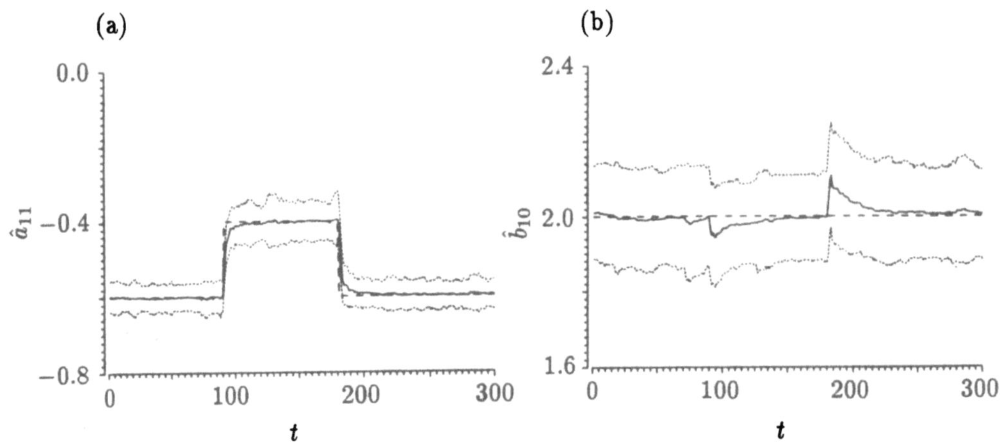


*Figure 19.4:* Recursive estimates of model I assuming a random walk model for the parameters; high signal-to-noise ratio.

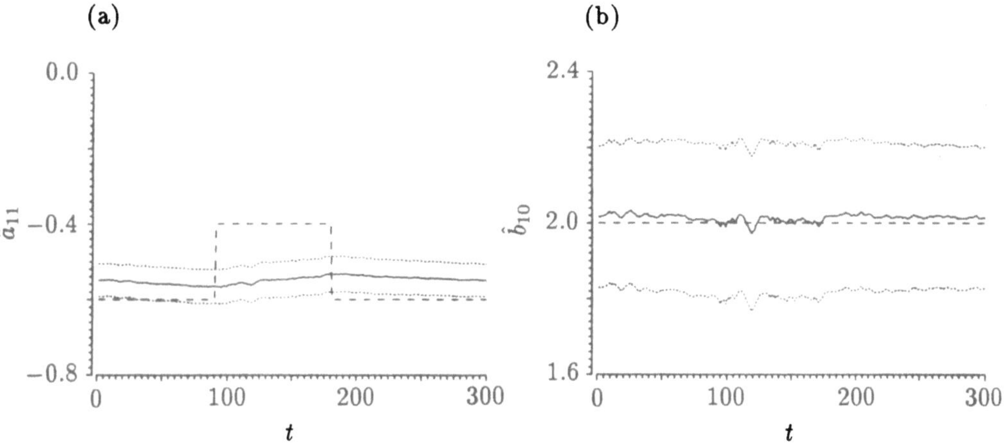


*Figure 19.5:* Recursive estimates of model I assuming constant parameters; low signal-to-noise ratio.

As we shall see in the following, the very appealing results for parameter $a_{11}$ in model I are somewhat coincidental (dependent on design).

When the same estimation procedures were applied to model II with high signal-to-noise ratio, the inherent weaknesses of the refined IV approach in connection with random walk modeling for parameter shifts were exposed.

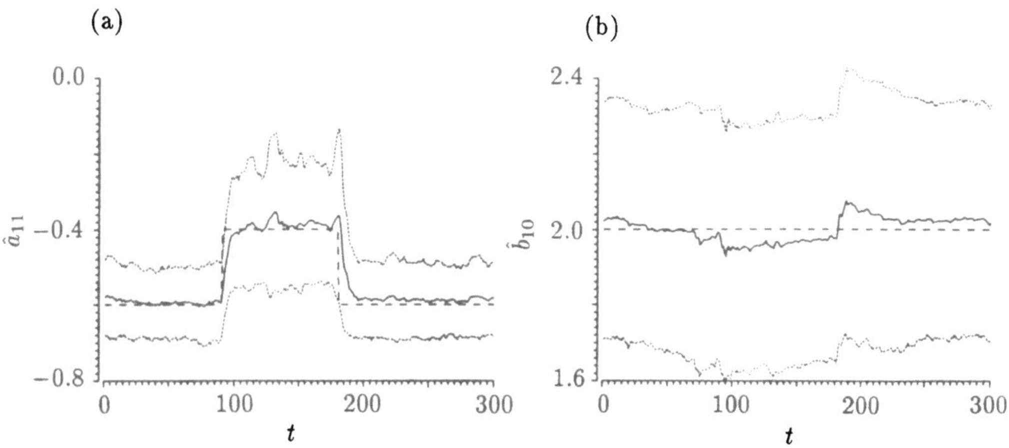


*Figure 19.6:* Recursive estimates of model I assuming a random walk model for the parameters; low signal-to-noise ratio.

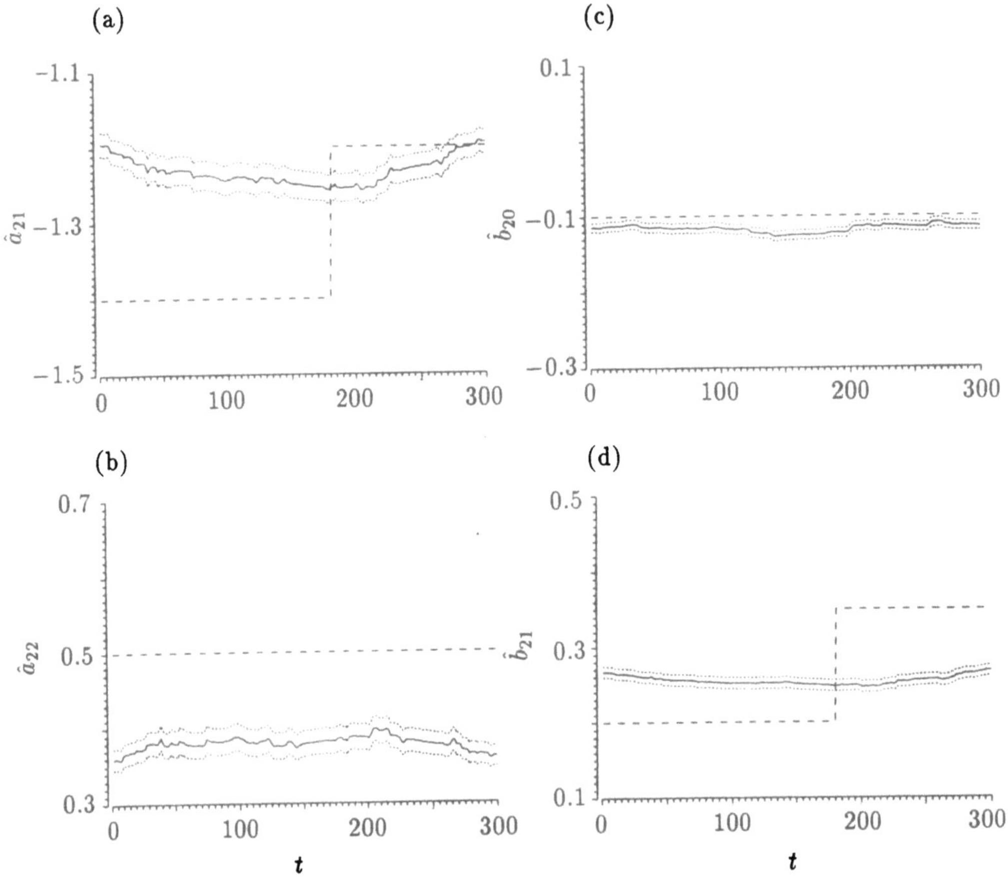


*Figure 19.7:* Recursive estimates of model II assuming constant parameters; high signal-to-noise ratio.

*Figure 19.7* shows the recursive estimates assuming constant parameters. One can note the rather peculiar estimates of the autoregressive coefficients.

*Figure 19.8* shows the recursive estimates using the random walk approach. The shifts are still detected, although it takes a while longer to reach the right level (on average) compared with the results for model I. The poor estimates in the beginning of the series are obvious. They have to do with the construction of the instruments (and to some extent

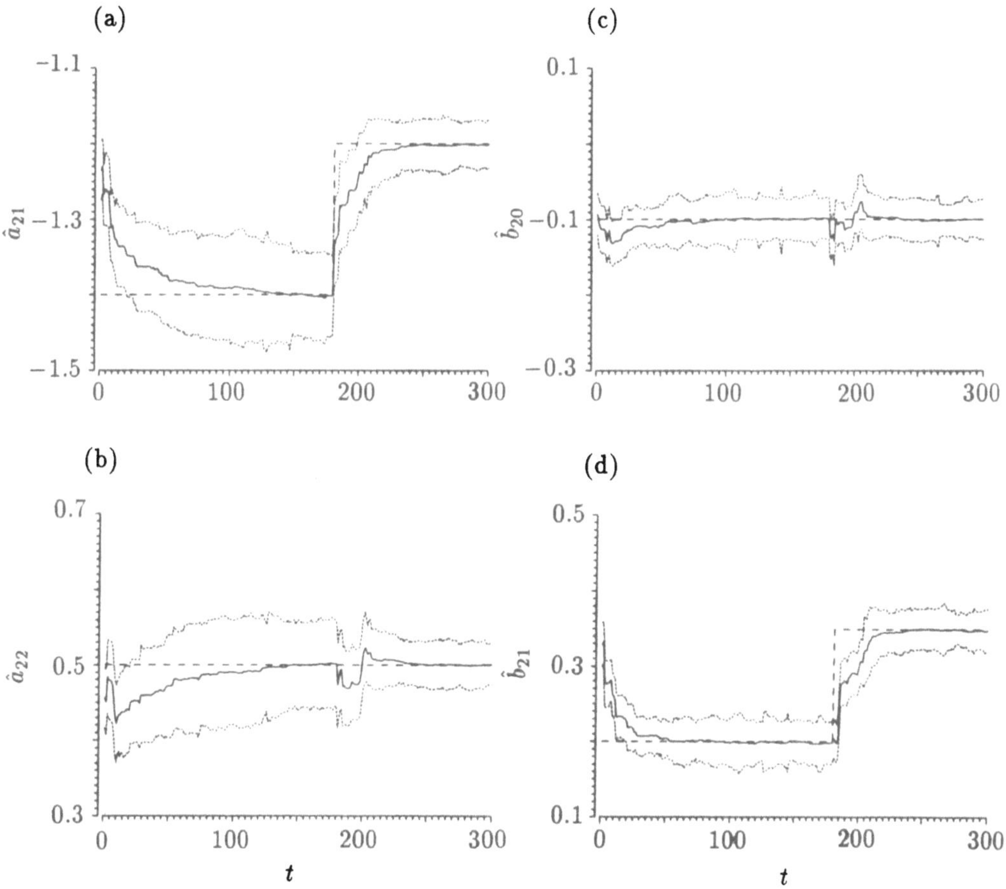


*Figure 19.8:* Recursive estimates of model II assuming a random walk model for the parameters; high signal-to-noise ratio.

the prefilters) and, furthermore, with the starting values of the algorithm, which all are determined by the final estimates (at recursion 300) in the previous iteration. The effect of this is that the parameter changes are not reflected in the IV series, and that explains why the first recursion gives estimates close to the final level of the coefficient. To avoid this, the construction of the IV series has to be made more adaptive. Note that this problem did not appear in model I, since parameter $a_{11}$ shifted back to its original level.

The need to make the instruments (and prefilters) more adaptive would be even more necessary when there are two or more inputs. Otherwise, the explicit effects of the parameter changes on the series $\hat{\delta}_{it}$ will be mixed up by effects from determining the other input series.

## 19.5 Empirical Observations on Transfer Function Variability

In Section 19.2 we briefly indicated that various economic theories explaining business-cycle phenomena include assumptions about dynamic relations. We also stated that such dynamic structures, for the purpose of business-cycle forecasting, often are measured by so-called leading indicators. We have here estimated a number of leading indicator relations modeled by transfer functions. By splitting the sample period in various ways we will get an idea about the degree ofstructural variability involved and also indications whether and, if so, to what extent dynamics will vary over time. Some typical results are given in this section. The variability thus observed clearly justifies the use of a recursive estimation approach, e.g., according to the principles described above. A few empirical results are given.

Analysis and forecasting of business-cycle behavior requires some reference series to represent the cyclical growth of the aggregate economy. Here we use an index of industrial production (covering mining, manufacturing, and public utilities) and concentrate on the Swedish case. The leading indicators selected by OECD for Sweden are listed in connection with *Figure 19.1*. They are chosen with respect to various criteria. Besides the above-mentioned need for consistency of the lead of the indicator over the reference series (especially at turning points), some other criteria are worthwhile mentioning. "Economic significance" is, of course, the basic criterion according to which an indicator is accepted. Furthermore, it is desirable that an indicator shows cyclical profiles highly correlated with that of the reference series, and that these profiles are smooth enough to allow turning points to be distinguished from random movements.

Transfer function models are identified and estimated for several relations between leading indicators and the reference series. Multiple-input as well as single-input models are determined. Furthermore, transfer function relations between some alternative international indicators (and foreign reference series) and the Swedish industrial production index are studied. Finally, the relation between the Swedish composite index (based on the set of leading indicators) and the Swedish industrial production index is analyzed. The total time series period (covering 1960:1–1988:11) is split up into numerous subperiods.

To analyze variability in structures (particularly in dynamics) different approaches can be used. One means is to identify a single-model structure, which is used over all the different sample sets. Varying parameter estimates indicate structural variability, subject to the particular model actually used. Another procedure allows for new identification as well as estimation for each new sample. This approach, of course, more efficiently indicates changes with respect to dynamics, but at the same time also makes it more difficult (or

*Table 19.2:* Swedish industrial production ($y_t$) = $f$[Swedish share prices ($x_{1t}$), Swedish yield of long-term government bonds ($x_{2t}$)] for 1963:10–1988:9, and subperiods thereof.

| Time period ($T$) | Diff. (order, degree) | | | Pure lag | | Estimated parameters Input lag | | Output lag | | Noise MA/AR | |
|---|---|---|---|---|---|---|---|---|---|---|---|
| | $y_t$ | $x_{1t}$ | $x_{2t}$ | $x_{1t}$ | $x_{2t}$ | $x_{1t}$ | $x_{2t}$ | $x_{1t}$ | $x_{2t}$ | | |
| 63:10–88:9 ($T = 300$) | Univariate (input variables are rejected) | | | | | | | | | | |
| 76:4–88:9 ($T = 150$) | (1,1) | (1,1) | – | 9 | – | 0/.02 | – | – | – | MA(1) | .65 |
| 63:10–76:3 ($T = 150$) | (1,1) | – | (1,1) | – | 7 | – | 0/–.003 | – | – | MA(1) | .53 |
| 82:7–88:9 ($T = 75$) | (1,1) | (1,1) | (1,1) | 9 | 1 | 0/.01 | 0/.01 | – | 1 / –.80 | MA(1) | .70 |
| 76:4–82:6 ($T = 75$) | Univariate (input variables are rejected) | | | | | | | | | | |
| 70:1–76:3 ($T = 75$) | (1,1) | – | (1,2) | – | 8 | – | 0/–.04 | – | – | MA(1) | .41 |
| 63:10–69:12 ($T = 75$) | Univariate (input variables are rejected) | | | | | | | | | | |

impossible) to compare estimates of different periods. As the identification procedure is extremely complicated and time-consuming, the second approach requires very efficient computer software. In the entire study we have applied AUTOBOXPLUS [see Shumway (1988)], which permits automatic transfer function identification.

The overall and very general results obtained through these extensive exercises are (i) most transfer function relations studied show high instability with respect to the dynamic structures involved and (ii) subject to a specific transfer function model, in most cases the parameter variability is significant. In conclusion, the above-mentioned requirements with respect to lead consistency is very rarely satisfied.

It is of course not possible to demonstrate all our results. We restrict ourselves to a few typical examples. These include Swedish industrial production as a function of (i) Swedish long-term interest rates and share prices; (ii) an OECD composite leading index; and (iii) industrial production indices in the UK, the United States, and West Germany.

*Tables 19.2–19.5* above summarize some results obtained when transfer function models are reidentified and re-estimated over various time periods.

Recursive estimation according to (19.10)–(19.11) is now used to determine various transfer function models. Some results are given in *Figures 19.9–19.11.*

## 19.6 Conclusion

Transfer function models are used here empirically to identify economic dynamic systems. As economic structures in general, dynamic relations (and, thus, transfer function

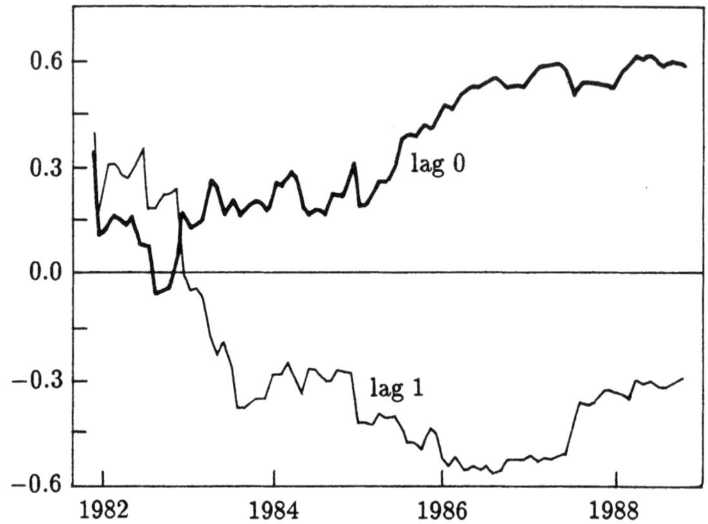


*Figure 19.9:* Recursive estimation: UK industrial production.

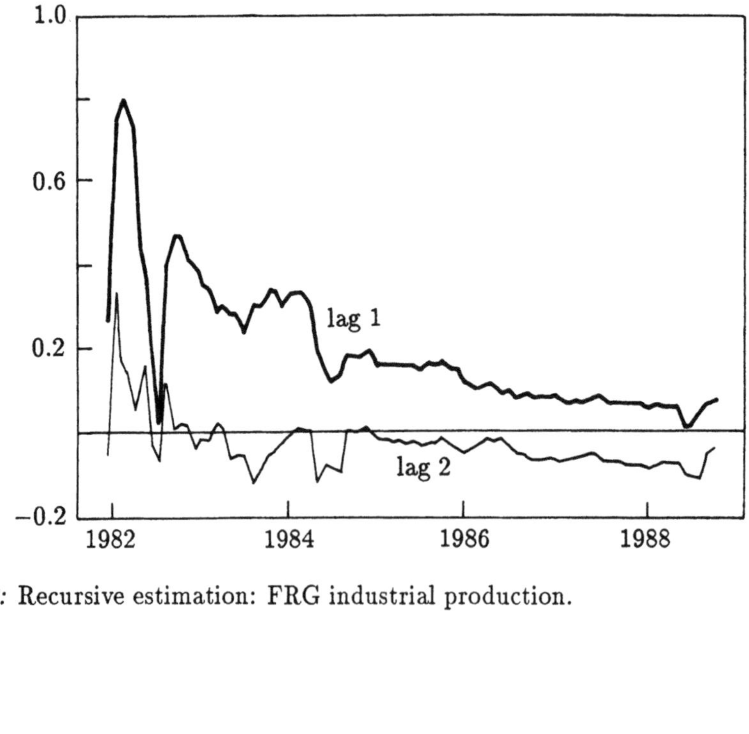


*Figure 19.10:* Recursive estimation: FRG industrial production.

*Table 19.3:* Swedish industrial production $(y_t)$ = $f$[Swedish share prices $(x_{1t})$] for 1963:10–1988:9, and subperiods thereof.

| Time period ($T$) | Diff. (order, degree) $y_t$ | $x_{1t}$ | Pure lag $x_{1t}$ | Estimated parameters Input lag $x_{1t}$ | Output lag $x_{1t}$ | Noise MA/AR | |
|---|---|---|---|---|---|---|---|
| 76:4–88:9 ($T = 150$) | (1,1) | (1,1) | 9 | 0 / .017 | – | MA(1) | .65 |
| 76:4–88:4 ($T = 145$) | Univariate (input variables are rejected) | | | | | | |
| 76:4–87:11 ($T = 140$) | Univariate (input variables are rejected) | | | | | | |
| 76:4–87:6 ($T = 135$) | Univariate (input variables are rejected) | | | | | | |
| 76:4–87:1 ($T = 130$) | (1,1) | (1,1) | 2 | 0 / .026 | 1 / –.39<br>2 / –.74 | MA(1) | .59 |
| 76:1–86:8 ($T = 125$) | (1,1) | (1,1) | 9 | 0 / .027 | – | MA(1)<br>MA(3) | .77<br>–.17 |
| 76:4–86:3 ($T = 120$) | Univariate (input variables are rejected) | | | | | | |
| 76:4–85:10 ($T = 115$) | – | (1,1) | 5 | 0 / .032 | – | MA(1)<br>MA(3) | .75<br>–.18 |
| 76:4–85:5 ($T = 110$) | – | (1,1) | 9 | 0 / .076<br>4 / –.053 | – | AR(1)<br>MA(3) | .49<br>.45 |
| 76:4–84:12 ($T = 105$) | Univariate (input variables are rejected) | | | | | | |

parameters) are often assumed to be time-varying. An iterative recursive instrumental variables estimator for transfer function estimation, originally suggested by Young [see, e.g., Young (1984)], is applied here in estimating structurally varying transfer function models describing leading indicator relations for business-cycle forecasting. A series of numerical examples clearly indicate the good capability of the estimator to discover parameter changes. At least for single-input models the estimator converges well enough toward the correct and time-varying parameter values within a few iterations, as soon as the signal-to-noise ratio is not too small. As a whole, the results are promising, but the estimator is still to be systematically evaluated. Besides further empirical studies, the only possible and meaningful strategy involves well-designed Monte Carlo studies. It is also suggested that various dynamic economic models should be re-estimated by this approach. For example, the study of lead stability through transfer function analysis by Koch and Rasche (1988) is expected to give us new insights when allowing for time-varying parameters, and after recursive parameter estimation.

## References

Athans, M. (1974), The importance of Kalman filtering methods for economic systems. *Annals of Economic and Social Measurement*, **3**, 49-64.

*Table 19.4:* Swedish Industrial Production ($y_t$) = $f$[OECD Composite Index ($x_{3t}$)] for 1963:10–1988:9, and subperiods thereof.

| Time period ($T$) | Diff. (order, degree) $y_t$ | $x_{3t}$ | Pure lag $x_{3t}$ | Estimated parameters Input lag $x_{3t}$ | Output lag $x_{3t}$ | Noise MA/AR | |
|---|---|---|---|---|---|---|---|
| 63:10–88:9 | (1,1) | (1,1) | 0 | 0 / .53 | – | AR(1) | –.66 |
| ($T = 300$) | | | | | | AR(2) | –.47 |
| | | | | | | AR(3) | –.24 |
| | | | | | | MA(24) | .19 |
| 76:4–88:9 | (1,1) | (1,1) | 0 | 0 / .43 | – | MA(1) | .77 |
| ($T = 150$) | | | | | | MA(3) | –.18 |
| 63:10–76:3 | (1,1) | (1,2) | 11 | 0 / 1.57 | – | MA(1) | .53 |
| ($T = 150$) | | | | 4 / –1.83 | | | |
| 82:7–88:9 | (1,1) | (1,1) | 10 | 0 / .63 | | MA(1) | .76 |
| ($T = 75$) | | | | | | | |
| 76:4–82:6 | | | Univariate (input variables are rejected) | | | | |
| ($T = 75$) | | | | | | | |
| 70:1–76:3 | | | Univariate (input variables are rejected) | | | | |
| ($T = 75$) | | | | | | | |
| 63:10–69:12 | (1,1) | (1,1) | 8 | 0 / –1.93 | | MA(1) | .75 |
| ($T = 75$) | | | | 1 / –2.77 | | MA(24) | .39 |

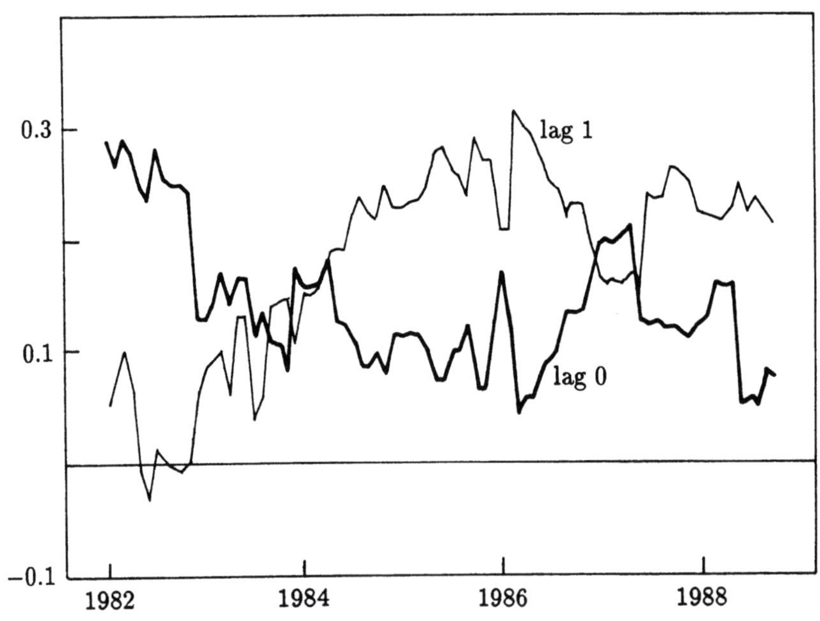


*Figure 19.11:* Recursive estimation: US industrial production.

*Table 19.5:* Swedish industrial production ($y_t$) = $f$[US industrial production ($x_{4t}$), West German industrial production ($x_{5t}$), UK industrial production ($x_{6t}$)] for 1970:1–1988:9, and subperiods thereof. Output variable always differentiated by (1,1).

| Time period ($T$) | Diff. (order, degree) $x_{4t}$ | $x_{5t}$ | $x_{6t}$ | Pure lag $x_{4t}$ | $x_{5t}$ | $x_{6t}$ | Estimated parameters Input lag $x_{4t}$ | $x_{5t}$ | $x_{6t}$ | Noise MA/AR | |
|---|---|---|---|---|---|---|---|---|---|---|---|
| 70:1–88:9 | – | (1,1)(1,1) | (1,1) | – | 1 | 9 | – | 0 / .26 | 0 / .26 | MA(1) | .57 |
| ($T$ = 225) | | | | | | | | – | 7 / –.51 | MA(3) | –.18 |
| | | | | | | | | – | 8 / .26 | MA(12) | .22 |
| 70:11–88:9 | – | (1,1)(1,1) | (1,1) | – | 1 | 9 | – | 0 / .26 | 0 / .29 | AR(12) | –.19 |
| ($T$ = 215) | | | | | | | – | – | 7 / –.49 | | |
| | | | | | | | – | – | 8 / .23 | MA(1) | .56 |
| 71:9–88:9 | – | (1,1)(1,1) | (1,1) | – | 1 | 9 | – | 0 / .26 | 0 / .27 | MA(1) | .54 |
| ($T$ = 205) | | | | | | | – | – | 7 / –.49 | MA(12) | .26 |
| | | | | | | | – | – | 8 / .25 | | |
| 72:7–88:9 ($T$ = 195) | Univariate (input variables are rejected) | | | | | | | | | | |
| 73:5–88:9 | – | (1,1)(1,1) | (1,1) | – | 1 | 0 | – | 0 / .23 | 0 / .30 | MA(1) | .56 |
| ($T$ = 185) | | | | | | | – | – | 9 / –.32 | | |
| | | | | | | | – | – | 16 / –.54 | | |
| | | | | | | | – | – | 17 / .30 | | |
| 74:3–88:9 | (1,1) | $(1,1)^2$ | (1,1) | 0 | 1 | 9 | 0 / .30 | 0 / .30 | 0 / .26 | MA(1) | .58 |
| ($T$ = 175) | | | | | | | – | 16 / .21 | 7 / –.59 | MA(12) | .28 |
| | | | | | | | – | – | 8 / .36 | AR(1) | .21 |
| 75:1–88:9 ($T$ = 165) | (1,1) | (1,1)(1,1) | (1,1) | – | 1 | 16 | – | 0 / .29 | 0 / .49 | MA(1) | .57 |
| 75:11–88:9 | (1,1) | (1,1)(1,1) | (1,1) | 0 | 1 | 16 | 0 / .44 | 0 / .26 | 0 / .64 | MA(1) | .66 |
| ($T$ = 155) | | | | | | | – | – | 1 / .49 | MA(12) | .34 |
| 76:9–88:9 | (1,1) | (1,1)(1,1) | (1,1) | 0 | 1 | 0 | 0 / .24 | 0 / .30 | 0 / .25 | MA(1) | .72 |
| ($T$ = 145) | | | | | | | – | – | 9 / –.31 | | |
| | | | | | | | – | – | 16 / –.38 | | |
| 77:7–88:9 | (1,1) | (1,1) | (1,1) | 0 | 2 | 0 | 0 / .29 | 0 / –.35 | 0 / .35 | MA(1) | .71 |
| ($T$ = 135) | | | | | | | – | – | 9 / –.36 | | |
| | | | | | | | – | – | 16 / –.49 | | |
| 78:5–88:5 | (1,2) | (1,2) | (1,2) | 0 | 1 | 9 | 0 / .67 | 0 / .36 | 0 / .43 | AR(1) | –.69 |
| ($T$ = 125) | | | | | | | | | | AR(2) | –.39 |
| | | | | | | | | | | MA(6) | –.28 |
| 79:3–88:5 ($T$ = 115) | (1,1) | (1,1) | (1,1) | 0 | 2 | 9 | 0 / .50 | 0 / –.40 | 0 / .39 | MA(1) | .80 |
| 80:1–88:9 ($T$ = 105) | (1,1) | (1,1) | – | 0 | 1 | – | 0 / .43 | 0 / .32 | – | MA(1) | .79 |
| 80:11-88:9 ($T$ = 95) | Univariate (input variables are rejected) | | | | | | | | | | |
| 81:9–88:9 | (1,1) | (1,1) | (1,1) | 0 | 2 | 0 | 0 / .60 | 0 / –.57 | 0 / .68 | MA(1) | .66 |
| ($T$ = 85) | | | | | | | – | 13 / .25 | 1 / .52 | | |
| 82:7–88:9 | (1,1) | (1,1) | (1,1) | 0 | 2 | 0 | 0 / .68 | 0 / –.61 | 0 / .74 | MA(1) | .66 |
| ($T$ = 75) | | | | | | | – | 13 / .27 | 1 / .55 | | |
| 83:5–88:9 | (1,2) | (1,1) | (1,2) | 0 | 0 | 0 | 0 / –.73 | 0 / .44 | 0 / .66 | MA(6) | –.97 |
| ($T$ = 65) | | | | | | | – | 13 / –.26 | 19 / –.48 | | |
| 84:3–88:9 ($T$ = 55) | – | – | (1,1) | 1 | 2 | – | 0 / .68 | 0 / –.34 | 0 / .74 | | |

In the case of 80:1–88:9, the output lag parameters for $x_{6t}$ were 1/ – .62. No other significant output lag parameters were observed for this or other cases.

Hackl, P. and Westlund, A.H. (1989), Statistical analysis of "Structural Change": An annotated bibliography. *Empirical Economics*, **14**, 167-192.

Hendry, D.F., Pagan, A.R., and Sargan, J.D. (1984), Dynamic specification, chapter 18 in Z. Griliches and M.D. Intriligator (eds.), *Handbook of Econometrics.* Amsterdam: North-Holland.

Jakeman, A.J., Steele, L.P., and Young, P.C. (1980), Instrumental variable algorithms for multiple input systems described by multiple transfer functions. *IEEE Transactions on Systems, Man, and Cybernetics*, **10**, 593-602.

Kalman, R.E. (1960), A new approach to linear filtering and prediction problems. *Journal of Basic Engineering*, **82**, 34-45.

Koch, P.D. and Rasche, R.H. (1988), An examination of the commerce department leading-indicator approach, *Journal of Business and Economic Statistics*, **6**, 167-187.

Laidler, D.E.W. (1975), *Essays on Money and Inflation.* Chicago: Univ. of Chicago Press.

Ledolter, J. (1981), Recursive estimation and adaptive forecasting in ARIMA models with time-varying coefficients, pp. 449-471 in D.F. Findley (ed.), *Applied Time Series Analysis.* New York: Academic Press.

Shumway, R.W. (1988), *Applied Statistical Time Series Analysis.* Englewood Cliffs: Prentice-Hall.

Young, P.C., Jakeman, A.J. and McMurtrie, R. (1984), An instrumental variable method for model order identification. *Automatica*, **16**, 281-299.

Young, P.C. (1984), *Recursive Estimation and Time Series Analysis.* Berlin: Springer-Verlag.

Zarnowitz, V. (1985), Recent work on business cycles in historical perspective: A review of theories and evidence. *Journal of Economic Literature*, **23**, 523-580.

CHAPTER 20

# Abrupt and Gradual Structural Shifts in State-Space Models

*Hiroki Tsurumi*

## Summary

A Bayesian procedure is proposed to estimate an unknownjoin point and the speed of transition from the old to new regimes in the state-space model with an autocorrelated and heteroscedastic error term. The procedure is applied to detect a stuctural shift in yen-US dollar foreign exchange rate models using monthly observations from January 1981 to September 1988. A join point around the Plaza Hotel agreement of late September of 1985 is identified.

## 20.1 Introduction

Assuming that there is one unknown join point in a sample and the change of regimes is either abrupt or gradual, Bacon and Watts (1971) introduced a class of transition functions to estimate the unknown join point and the speed of transition from the old to new regime. Tsurumi (1980) modified Bacon and Watts' parametric transition function.

In this chapter I propose a Bayesian procedure to estimate an unknown join point and speed of transition in the state-space model whose error term is autocorrelated and heteroscedastic. This Bayesian procedure is applied to detect a structural shift in yen-US dollar foreign exchange rate models.

The organization of the chapter is as follows. In Section 20.2 I derive the Bayesian procedure, and in Section 20.3 after a brief survey of empirical models of foreign exchange rates I apply the Bayesian procedure to regression models of the yen-US dollar exchange rate determination using monthly observations from January 1981 to September 1988. Concluding remarks are given in Section 20.4.

## 20.2 Bayesian Estimation of a Structural Shift in the State Space Model with an Autocorrelated and Heteroscedastic Error Term

Let the state-space model with a shift be given by

$$\begin{aligned} y_t &= x_t\beta_t + u_t && (20.1) \\ \beta_t &= \beta_{t-1} + \alpha\eta_{t^*,t} + v_t\,, && (20.2) \end{aligned}$$

where $y_t$ is the $t$-th observation on the dependent variable; $x_t$ is a $(1 \times k)$ vector of regressors; $\beta_t$ is a $(k\times 1)$ vector of time-varying regression coefficients; $\alpha$ is a $(k\times 1)$ vector of unknown jump parameters; $\eta_{t^*,t}$ is a scalar to detect a jump point. The exact value of $\eta_{t^*,t}$ is given in equation (20.3). The variable $u_t$ is the $t$-th error term with $u_t = \rho u_{t-1} + \epsilon_t$, and $E\epsilon_t\epsilon_s = \sigma^2\delta_{ts}$, where $\sigma^2$ follows

$$\sigma^2 = \begin{cases} \sigma_1^2 & \text{for } t \le t^* \\ \sigma_2^2 & \text{for } t > t^* \end{cases}$$

and $\delta_{ts}$ is the Kronecker delta with 1 for $t = s$ and 0 otherwise. The variable $v_t$ is a $(k \times 1)$ vector of disturbances with $Ev_t = 0$; $Ev_tv_s' = \sigma^2R\delta_{ts}$; $u_t$ is uncorrelated with $v_s$; $Eu_tv_s = 0$ for all $t$ and $s$. Since we have from (20.2)

$$\begin{aligned} \beta_1 &= \beta_0 + \alpha\eta_{t^*,1} + v_1 \\ \beta_2 &= \beta_0 + \alpha\eta_{t^*,1} + \alpha\eta_{t^*,2} + v_1 + v_2 \\ &\vdots \\ \beta_t &= \beta_0 + \alpha(\eta_{t^*,1} + \ldots + \eta_{t^*,t}) + v_1 + \ldots + v_t\,. \end{aligned}$$

Let

$$\eta_{t^*,1} + \ldots + \eta_{t^*,t} = \operatorname{trn}(s_t\zeta)\,, \qquad (20.3)$$

where $\operatorname{trn}(s_t\zeta)$ satisfies

$$\begin{aligned} &(i) && \lim_{s_t\to\infty} \operatorname{trn}(s_t\zeta) = 1 \\ &(ii) && \operatorname{trn}(0) = 0 \\ &(iii) && \lim_{\zeta\to\infty} \operatorname{trn}(s_t\zeta) = 1 \end{aligned}$$

and

$$s_t = \begin{cases} 0 & \text{for } t \le t^* \\ t - t^* & \text{for } t > t^*\,. \end{cases}$$

Given the initial value, $y_0$, equation (20.1) may be expressed as

$$\begin{aligned} y_t - \rho y_{t-1} &= x_t\beta_t - \rho x_{t-1}\beta_{t-1} + \epsilon_t \\ &= (x_t - \rho x_{t-1})\beta_0 + [x_t \mathrm{trn}(s_t\zeta) - \rho x_{t-1}\mathrm{trn}(s_{t-1}\zeta)]\alpha \\ &\quad + (x_t - \rho x_{t-1})v_1 + \ldots + (x_t - \rho x_{t-1})v_{t-1} + x_t v_t + \epsilon_t \end{aligned}$$

for $t = 1, \ldots, n$. Let

$$Y_1 = \begin{bmatrix} y_1 - \rho y_0 \\ \vdots \\ y_{t^*} - \rho y_{t^*-1} \end{bmatrix}, \quad Y_2 = \begin{bmatrix} y_{t^*+1} - \rho y_{t^*} \\ \vdots \\ y_n - \rho y_{n-1} \end{bmatrix}, \quad X_1 = \begin{bmatrix} x_1 - \rho x_0 \\ \vdots \\ x_{t^*} - \rho x_{t^*-1} \end{bmatrix}$$

$$X_2 = \begin{bmatrix} x_{t^*+1} - \rho x_{t^*} \\ \vdots \\ x_n - \rho x_{n-1} \end{bmatrix}, \quad \bar{X}_2 = \begin{bmatrix} \mathrm{trn}(s_{t^*+1}\zeta)x_{t^*+1} \\ \mathrm{trn}(s_{t^*+2}\zeta)x_{t^*+2} - \rho\ \mathrm{trn}(s_{t^*+1}\zeta)x_{t^*+1} \\ \vdots \\ \mathrm{trn}(s_n\zeta)x_n - \rho\ \mathrm{trn}(s_{n-1}\zeta)x_{n-1} \end{bmatrix}$$

$$\epsilon_1 = \begin{bmatrix} u_1 - \rho u_0 \\ \vdots \\ u_{t^*} - \rho u_{t^*-1} \end{bmatrix}, \quad \epsilon_2 = \begin{bmatrix} u_{t^*+1} - \rho u_{t^*} \\ \vdots \\ u_n - \rho u_{n-1} \end{bmatrix}$$

then equations (20.1) and (20.2) may be expressed as

$$\begin{aligned} Z &= X\beta_0 + V\alpha + \epsilon \\ &= W\xi + \epsilon, \end{aligned}$$

where

$$Z = \begin{pmatrix} Y_1 \\ \omega Y_2 \end{pmatrix}, \quad X = \begin{pmatrix} X_1 \\ \omega X_2 \end{pmatrix}, \quad V = \begin{pmatrix} 0 \\ \omega \bar{X}_2 \end{pmatrix}$$

$$\xi = \begin{pmatrix} \beta_0 \\ \alpha \end{pmatrix}, \quad \epsilon = \begin{pmatrix} \epsilon_1 \\ \omega\epsilon_2 \end{pmatrix} + AU, \quad \omega = \frac{\sigma_1}{\sigma_2}$$

$$A = \begin{bmatrix} x_1 & 0 & 0 & \cdots & \cdots & 0 & \cdots & \cdots & 0 \\ x_2(\rho) & x_2 & 0 & \cdots & \cdots & \cdots & \cdots & \cdots & 0 \\ \vdots & \vdots & \vdots & \vdots & \vdots & \vdots & \vdots & \vdots & \vdots \\ x_{t^*}(\rho) & x_{t^*}(\rho) & \cdots & x_{t^*}(\rho) & x_{t^*} & 0 & 0 & \cdots & 0 \\ x_{t^*+1}(\rho) & \cdots & \cdots & \cdots & \omega x_{t^*+1}(\rho) & \omega x_{t^*+1} & 0 & \cdots & 0 \\ \vdots & \vdots & \vdots & \vdots & \vdots & \vdots & \vdots & \vdots & \vdots \\ x_n(\rho) & x_n(\rho) & \cdots & \cdots & \cdots & \cdots & \cdots & \omega x_n(\rho) & \omega x_n \end{bmatrix}$$

$$U = \begin{bmatrix} v_1 \\ \vdots \\ v_n \end{bmatrix}, \quad \Omega = [I_n + A(I \otimes R)A'],$$

where $x_t(\rho) = x_t - \rho x_{t-1}$.

Let $\epsilon$ be given by the multiple student (MS) $t$-distribution

$$p(\epsilon \mid \nu_0, \sigma^2, R) = \frac{g(\nu_0)}{\sigma^n}\left[\nu_0 + \frac{\epsilon'\Omega^{-1}\epsilon}{\sigma^2}\right]^{-(n+\nu_0)/2}, \tag{20.4}$$

where $g(\nu_0)$ is a scaling constant. If we assume that the value of $\nu_0$ is given and the prior pdf of $\beta_0$, $\alpha$, $\sigma^2$, $t^*, \zeta, \omega, \rho$, and $R$ is given by

$$p(\beta_0, \alpha, \sigma^2, t^*, \zeta, \omega, \rho, R) \propto \sigma^{-2} \mid W'\Omega^{-1}W \mid^{1/2}, \tag{20.5}$$

then the posterior pdf for $t^*, \zeta, \omega, \rho$, and $R$, after integrating out $\beta_0, \alpha$, and $\sigma^2$ is given by

$$p(t^*, \zeta, \omega, \rho, R \mid \text{data}) \propto (\nu\, s^2)^{-(n-2k)/2}, \tag{20.6}$$

where $\nu\, s^2 = Z'[\Omega^{-1} - \Omega^{-1}W(W'\Omega^{-1}W)^{-1}W'\Omega^{-1}]Z$, and $\nu = n - 2k$.

**Remark 20.1** We assumed that the vector of error terms, $\epsilon$, has a MS $t$-distribution. In the regression analysis, the MS $t$-error vector was first introduced by Zellner (1976). Instead of the MS $t$-distribution we could use an elliptically symmetric distribution, $h(\epsilon) = h_1(\epsilon'\Omega^{-1}\epsilon)$. In equation (20.5), $\mid W'\Omega^{-1}W \mid^{1/2}$ is used to make sure that the prior pdf is zero for $t^* \leq 1$ or $t^* \geq n$.

**Remark 20.2** The posterior pdf in equation (20.6) is independent of $\nu_0$. We obtain equation (20.6) if we assume that $\epsilon$ is normal. This is because we used the diffuse prior pdf in equation (20.5). If we employ natural conjugate priors for $\sigma$ and $\beta_0$, the posterior pdf will contain an auxiliary parameter that needs to be integrated out. If this auxiliary parameter is equal to 1, the posterior pdf is identical to that which is derived when the error term, $\epsilon$, is normal. Consequently, we may examine the validity of assuming normally distributed error terms by testing whether this auxiliary parameter is unity or not. For more on this point, see Tsurumi (1977).

**Remark 20.3** $\text{Cov}(v_t) = R$ is assumed to be constant over time, and thus $v_t$ is stationary. One could make $v_t$ nonstationary by letting $\text{Cov}(v_t)$ change over time. However, we shall focus our attention on the nonstationary state-space process of $\beta_t$ rather than on a time varying $\text{Cov}(v_t)$ since our primary interest is in estimating $t^*$.

## 20.3 A Shift in Yen-US Dollar Exchange Rate Models

We shall apply the Bayesian methodology given in the previous section to the yen-US dollar foreign exchange rate models using monthly observations from 1981(i) to 1988(ix). The starting point, 1981(i), is chosen since in 1980(xii) a new foreign exchange law came into effect in Japan to lift various controls on foreign exchange transactions.

Empirical studies of foreign exchange rates may be classified into three groups. The first group consists of the studies that analyze time series data on foreign exchange rates either to fit distribution functions or to test random walk hypotheses. The main point of interest is to analyze time series data *per se* without looking at the relationship between a set of variables and a foreign exchange rate. Boothe and Glassman (1987), for example, fit three non-normal pdf's to daily data on the changes in logarithms of exchange rates and conclude that there is evidence that distribution parameters may vary over time. Westerfield (1977) and So (1987) also fit time series data on foreign exchange rates to some pdf's. Hakkio (1986) argues that one reason why the tests of random walk of foreign exchange rates have yielded mixed results is that the tests on a random walk have low power. Kariya and Matsue (1988) test to see whether daily and weekly foreign exchange rates follow a random walk. They use nonlinear conditional heteroscedastic variance models of Taylor (1986) and find that daily exchange rates follow a stochastic trend, and that Monday rates follow a random walk while other weekly rates follow autoregressive processes.

The second group of studies consists of regression models of foreign exchange rates. Based on economic theory a relationship between a set of regressors and a foreign exchange rate is specified and estimated. Comprehensive surveys are given in Levich (1985) and Isard (1988). Many regression models of foreign exchange rates use some form of monetary theory, and depending on how it is used they may be labeled as the *flexible-price monetary models* [Frankel (1979), Bilson (1978, 1979)], *sticky-price monetary models* [Dornbusch (1983)], and *sticky-price asset models* [Frenkel (1983), Hooper and Morton (1982)]. Kariya and Fukao (1988) develop a *rational expectation model* based on a simultaneous equations system.

The third group of studies is the combination of the first two groups: Within regression framework random walk or cointegration tests are carried out or regression models are specified in state-space form in which regression coefficients follow a random walk. Enders (1988) tested the purchasing power parity under fixed and flexible exchange rate regimes and showed that the regression error processes have unit roots. Baille and Selover (1987) estimated the flexible monetary model of Frankel (1979) and sticky-price monetary model of Dornbusch (1983) in addition to other monetary models and concluded that many of the regression coefficients are significant but the sign and magnitude are different from what one expects from theory and that the error terms tend to have unit roots. They find that the regressor possesses different orders of cointegrability, implying that the regression models do not exist as a stable long-run relationship. State space models of foreign exchange rates have been made by Wolff (1987) and Fukao (1988), among others. Wolff used the Kalman filter to improve the predictive performance of certain monetary exchange rate models. Fukao, on the other hand, used the Kalman filter to trace the changes in the regression parameters associated with interest rate differentials and trade balances. He concludes that the importance of the interest rate differential has increased while the influence of the trade balances on the foreign exchange rate has been reduced in recent years reflecting the internationalization of the major financial markets of the world.

Common findings among the three groups of studies appear to be that the parameters of the distributions of the foreign exchange rates are not stable, and they appear to change

*Table 20.1:* Various foreign exchange rate equations.

| | Flexible-price monetary model | Sticky-price monetary model | Sticky-price asset model | Risk-premium[a] model |
|---|---|---|---|---|
| $\beta_1$ | NC | NC | NC | NC |
| $\beta_2$ | NC[b] | NC[b] | NC[b] | 0 |
| $\beta_3$ | NC | NC | NC | 0 |
| $\beta_4$ | NC | NC | NC | NC |
| $\beta_5$ | 0 | NC | NC | 0 |
| $\beta_6$ | 0 | 0 | NC | NC |

NC = no constraints.

[a] The risk-premium model is given in Fukao (1988).

[b] Flexible-price monetary model, sticky-price monetary model, and sticky-price asset model all assume that the exchange rate is homogeneous of degree 1 in relative money supply ($\beta_2$=1), but in this study this assumption is relaxed.

over time. In short, the foreign exchange rates exhibit characteristics of nonstationarity. One way of modeling a nonstationary process is to use, after confirming unit roots, cointegrated and error-corrected models or to specify a regression model with changes in means (or regression coefficients) over time or to combine both approaches in a regression model as done in this chapter.

I shall use single equation models of foreign exchange determination with state-space regression coefficients and with autocorrelated and heteroscedastic error. Equation (20.1) becomes

$$\begin{aligned} s_t &= \beta_1 + \beta_2(m_t - m_t^*) + \beta_3(y_t - y_t^*) + \beta_4(r_{s,t} - r_{s,t}^*) \\ &\quad + \beta_5(\dot{p}_t - \dot{p}_t^*) + \beta_6 t_b + u_t , \end{aligned} \tag{20.7}$$

where $s_t$ is the logarithm of the domestic currency price of foreign exchange currency (spot exchange rate); $m - m^*$ is the logarithm of the ratio of domestic to foreign money; $r_s - r_s^*$ is the short-term interest rate differential; $\dot{p}_t - \dot{p}_t^*$ is the differential of the expected inflation rates, and $t_b$ is the domestic trade balance. *Table 20.1* presents the constraints on parameters to obtain various exchange rate regression equations.

Using data from 1981(i) to 1988(ix), we estimated the join point $t^*$, the speed of transition $\zeta$, autocorrelation coefficient $\rho$, and the ratio of standard deviations $\omega$, for each exchange rate model. (Data are explained in the Appendix.) *Table 20.2* presents the results for each exchange rate model given in *Table 20.1* and for two values of $Ev_tv_t'$. We used the hyperbolic tangent, $\tanh(s_t\zeta)$, as the transition function.

From *Table 20.2* we may obtain the following observations:

(1) Regardless of the models and of assumed values for $Ev_tv_t'$, the point estimates given by the posterior means and posterior modes for $t^*$, $\zeta$, $\omega$, and $\rho$ are quite similar.

(2) The join point, $t^*$, is estimated at 1985(v); the speed of transition, $\zeta$, is about .03;

*Table 20.2:* Estimation of join points for foreign exchange rate equations.

| Flexible-price monetary model | | | | | | |
|---|---|---|---|---|---|---|
| | $R = \hat{\Sigma}$ | | | $R = 0$ | | |
| | P-Mode | P-Mean | PSD | P-Mode | P-Mean | PSD |
| $t^*$ | 1985(v) | 1985(v) | 1.67 | 1985(v) | 1985(v) | 1.68 |
| $\zeta$ | .030 | .035 | .017 | .050 | .045 | .019 |
| $\omega$ | 1.15 | 1.18 | .13 | 1.14 | 1.15 | .087 |
| $\rho$ | .838 | .803 | .064 | .840 | .808 | .060 |
| | AIC=−1060.33 | | | AIC=−997.45 | | |
| Sticky-price monetary model | | | | | | |
| | $R = \hat{\Sigma}$ | | | $R = 0$ | | |
| | P-Mode | P-Mean | PSD | P-Mode | P-Mean | PSD |
| $t^*$ | 1985(v) | 1985(v) | 1.72 | 1985(v) | 1985(v) | 1.72 |
| $\zeta$ | .026 | .035 | .017 | .025 | .035 | .017 |
| $\omega$ | 1.15 | 1.19 | .14 | 1.15 | 1.18 | .13 |
| $\rho$ | .838 | .805 | .060 | .840 | .805 | .063 |
| | AIC=−1049.83 | | | AIC=−1002.87 | | |
| Sticky-price asset model | | | | | | |
| | $R = \hat{\Sigma}$ | | | $R = 0$ | | |
| | P-Mode | P-Mean | PSD | P-Mode | P-Mean | PSD |
| $t^*$ | 1985(v) | 1985(v) | 1.63 | 1985(v) | 1985(v) | 1.62 |
| $\zeta$ | .025 | .039 | .027 | .025 | .039 | .027 |
| $\omega$ | 1.15 | 1.24 | .20 | 1.16 | 1.24 | .20 |
| $\rho$ | .725 | .756 | .088 | .725 | .756 | .088 |
| | AIC=−1120.69 | | | AIC=−1111.29 | | |
| Risk-premium model | | | | | | |
| | $R = \hat{\Sigma}$ | | | $R = 0$ | | |
| | P-Mode | P-Mean | PSD | P-Mode | P-Mean | PSD |
| $t^*$ | 1985(v) | 1985(vi) | 1.39 | 1985(v) | 1985(vi) | 1.39 |
| $\zeta$ | .025 | .039 | .025 | .025 | .039 | .025 |
| $\omega$ | 1.15 | 1.16 | .089 | 1.15 | 1.16 | .088 |
| $\rho$ | .838 | .801 | .064 | .838 | .801 | .064 |
| | AIC=−1113.10 | | | AIC=−1114.54 | | |

$\hat{t}^*$ = join point; $\zeta$ = speed of transition; $\omega = \sigma_1/\sigma_2$; $\rho$ = autocorrelation coefficient; P-Mode = posterior mode; P-Mean = posterior mean; and PSD = posterior standard deviation.
The variance-covariance matrix $Ev_tv_t'$ is specified by $\hat{\Sigma}$ and 0, where $\hat{\Sigma}$ is the estimate of the variance-covariance matrix of the OLS regression parameters assuming no jumps for the entire sample period, 1981(i)-1988(ix).
AIC is the Akaike information criterion: $n \log \hat{\sigma}^2(R) + 2p$ where $p$ is the number of regressors.

*Table 20.3:* Values of the transition function with $\zeta = 0.03$, $\tanh(0.3s_t)$.

| month | $\tanh(0.3s_t)$ | month | $\tanh(0.3s_t)$ | month | $\tanh(0.3s_t)$ |
|---|---|---|---|---|---|
| 1985(v) | 0 | 1986(vii) | .336 | 1987(ix) | .605 |
| 1985(vi) | .025 | 1986(viii) | .358 | 1987(x) | .620 |
| 1985(vii) | .050 | 1986(ix) | .380 | 1987(xi) | .635 |
| 1985(viii) | .075 | 1986(x) | .401 | 1987(xii) | .650 |
| 1985(ix) | .100 | 1986(xi) | .422 | 1988(i) | .664 |
| 1985(x) | .124 | 1986(xii) | .442 | 1988(ii) | .678 |
| 1985(xi) | .149 | 1987(i) | .462 | 1988(iii) | .691 |
| 1985(xii) | .173 | 1987(ii) | .482 | 1988(iv) | .704 |
| 1986(i) | .197 | 1987(iii) | .501 | 1988(v) | .716 |
| 1986(ii) | .221 | 1987(iv) | .519 | 1988(vi) | .728 |
| 1986(iii) | .245 | 1987(v) | .537 | 1988(vii) | .740 |
| 1986(iv) | .268 | 1987(vi) | .555 | 1988(viii) | .751 |
| 1986(v) | .291 | 1987(vii) | .572 | 1988(ix) | .761 |
| 1986(vi) | .314 | 1987(viii) | .588 | | |

the ratio of standard deviations, $\omega = \sigma_1/\sigma_2$, is about 1.2; and the autocorrelation coefficient is between .73 and .84.

Although in *Table 20.2* we only report two cases of $Ev_tv_t'$, we have varied the values of $Ev_tv_t' = \gamma\hat{\Sigma}$ by setting $\gamma$ at .7, .4, and .1. The results are virtually the same regardless of the values of $\gamma$.

Among the four models, which one should one choose? If we use the AIC as a model selection criterion, we may choose the sticky-price asset model with $Ev_tv_t' = \hat{\Sigma}$ on the grounds that the minimum AIC is obtained for this model. We use the AIC as the model selection criterion, since it is easy to use, and, compared with other model selection criteria such as Cox's tests of separate families, the AIC tends to have a better power of choosing the correct model. [See Tsurumi and Wago (1988).]

Regardless of model specification, it appears that the yen-US dollar exchange rate experienced a structural shift at around 1985(v). The speed of transition, $\zeta = 0.03$, indicates a very slow adjustment to the new regime. The transition function is given by the hyperbolic tangent. The values of the transition function are presented in *Table 20.3* for the period of 1985(v) to 1988(ix). Although the regime change started from 1985(vi), at least for a few months the transition is virtually negligible; thus for practical purposes, we may say that the transition to the new regime started in the last quarter of 1985, and the transition to the new regime is only 76.1% complete in 1988(ix), the end of the sample period.

The transition point around the last quarter of 1985 may be explained by the Plaza Hotel agreement of 1985(ix). Since the agreement, the five industrialized countries (the USA, France, West Germany, the UK, and Japan) have intervened in the foreign exchange markets with coordinated purchases and sales of dollars and other currencies, and they

*Table 20.4:* Sticky-price asset model with $Ev_tv_t' = \hat{\Sigma}$: Estimated coefficients.

| | Estimate of $\beta_0$ | Estimate of $\alpha$ | $\beta_0 + \alpha$ |
|---|---|---|---|
| Intercept | 5.197 | −.808 | 4.389 |
| | (7.60) | (.57) | |
| $m - m^*$ | .045 | −.009 | .036 |
| | (.34) | (.03) | |
| $y - y^*$ | −.041 | .097 | .056 |
| | (.41) | (.51) | |
| $r_s - r_s^*$ | .004 | −.054 | −.050 |
| | (.77) | (2.08) | |
| $\dot{p} - \dot{p}^*$ | −7.514 | 9.698 | 2.184 |
| | (1.49) | (1.01) | |
| $t_b$ | .001 | −.001 | 0 |
| | (1.90) | (1.80) | |

Figures in parentheses just below the estimated coefficients are the $t$-ratios.
The regression coefficients are the posterior means conditioned on the posterior means of $t^*$, $\zeta$, $\omega$, and $\rho$. The $t$-ratios are the posterior means divided by posterior standard deviations.

have backed up their actions with economic cooperation. The fact that the ratio of standard deviations, $\omega = \sigma_1/\sigma_2$, is 1.2 means that the standard deviation of the regression term is smaller in the second regime, perhaps indicating that the economic cooperation among the industrialized countries is responsible for this reduction in the variance.

The direction of jumps of regression parameters is given in *Table 20.4*. The first set of regression parameters is the estimates of the random parameters $\beta_0$. The second set is the estimate of $\alpha$, and the third is $\beta_0 + \alpha$. Examining the values of $\beta_0 + \alpha$ one sees that these results support Fukao's (1988) findings that the importance of the interest rate differential has increased while the influence of the trade balances has decreased in recent years.

What can one say about a unit root in the regression model in (20.7)? In the Bayesian analysis we may test the null hypothesis of whether the error term, $u_t$, which follows an autoregressive error of order 1, AR(1), has a unit root or not (i.e., the null hypothesis of H: $\rho = 1$) by checking whether a highest posterior density interval (HPDI) contains $\rho = 1$ or not. This Bayesian test has an advantage over the sampling tests of a unit root such as those proposed by Dickey and Fuller (1979) and by Phillips and Perron (1988) at least on two accounts: (i) the Bayesian HPDI test is a test with a finite sample size rather than an asymptotic test and (ii) the HPDI test can allow for inclusion of regressors other than a time trend. As is well known the asymptotic sampling distributions of unit roots are sensitive to model specification [see, for example, Dickey and Fuller (1979)]. Sampling experiments in Wago and Tsurumi (1990) show that the HPDI test is robust, and it has in general more power than the Dickey-Fuller or Phillips-Perron tests.

*Table 20.5* presents 95% HPDI's for the sticky-price asset model for the two cases:

*Table 20.5:* 95% Highest posterior density intervals (HPDI) for $\rho$: Sticky-price asset model.

| | Regression model with gradual switching | without switching |
|---|---|---|
| $R = \widehat{\Sigma}$ | (0.562, 0.950) | (0.810, 1.02) |
| $R = 0$ | (0.561, 0.948) | (0.811, 1.03) |

The 95% HPDI's are computed from the marginal posterior pdf's of $\rho$ given $R$.
Model with gradual switching has $t^*$, $\zeta$, and $\omega$. The model without switching is equation (20.7) without the transition function and with $\omega = 1$.
Sample period is 1981(i)–1988(ix).

(i) the gradual switching regression model with $t^*$, $\zeta$, and $\omega$ and (ii) the regression model with no switching and heteroscedasticity (i.e., $\omega = 1$). The values of $R$ are set at $\widehat{\Sigma}$ and 0. The results show that the regression model without switching and heteroscedasticity exhibits the possibility of a unit root since the 95% HPDI's for $R = \widehat{\Sigma}$ and for $R = 0$ both include $\rho = 1$, whereas the gradual switching regression with heteroscedasticity leads one to say that the error process is stationary, since the 95% HPDI's are between .56 and .95. This result indicates that if one neglects a structural change and heteroscedasticity, then one tends to accept the unit-root hypothesis.

## 20.4 Concluding Remarks

In this chapter I proposed a Bayesian procedure to detect join points with an abrupt or gradual shift in the state-space model when the regression error term is autocorrelated and heteroscedastic, and I applied the procedure to single equation models of the yen-US dollar exchange rate determination using monthly data from 1981(i) to 1988(ix). The estimates of a join point, speed of transition, the ratio of standard deviations, and autocorrelation coefficient are insensitive to model specification. I identified a join point at around 1985(vi). This shift may be due to the Plaza Hotel agreement of late September of 1985.

In this chapter, I assumed that the number of join points in a sample is known. If it is not known, one needs to make inferences on the number and timing of shifts. Tsurumi (1989) proposes a Bayesian sequential test procedure when one does not know the number and timing of shifts. In the control literature, shifts in state-space models are often tested by the generalized likelihood ratio test (GLRT) that was proposed by Willsky and Jones (1974, 1976), but the Bayesian sequential procedure tends to be more robust to the specification of the covariance matrix for the state-space equation error terms. A critique of the GLRT is given in Kerr (1983) and Isermann (1984).

## References


Bacon, D.W. and Watts, D.G. (1971), Estimating the transition between two intersecting lines. *Biometrica*, **58**, 525-534.

Baille, R.T. and Selover, D.D. (1987), Cointegration and model of exchange rate determination. *International Journal of Forecasting*, **3**, 43-51.

Bilson, J.F.O. (1978), Rational expectations and the exchange rates, pp. 75-96 in J. Frankel and H. Johnson (eds.), *The Economics of Exchange Rates.* Reading: Addison-Wesley.

Bilson, J.F.O. (1979), The Deutsche mark/dollar rate: A monetary analysis, pp. 59-101 in K. Brunner and A.H. Meltzer (eds.), *Policies for Employment, Prices and Exchange Rates.* Amsterdam: North-Holland.

Boothe, P. and Glassman, D. (1987), The statistical distribution of exchange rates. *Journal of International Economics*, **22**, 297-319.

Dickey, D.A. and Fuller, W.A. (1979), Distribution of the estimators for autoregressive time series with a unit root. *Journal of the American Statistical Association*, **74**, 427-431.

Dornbusch, R. (1983), Flexible exchange rates and interdependence. *IMF Staff Papers*, **30**, 3-30.

Enders, W. (1988), ARIMA and cointegration tests of PPP under fixed and flexible exchange rate regimes. *Review of Economics and Statistics*, **55**, 251-276.

Frankel, J.A. (1979), On the mark: A theory of floating exchange rates based on real interest differentials. *American Economic Review*, **69**, 610-622.

Frenkel, J.A. (1983), Monetary and portfolio-balance models of exchange rate determination, pp. 3-27 in J.S. Bhandari and B.H. Putnam (eds.), *Economic Interdependence and Flexible Exchange Rates.* Cambridge: MIT Press.

Fukao, M. (1988), Kinyu no kokusaika ga kawase rate no hendo oyobi kokusai shushi ni ataeru eikyo ni tsuite (Influences of liberalization of financial markets on the foreign exchange rates and on balances of payments). *Kinyu Kenkyu*, **7**, 1-42.

Hakkio, C.S. (1986), Does the exchange rate follow a random walk? A Monte Carlo study of four tests for a random walk. *Journal of International Money and Finance*, **5**, 221-229.

Hooper, P. and Morton, J. (1982), Fluctuations in the dollar: A model of nominal and real exchange rate determination. *Journal of International Money and Finance*, **1**, 39-56.

Isard, P. (1988), Exchange rate modeling: An assessment of alternative approaches, pp. 183-201 in R.C. Bryant, D.W. Henderson, G. Holthar, P. Hooper, and S.A. Symansky (eds.), *Empirical Macroeconomics for Interdependent Economies. Volume 1.* Washington, D.C.: Brookings Institution.

Isermann, R. (1984), Process fault detection based on modeling and estimation: A survey. *Automatica*, **20**, 387-404.

Kariya, T. and Fukao, K. (1988), *Goriteki yosokeisei ni yoru inflation-kawase bunseki* (Rational expectation analysis of foreign exchange rates). Tokyo: Yuhikaku.

Kariya, T. and Matsue, Y. (1988), Hisenkei bunsan hendo model ni yoru nichiji shuji kawase rate hendo bunseki (Analysis of daily and weekly foreign exchange rates within the framework of nonlinear heteroscedastic variance models). Mimeograph.

Kerr, T.H. (1983), The controversy over use of SPRT and GLR techniques and other loose ends in failure detection, pp. 966-977 in *Proceedings of the American Control Conference.*

Levich, R.M. (1985), Empirical studies of foreign exchange rates: Price behavior, rate

determination and market efficiency, pp. 979-1040 in R.W. Jones and P.B. Kenen (eds.), *Handbook of International Economics II.* Amsterdam: North-Holland.
Phillips, P.C.B. and Perron, P. (1988), Testing for a unit root in time series regression. *Biometrika*, **75**, 335-346.
So, J.C. (1987), The sub-Gaussian distribution of currency futures: Stable Paretian or nonstationary? *Review of Economics and Statistics*, **69**, 100-107.
Taylor, S.J. (1986), *Modeling Financial Time Series.* New York: John Wiley.
Tsurumi, H. (1977), A Bayesian test of a parameter shift with an application. *Journal of Econometrics*, **6**, 371-380.
Tsurumi, H. (1980), Bayesian estimation of structural shifts by gradual switching regressions with an application to the US gasoline market, pp. 212-240 in A. Zellner (ed.), *Bayesian Analysis in Econometrics and Statistics: Essays in Honor of Harold Jeffreys.* Amsterdam: North-Holland.
Tsurumi, H. (1990), Non-Gaussian fault detection and intervention analysis: A comparison of Bayesian with generalized likelihood ratio procedures. Paper in preparation.
Tsurumi, H. and Wago, H. (1988), Mean squared errors of forecast for selecting non-nested linear models and comparison with other criteria. Forthcoming in *Journal of Econometrics.*
Wago, H. and Tsurumi, H.(1990), Bayesian tests of unit roots and stationarity. Mimeograph.
Westerfield, J.M. (1977), An examination of foreign exchange risk under fixed and floating rate regimes. *Journal of International Economics*, **7**, 181-200.
Willsky, A.S. and Jones, H.L. (1974), A generalized likelihood ratio approach to estimation in linear systems subject to abrupt changes, pp. 34-47 in *Proceedings of 1974 IEEE Decision and Control.*
Willsky, A.S. and Jones, H.L. (1976), A generalized likelihood ratio approach to the detection and estimation of jumps in linear systems. *IEEE Transactions on Automation and Control*, **AC-21**, 108-112.
Wolff, C.C.P. (1987), Time-varying parameters and the out-of-sample forecasting performance of structural exchange models. *Journal of Business and Economic Statistics*, **5**, 87-97.
Zellner, A. (1976), Bayesian and non-Bayesian analysis of the regression model with multivariate student-$t$ error terms. *Journal of the American Statistical Association*, **71**, 400-405.

# Appendix: Data Sources

$s_t$ is the logarithm of the yen price of the US dollar (spot rate), taken from *International Financial Statistics* (IFS).

$m - m^*$ is the logarithm of the ratio of the Japanese money supply to the US money supply. Money supplies are seasonally unadjusted *M1* figures, taken from *Main Economic Indicators* (MEI), published by the Organisation for Economic Co-operation and Development for the US, and from IFS for Japan.

$y - y^*$ is the logarithm of the ratio of Japanese to US real income. Seasonally unadjusted industrial production indexes are used for real income, taken from the MEI.

$r_s - r_s^*$ is the short-term interest rate differential: $r_s$ is the call rate and $r_s^*$ is the US treasury bill rate, both taken from IFS.

$p - p^*$ is the expected inflation rate differential; 12-month moving averages of consumer price indexes over the preceding 12-month period are used as proxies for expected price levels. Consumer price indexes are taken from IFS.

$t_b$ are the accumulated Japanese current accounts: $t_{b,t} = \sum_{s=1972(xii)}^{t-1} c_s$, where $c_s$ is the current accounts at time $s$ deflated by the consumer price index at time $s$.

CHAPTER 21

# Structural Changes in the Real GNP Interdependence of the USA, West Germany, and Japan: 1970–1986

*Masanao Aoki*

## Summary

This chapter first locates quarters in the early 1970s at which the cUovariance matrices of the innovation vectors have shifted for the real GNPs of the USA, West Germany, and Japan treated as univariate series. The chapter then exhibits differences in the impulse response time profiles of the two models estimated from the data primarily before and after the break as a concise summary of the changes in dynamic interactions of the three real GNPs.

## 21.1 Introduction

This chapter uses the state space modeling method for "trending" time series, i.e., time series with roots close to one, to locate possible breaks in the real GNPs and money stock series of the USA, West Germany, and Japan by the likelihood ratio test. The real GNP series are first treated separately as univariate series to locate a likely quarter at which changes in some characteristics of the data-generating process have occurred.

The chapter then jointly treats the three real GNP series as a trivariate series. The focus in this part of the chapter is to examine the differences in the structure of dynamic interdependence by the time profiles of impulse responses (dynamic multipliers), rather than pinpointing the quarter of structural shifts.

Several methods have been proposed in literature to detect sudden (or gradual) changes in parameters of data-generating processes [Andrews and Fair (1988), Goldfeld and Quandt (1976), Lo and Newey (1985) in the econometric literature and Willsky and Jones (1976) and Basseville *et al.* (1987) in the systems literature]. We are interested in detecting structural changes in vector-valued macroeconomic time series, such as money stocks and real GNP. Most of the methods are for univariate series although some extension for vector-valued series are available.

In the context of state space innovation modeling of time series, the innovation vectors, $e_t = y_t - y_{t|t-1}$, where $y_{t|t-1}$ is the orthogonal projection of $y_t$ onto the manifold spanned by its own past data, are modeled as approximately normally distributed with mean zero and sample covariance matrix $\Delta$. The joint probability distribution of $y_1, \ldots, y_T$ has only $\Delta$ as the parameters when the innovation representation is used.

A parameter shift in the data-generating process manifests itself then as changes in the covariance matrix $\Delta$ of the innovation vector. [We need not adopt ad hoc assumptions on the breaks of "slope" or intercept points of the time series.] Given that a single shift in the covariance matrix has occurred in a sample period, we can adopt the method of Goldfeld and Quandt (1976) to locate the time instant which is the most likely instant of the parameter shift by maximizing the joint likelihood function over the sample period.

Suppose that $t_c$ is the instant of the parameter shift so that $\mathrm{Cov}\{e_t\} = \Delta_1$ for $t \leq t_c$ but $\mathrm{Cov}\{e_t\} = \Delta_2$ for $t > t_c$.

The joint likelihood function is

$$L(y_1, \ldots, y_T | t_c) = \text{const } |\Delta_1|^{-t_c/2} |\Delta_2|^{(T-t_c)/2} \exp[-1/2 \operatorname{tr}(\Delta_1 S_1 + \Delta_2 S_2)] \, ,$$

where

$$S_1 = \sum_{t=1}^{t_c} e_t e_t'$$

and

$$S_2 = \sum_{t=t_c+1}^{T} e_t e_t' \, .$$

The regime shift is identified with the $t_c$, which maximizes the joint likelihood function.

## 21.2 Univariate Series

Episodes in the late 1960s and early 1970s such as the demise of the Brettonwood accord and the oil shocks tell us that a regime shift is likely during a period spanning from the late 1960s to the early 1970s. The procedure outlined above is applied to the quarterly US money stock data from the first quarter 1947 [1947(I)] to 1982(II). The total of 141 data points is split into two periods, and separate modes are estimated for each subperiod to calculate the joint likelihood function. This data produces 1970(IV) as the most likely quarter in which the US monetary regime has shifted. The method is then applied to the US real GNP series. It produces 1971(I) as the most likely shift point for the quarterly US real GNP series based on 152 data points from 1949(I). The shift in the real GNP series occurred one quarter later than that for the *M*2 series.

Since both money stock and real GNP series are apparently trending with (near) unit roots, we apply the two-step modeling procedure outlined in Aoki (1989, 1990a, 1990b) to separately model the largest eigenvalue, which is near one (slowest dynamic model). For example, for the US real GNP series, the model before the break is

$$y_t = 0.482\,\tau_t + (0.779, 0.144)z_t + e_t\,,$$

where

$$\begin{bmatrix} \tau_{t+1} \\ z_{t+1} \end{bmatrix} = \begin{bmatrix} 0.984 & 1.606 & 0.296 \\ 0 & 0.429 & -0.569 \\ 0 & 0.569 & 0.484 \end{bmatrix} \begin{bmatrix} \tau_t \\ z_t \end{bmatrix} + \begin{bmatrix} 2.063 \\ -0.585 \\ 0.402 \end{bmatrix} e_t$$

with $\mathrm{Cov}\{e_t\} = 0.931 \times 10^{-4}$. The model after the break is

$$y_t = 0.238\,\tau_t + (0.598, 0.021)z_t + e_t\,,$$

where

$$\begin{bmatrix} \tau_{t+1} \\ z_{t+1} \end{bmatrix} = \begin{bmatrix} 0.974 & 2.512 & 0.087 \\ 0 & 0.612 & -0.337 \\ 0 & 0.337 & 0.956 \end{bmatrix} \begin{bmatrix} \tau_t \\ z_t \end{bmatrix} + \begin{bmatrix} 4.200 \\ -0.402 \\ 0.064 \end{bmatrix} e_t$$

with $\mathrm{Cov}\{e_t\} = 0.114 \times 10^{-3}$. The statistic $T(\hat{\rho}-1)$ is about $-1.5$ for the largest eigenvalue, and the largest eigenvalue may or may not equal 1. We treat both series not as random walks, but rather as nearly integrated series as in Chan (1988).

When we examine the real GNP and money stock series of Japan and West Germany on the assumption that there is a simple break, the method of Goldfeld and Quandt applied to the estimated state space innovation models place the breaks during 1971 well before the episode of the first oil shock.

## 21.3 Three Real GNPs

Our preliminary analysis of the three real GNP series of the USA, West Germany, and Japan as univariate series indicates that they are likely to have individually experienced shifts in the paramerters of the data-generating process somewhere in 1970-1971. This section treats them as a trivariate series. It would be useful to break the data at 1972(I).

Since the data set at our disposal covers the period from 1965(II) to 1985(IV), a total of 83 data points, this choice would leave only about 30 data points for the initial period. To increase the data points of the first subperiod to about 40 to balance the magnitudes of statistical errors in the two models for the two subperiods, we break the data at 1974 and estimate two models for the two subperiods 1965(II)-1974(II) and 1974(I)-1985(IV). We are now more interested in learning what differences if any are exhibited by the two models rather than locating the break-point exactly. To this end, we evaluate the differences in the time paths of the impulse responses, i.e., dynamic multiplier profiles implied by the two models. The greater the discrepancies measured, the more significant are the consequences of the structural shift.

To calculate impulse responses we need to identify the matrix $D$, which relates the time series innnovation vector, $e_t$, with the shocks, $n_t$, in the structural model

$$e_t = Dn_t\,.$$

[See Bernanke (1986) or Sims (1986) for several ways for identification.] As explained in these references, the matrix $D$ may be thought of as $\phi_0^{-1}\theta_0$ in the structural model $\phi(L)y_t = \theta(L)n_t$, where $y_t$ is the three-dimensional vector composed of the real GNP of the USA, West Germany, and Japan in that order, and $\phi_0$ and $\theta_0$ are the $3 \times 3$ constant matrices in the lag polynomination matrices $\phi(L)$ and $\theta(L)$, respectively. The covariance matrix $\mathrm{Cov}\{n_t\}$ is normalized to be $I_3 \times 10^{-4}$. Thus

$$\mathrm{Cov}\{e_t\} = \Delta = DD' \times 10^{-4}\,. \tag{21.1}$$

There are nine elements in $D$, and there are six elements in $\Delta$ that are estimated. Thus, we need three additional conditions to uniquely specify the matrix $D$. Once the matrix $D$ is specified, the multiplier profiles are given by $\theta\Phi^k\Psi D$, $k = 0, 1, \ldots$, where the matrices appear in the innovation model

$$\begin{aligned} y_t &= \theta\chi_t + e_t \\ \chi_{t+1} &= \Phi\chi_t + \Psi e_t\,, \end{aligned}$$

which are estimated as in Aoki (1989) by the two-step procedure series since $y_t$ is "trending".

One way to identify the matrix $D$ is to use the Wold causal chain structure, i.e., to use the Cholesky decomposition. Instead, we use the decomposition into a common shock and uncorrelated country specific shocks, i.e., we model $e_t$ by

$$e_t = \mu\nu_t + \begin{bmatrix} d_1 n_U \\ d_2 n_G \\ d_3 n_J \end{bmatrix}_t,$$

where $\nu$, $n_U$, $n_G$, and $n_J$ all have variance $1 \times 10^{-4}$ and are mutually uncorrelated, i.e.,

$$\Delta = \left[\mu\mu' + \mathrm{diag}(d_1^2, d_2^2, d_3^2)\right] \times 10^{-4}\,.$$

The elements of the vector $\mu$ indicate how the common shock impinges on the three countries. The set of six algebraic relations in (21.1) can be solved for the three components of $\mu$ and $d_i$, $i = 1, 2, 3$, uniquely in general. [A related work by Gerlach and Klock (1988) covers a period roughly comparable with the second subperiod. They jointly model six real GNPs with VAR. Our model is equivalent to vector ARMA.]

Then the multiplier profiles $\theta\Phi^k\Psi\mu$, $k = 0, 1, \ldots$, show how the three real GNP respond to a common shock and the profiles $\theta\Phi^k d_i\psi_i$, $i = 0, 1, 2, 3$, where $\psi_i$ is the $i$-th column vector of the matrix $\Psi$, show how the three real GNP respond to a shock originating in one country only. Note that $d_1$, $d_2$, and $d_3$ affect the relative magnitudes but not the shapes nor the timings of peaks and troughs in the multiplier time profiles, if any.

The parameters of the estimated models are as follows:

*The first period model*

$$\theta = \begin{bmatrix} 0.182 & 0.084 & -0.065 \\ 0.207 & -0.280 & -0.105 \\ 0.502 & -0.120 & -0.106 \end{bmatrix}, \quad \Psi = \begin{bmatrix} 1.898 & -1.038 & 1.372 \\ 1.038 & -2.279 & 0.884 \\ -4.376 & 0.427 & 1.381 \end{bmatrix}$$

$$\Phi = \begin{bmatrix} 0.971 & 0.000 & 0.010 \\ 0.000 & 0.941 & -0.132 \\ 0.000 & 0.147 & 0.869 \end{bmatrix}, \quad \mu = \begin{bmatrix} 0.267 \\ 0.995 \\ 0.483 \end{bmatrix}$$

$$d_1 = 1.078, \quad d_2 = 0.660, \quad d_3 = 1.093.$$

*The second period model*

$$\theta = \begin{bmatrix} 0.148 & 0.212 & -0.051 \\ 0.120 & 0.157 & 0.070 \\ 0.282 & -0.029 & -0.061 \end{bmatrix}, \quad \Psi = \begin{bmatrix} -0.233 & 1.413 & 3.017 \\ 5.266 & 1.142 & -1.155 \\ -6.145 & 8.435 & -3.181 \end{bmatrix}$$

$$\Phi = \begin{bmatrix} 0.966 & 0.009 & -0.007 \\ 0.000 & 0.805 & -0.052 \\ 0.000 & 0.213 & 0.783 \end{bmatrix}, \quad \mu = \begin{bmatrix} 0.676 \\ 0.671 \\ 0.356 \end{bmatrix}$$

$$d_1 = 0.655, \quad d_2 = 0.504, \quad d_3 = 0.604.$$

(When the whole period is modeled jointly, then $\mu = (0.485, 0.891, 0.522)'$ and $d_1 = 0.896$, $d_2 = 0.894$, and $d_3 = 0.904$.)

Even before we examine the multiplier profiles, some differences are clearly evident. The eigenvalues of $\Phi$ of the second model is slower, for example. The common shock affects West Germany more than the other two in each period, but less so in the second period. The country-specific shocks of the USA and Japan are smaller in the second period. The USA economy is more exposed to a common shock in the latter period, while the Japanese exposure remains about the same.

The differences in the dynamic interactions in the two periods are clearly visible from the multiplier in *Figures 21.1–21.8*. *Figures 21.1* and *21.2* show responses to a common shock in the first and second period. The model dimension is three in both periods, i.e., $\dim \tau_t = 1$, $\dim z_t = 2$ where $\tau_t$ is the state variable with the slowest decay, and the vector

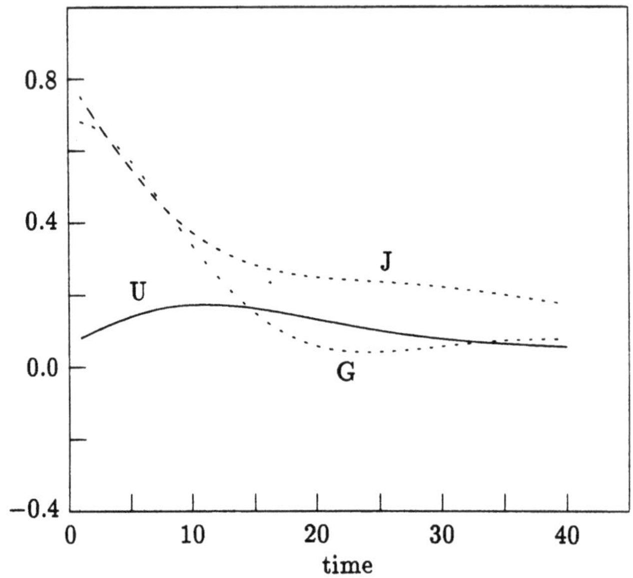


*Figure 21.1:* Response to a common shock of the first model from 1965(II).

$z_t$ is for the shorter-run dynamics. *Figures 21.3–21.5* are for the first period and *Figures 21.6–21.8* are for the second. The solid lines are for the US responses. The US shocks affect West Germany with opposite signs in the two periods. Japan is less affected by the German shock in the second period, while the opposite is true for the USA. The Japanese shock affects the USA and West Germany less in the second period than in the first.

## 21.4 Concluding Remarks

A state space modeling method for apparently nonstationary time series and the resulting impulse responses are used to portray concisely the qualitative differences in the interaction characteristics of the three real GNPs that apparently took place in the early 1970s. See Aoki (1990b, chapter 7) for results with a different identification scheme.

### Acknowledgments

The author gratefully acknowledges the comments by C.A. Sims, which were instrumental in shaping the chapter. This material is based on work supported by the National Science Foundation under Grant No. SES-8733451.

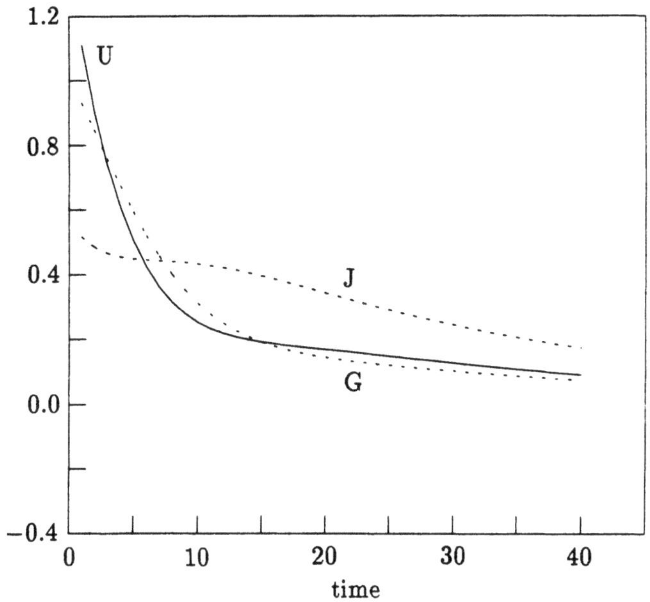


*Figure 21.2:* Response to a common shock of the second model from 1974(I).

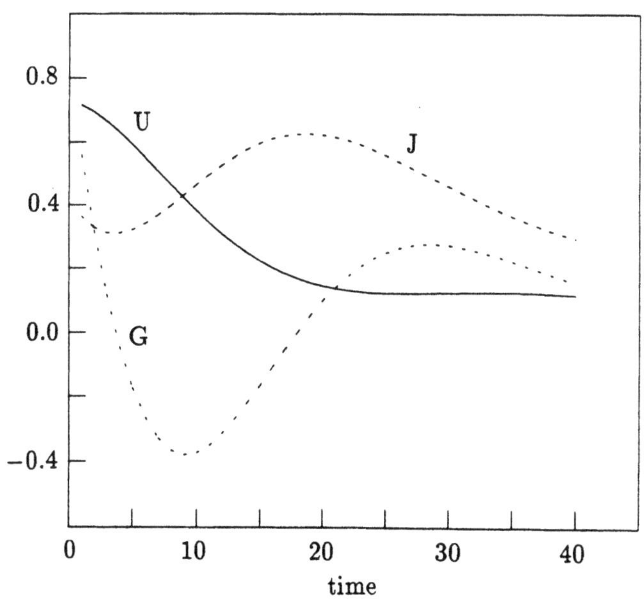


*Figure 21.3:* Response to US shock, first period.

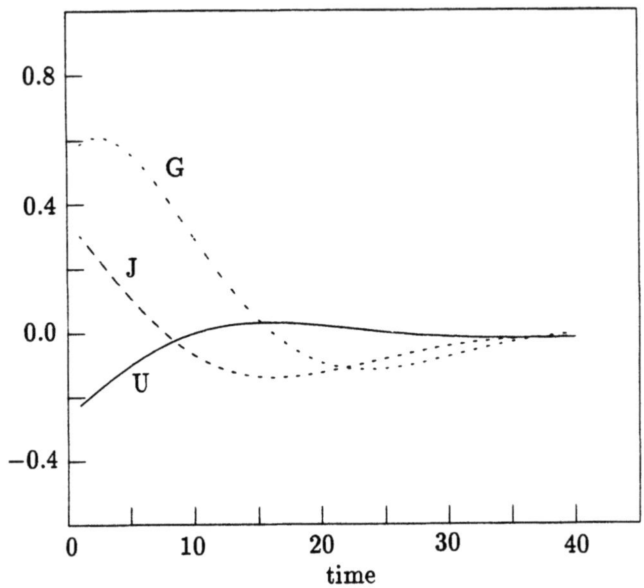


*Figure 21.4:* Response to West German shock, first period.

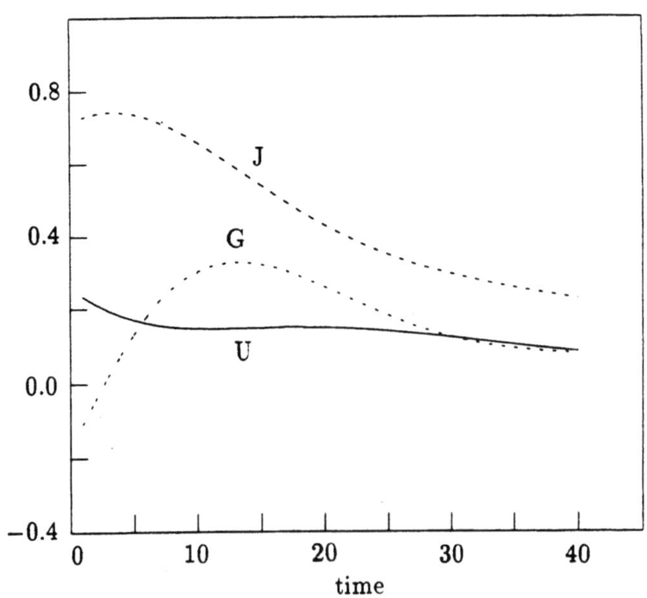


*Figure 21.5:* Response to Japanese shock, first period.

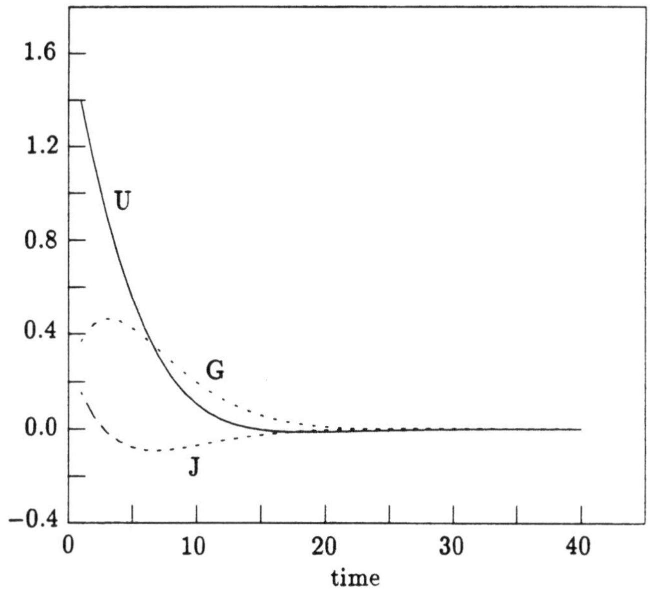


*Figure 21.6:* Response to US shock, second period.

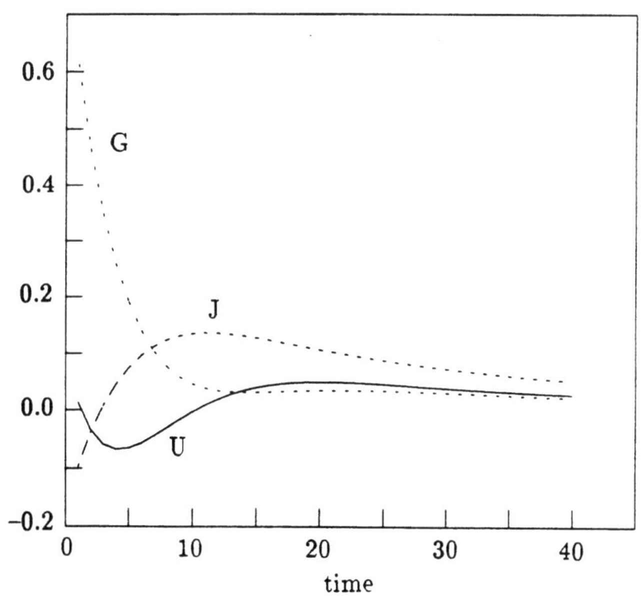


*Figure 21.7:* Response to West German shock, second period.

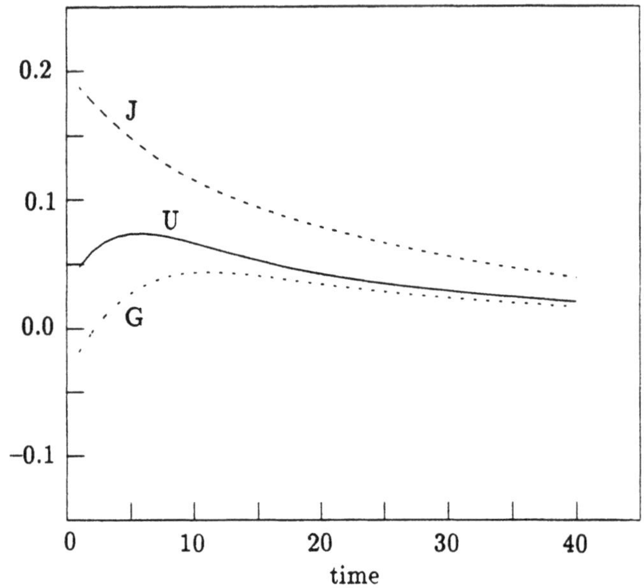


*Figure 21.8:* Response to Japanese shock, second period.

## References


Andrews, D.W.K. and Fair, R.C. (1988), Inference in nonlinear econometric model with structural change. *Review of Economic Studies*, **40**, 615-640.

Aoki, M. (1989), A two-step state space time series modeling method. *Computers and Mathematics with Applications*, **17**, 1165-1176.

Aoki, M. (1990a), A state space time series modeling method without individual detrending. Forthcoming in *Journal of Time Series Analysis.*

Aoki, M. (1990b), *State Space Modeling of Time Series, Second edition.* Heidelberg: Springer-Verlag.

Basseville, M., Benveniste, A., Moustakides, G., and Rougee, A. (1987), Detection and diagnosis of changes in the eigenstructure of nonstationary multivariate systems. *Automatica*, **21**, 479-489.

Bernanke, B. (1986), Alternative explanations of the money-income correlation. *Carnegie-Rochester Conference on Public Policy*, **25**, 49-100.

Chan, N.H. (1988), The parameter inference for nearly nonstationary time series. *Journal of the American Statistical Association*, **83**, 857-862.

Gerlach, S. and Klock, J. (1988), State-space estimates of international business cycle. *Economics Letters*, **28**, 231-234.

Goldfeld, S.M. and Quandt, R.E. (1976), *Studies in Nonlinear Estimation.* Cambridge: Ballinger.

Lo, A.W. and Newey, W.K. (1985), A large-sample Chow test for linear simultaneous equations. *Economics Letters*, **18**, 301-353.

Sims, C.A. (1986), Are forecasting models usable for policy analysis? *Federal Reserve Bank of Minneapolis Quarterly Review*, Winter, 2-16.

Willsky, A.S. and Jones, M.L. (1976), A generalized likelihood ratio approach to the detection and estimation of jumps in linear systems. *IEEE Transactions on Automation and Control*, **AC-21**, 108-112.

CHAPTER 22

# Interactive Poverty and Affluence of a Society: Change-Point Models for Structural Changes

*Pranab K. Sen*


## Summary

For any society or community, there is ample room for a prismatic view of its *income distribution*. Conventional measures of income *(distributional) inequalities*, such as the *Gini coefficient*, may not depict a realistic picture across the individual sectors of an income distribution. Stratification into *poor*, *middle-class*, and *affluent* sectors generally provides a better understanding of the socioeconomic intricacies of such income distributional inequalities. Combining such component measures into a single index requires careful statistical considerations and entails a detailed analysis of the entire income distributional data. *Economic structural changes* may occur within each sector and in plausibly rather diverse directions, so that the usual *linear models* may fail to be very appropriate for a composite analysis. A formulation of a *change-point model* in a setup of *constancy of regression surfaces* is therefore incorporated in the development of methodology for studying structural changes for such income distributions. Proper emphasis is placed on *nonparametric* as well as *robustness considerations* underlying such a nonstandard analysis. Such considerations also play a vital role in *forecasting* of economic structural changes with respect to some income inequalities.

## 22.1 Introduction

In any society or community, the distribution of *real income (or wealth)* is characterized by distinct *heterogeneity* even within individual sectors of relatively homogeneous groups. The extent of this divergence generally depends on various *socioeconomic factors* that may affect the individual sectors in a rather different manner. For example, a drought year may have serious impact on the income distribution of the agricultural households sector, quite perceptible by the lower income and the poor classes, but not so much by affluent people. There are literally thousands of such socioeconomic factors influencing the *income pattern and distribution* of any society, and their impacts are generally perceived rather differently across the different strata. Thus, for a better understanding of the intricacies of various socioeconomic factors affecting an income distribution and to relate them adequately to plausible *structural changes*, it may be wiser to encompass a rational stratification of an income distribution. This stratification may be accomplished by identifying the relative levels of income (viz., *poor, middle-class, and affluent people*) resulting in non-overlapping strata, or by the conventional *agricultural, industrial, and professional* sectors resulting in possibly overlapping income distributions. A combination of the two is also a possibility. With such a stratification, for drawing a neat picture for the entire spectrum, one needs to take into account both the *intra-strata* and *inter-strata variations*. In this chapter, we shall mainly confine ourselves to the interactive features of poverty and affluence of a society, and incorporate them in the change-point models for a better understanding of some structural changes that may arise in this context.

In Section 22.2, we outline the general statistical considerations underlying the affluence and poverty indexes, and based on these findings, we consider a general breakdown of an overall index in terms of some component indexes. It turns out that in the case of nonoverlapping strata, the Gini coefficient for the composite income distribution is explicitly expressible in terms of the component Gini coefficients, the relative proportion of the people in these strata, and their relative total incomes. Some additional parameters enter into the picture when the strata income distributions are possibly overlapping. This gives us a strong motivation to incorporate these component Gini coefficients and related income inequality indexes in the formulation of a so-called *response surface model*, which may be more conveniently used to study suitable structural changes for such income distributions. This is considered in Section 22.3. The *change-point model* is then introduced in Section 22.4 with a view to studying plausible structural changes over a span of time. The model is generally more complicated than the classical change-point model as here progressive realignment of the strata may be necessary to cope with the chronological changes in life-styles. These findings are finally incorporated in the last section in the forecasting of economic structural changes pertinent to the usual income distributions.

## 22.2 Income Inequality Indexes: A Prismatic View

The *Lorenz curve* for an income distribution provides a clear graphical picture of the overall concentration and dispersion of incomes of individuals or families in a society. However, as is generally the case, part of the income variation (or inequality) may be explained better in the light of between-sector variation when a suitable system of sectors is brought in the picture. For example, a stratification into three strata (poor, middle-class, and affluent sectors) may explain some of the variations as due to inter-strata variations while the rest are confined to the intra-strata ones. Hence, to study the income inequality picture more extensively (for the individual sectors as well), it may be wiser to look into the Lorenz curve a little bit more thoroughly with a view to depicting the picture for the component as well as the overall income inequality indexes. The sectors to be considered here are based on the relative levels of real income leading to a system of nonoverlapping strata. For simplicity (and practical relevance too), we confine ourselves to the case of three basic strata: *poor, middle-class, and affluent people.* A brief treatment of the general case of $k$ ($\geq 2$) (possibly overlapping) strata is also appended. In this context, there is a genuine need to pay attention to the following:

(i) For the correct labeling of the strata and real income, an assessment of real income of individuals or families in terms of a single *quantitative criterion* is needed. Once the real income is quantified, the drawing of the Lorenz curve and the associated indexes may not pose a serious problem. However, the issue of *robustness* remains as a pertinent one in such a quantification scheme.

(ii) Demarcation of the three strata rests on the proper fixation of the *line of poverty* and the *affluence line.* Poverty is usually defined as the extent to which individuals in a society or community fall below a *minimal acceptable standard of living.* Thus, quantified in terms of real income, the poverty line cuts off the lower tail of the income distribution: The truncated income distribution over this left-hand tail is often called the *income distribution of the poor.* Affluence of a society or community is similarly quantified by the proportion of its rich (or affluent) people and by the concentration of their wealth or real income. Again, in terms of real income, this amounts to the right-hand tail of the income distribution where the cutoff point (the affluence line) is determined by various socioeconomic factors. The truncated income distribution over this right-hand tail is often called the *income distribution of the rich.* The income distribution truncated from below and above by the poverty and affluence line, respectively, is termed the *middle-class income distribution.* It is quite clear that in the determination of poverty and affluence lines, various socioeconomic and related monetary utility functions play a basic role. Statistical considerations are very important in this respect too.

(iii) For each sector, some measures of concentration or inequality of wealth (or real income) need to be developed. Statistical considerations are very pertinent in this context too.

Arbitration of affluence and poverty lines is generally a very delicate task. The criteria may differ considerably from a socialistic to a capitalistic society. Even for the same society, they may vary progressively over time. These criteria appearing as deterministic in this context are also very relevant in the study of plausible structural changes of income indexes. In addition, there may be other important factors that should be taken into account in a proper formulation of a suitable response surface model for income inequality measures. Moreover, quantifications of all such basic factors are important for a proper formulation of real income on which everything is based.

With due considerations to this basic quantification of real income, we denote the income distribution (of a society or community in a given time period) by $F = \{F(x), x \in R^+\}$, $R^+ = [0, \infty)$. Also, with a proper arbitration of the poverty line ($\omega$) and the affluence line ($\rho$), we have two positive number $(\omega, \rho) : 0 < \omega < \rho < \infty$, such that the income in the ranges $[0, \omega]$, $(\omega, \rho)$, and $[\rho, \infty)$ characterizes the poor, middle-class, and affluent sectors, respectively. Thus, the income distribution of the poor is given by

$$F_P = \left\{ F_P(x) = \left\{ \begin{array}{ll} F(x)/F(\omega) & 0 \leq x \leq \omega \\ 1 & x > \omega \end{array} \right\} \right\} \tag{22.1}$$

and the proportion of the poor people is denoted by

$$\alpha_P = F(\omega) . \tag{22.2}$$

Similarly, the income distribution of the affluent people is given by

$$F_R = \left\{ F_R(x) = \left\{ \begin{array}{ll} 0 & x < \rho \\ [F(x) - F(\rho)]/[1 - F(\rho)] & x \geq \rho \end{array} \right\} \right\} , \tag{22.3}$$

and the proportion of affluent people is given by

$$\alpha_R = 1 - F(\rho) . \tag{22.4}$$

Finally, the middle-class income distribution is

$$F_M = \left\{ F_M(x) = \left\{ \begin{array}{ll} 0 & x \leq \omega \\ [F(x) - F(\omega)]/[F(\rho) - F(\omega)] & \omega \leq x \leq \rho \\ 1 & x > \rho \end{array} \right\} \right\} , \tag{22.5}$$

and the proportion of the middle-class people is

$$\alpha_M = F(\rho) - F(\omega) = 1 - \alpha_R - \alpha_P . \tag{22.6}$$

Assume that for the entire income distribution $F$, the *mean real income* $\mu = \int_{R^+} y dF(y)$ is finite and positive. Also, let $F^{-1}(t) = \inf\{x\colon\ F(x) \geq t\}$, $0 \leq t \leq 1$. Define then $\xi = \{\xi(t); t \in [0, 1]\}$ by letting

$$\xi(t) = \mu^{-1}\{\int_0^{F^{-1}(t)} y dF(y)\}, \quad t \in [0, 1] . \tag{22.7}$$

Thus, the relative contributions of the poor, middle-class, and the affluent people to the total income are given by $\gamma_P$, $\gamma_M$ and $\gamma_R$, respectively, where

$$\gamma_P = \xi(\alpha_P), \quad \gamma_M = \xi(1-\alpha_R) - \xi(\alpha_P), \quad \text{and } \gamma_R = 1 - \xi(1-\alpha_R). \tag{22.8}$$

Hence, the relative mean incomes of the poor, middle-class, and the rich are given by

$$\nu_P = \xi(\alpha_P)/\alpha_P \quad (\leq 1) \tag{22.9}$$
$$\nu_M = \gamma_M/\alpha_M = [\xi(1-\alpha_R) - \xi(\alpha_P)]/\alpha_M \tag{22.10}$$
$$\nu_R = \gamma_R/\alpha_R = [1 - \xi(1-\alpha_R)]/\alpha_R \quad (\geq 1), \tag{22.11}$$

where $\nu_M$ may be $<$, $=$, or $>$ 1 depending on $(\omega, \rho)$ and the income distribution $F$. Recall that

$$\nu_P = \mu^{-1}\{\int_0^\omega y dF(y)\}/F(\omega)$$
$$= \{\text{average income of the poor}\}/\{\text{mean income of all}\} \tag{22.12}$$
$$\nu_M = \{\text{average income of the middle-class}\}/\{\text{mean income of all}\} \tag{22.13}$$
$$\nu_R = \{\text{average income of the rich}\}/\{\text{mean income of all}\} \tag{22.14}$$

and these reflect the between-sector dispersion of the relative mean incomes; we must have $\nu_P \leq \nu_M \leq \nu_R$.

In a conventional setup, one plots $\xi(t)$ against $t$ $(0 \leq t \leq 1)$ and obtains the classical Lorenz curve for the income distribution $F$. To obtain the Lorenz curves for the individual sectors as well as for the entire distribution, we consider the decomposition of the classical Lorenz curve shown in *Figure 22.1*.

The entire picture is dictated by the proportions $\alpha_P$, $\alpha_M$, $\alpha_R$, $\nu_P$, $\nu_M$, $\nu_R$, and the individual sector Gini coefficients $G_P$, $G_M$, and $G_R$. In terms of the shaded areas $A$, $B$, $C$, and $D$, the Gini coefficient $(G)$ for the entire income distribution is given by

$$G = 2(A + B + C + D). \tag{22.15}$$

Similarly, the Gini coefficients of the income distributions for the poor, middle-class, and the rich are given by

$$G_P = (2A)/(\alpha_P\gamma_P) \tag{22.16}$$
$$G_M = (2B)/(\alpha_M\gamma_M) \tag{22.17}$$
$$G_R = (2C)/(\alpha_R\gamma_R), \tag{22.18}$$

respectively. Moreover, $D$ is a polygon whose area can easily be determined by some standard manipulations. Thus, we have

$$\begin{aligned}
2D &= \alpha_R\gamma_R - \alpha_R^2 + \alpha_P\gamma_P - \gamma_P^2 + (1-\alpha_R-\gamma_P)^2 - \alpha_M\gamma_M \\
&= (\gamma_R - \alpha_R) + \gamma_M(\alpha_P - \gamma_P) + \gamma_P(\gamma_M - \alpha_M) \\
&= \alpha_R(\gamma_R/\alpha_R - 1) + \alpha_P\gamma_M(1 - \gamma_P/\alpha_P) + \alpha_M\gamma_P(\gamma_M\alpha_M - 1) \\
&= \alpha_R(\gamma_R/\alpha_R - 1) + \alpha_P\alpha_M(\gamma_M/\alpha_M - \gamma_P/\alpha_P) \\
&= (\nu_R - 1)\alpha_R + (\nu_M - \nu_P)\alpha_P\alpha_M\,.
\end{aligned} \tag{22.19}$$

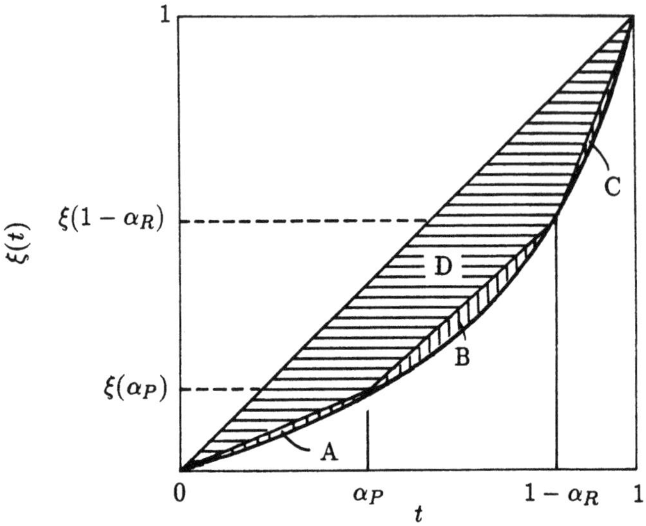


*Figure 22.1:* A prismatic decomposition of the classical Lorenz curve.

Also, note that from (22.9) through (22.14), we have

$$\nu_P \leq \nu_M \leq \nu_R, \quad \nu_P \leq 1, \quad \text{and} \quad \nu_R \geq 1. \tag{22.20}$$

Thus, the first term on the right-hand side of (22.19) represents the contributions of the affluent people [through their excessive relative income ($\nu_R - 1$) and their proportional representation ($\alpha_R$)], while the second term depicts the differential picture of relative mean incomes of the middle-class and the poor people, adjusted by their relative proportions too. From (22.15) through (22.19), we have

$$\begin{aligned} G &= \alpha_P\gamma_P G_P + \alpha_M\gamma_M G_M + \alpha_R\gamma_R G_R + \alpha_R(\nu_R - 1) + \alpha_M\alpha_P(\nu_M - \nu_P) \\ &= \alpha_P^2\nu_P G_P + \alpha_M^2\nu_M G_M + \alpha_R^2\nu_R G_R + \alpha_R(\nu_R - 1) + \alpha_M\alpha_P(\nu_M - \nu_P) \\ &= G(\alpha_P, \alpha_M, \alpha_R; \nu_P, \nu_M, \nu_R; G_P, G_M, G_R), \quad \text{say.} \end{aligned} \tag{22.21}$$

This clearly shows the structural dependence of the overall Gini coefficient, $G$, on the individual sector Gini coefficients ($G_P$, $G_M$, and $G_R$), the relative proportions ($\alpha_P$, $\alpha_M$, and $\alpha_R$) and the relative mean incomes ($\nu_P$, $\nu_M$, and $\nu_R$). Thus, for a better understanding of any plausible structural change in $G$ [due to a complex interplay of a (usually large) number of socioeconomic factors], it may be better to look back into the vectors $\boldsymbol{\alpha} = (\alpha_P, \alpha_M, \alpha_R)$, $\boldsymbol{\nu} = (\nu_P, \nu_M, \nu_R)$, and $\mathbf{G} = (G_P, G_M, G_R)$, and to examine the extent of coherence of such a change across the three sectors. We shall consider this model in greater details in the next section.

Let us next sketch the general case of $m$ ($\geq 1$) sectors with individual income distributions $F_1, \ldots, F_m$ and the relative proportions $\alpha_1, \ldots, \alpha_m$, respectively. Then

$$\sum_{i=1}^{m} \alpha_i = 1 \quad \text{and} \quad F(x) = \sum_{i=1}^{m} \alpha_i F_i(x), \quad x \in R^+ . \tag{22.22}$$

Note that these $F_j$ need not be all nonoverlapping. Then, we have

$$\mu = \int_0^\infty y dF(y) = \sum_{i=1}^{m} \alpha_i \int_0^\infty y dF_i(y) = \sum_{i=1}^{m} \alpha_i \mu_i , \tag{22.23}$$

where the $\mu_i$ are the individual sector mean incomes. Also let

$$\nu_i = \mu_i / \mu, \quad i = 1, \ldots, m \quad (\text{so that} \sum_{i=1}^{m} \alpha_i \nu_i = 1) . \tag{22.24}$$

Further, we let

$$\omega_{ij} = \mu^{-1}\{\int_0^\infty \int_0^\infty |x - y| \, dF_i(x) \, dF_j(y)\}, \quad \text{for } i, j = 1, \ldots, m . \tag{22.25}$$

Then, note that

$$\omega_{ii} = 2\mu^{-1}\mu_i G_i = 2\nu_i G_i, \quad i = 1, \ldots, m , \tag{22.26}$$

where the $G_i$ is the usual Gini coefficient for the income distribution $F_i$ ($i = 1, \ldots, m$); for this definition of the $G_i$, we may refer to Sen (1986). For $i \neq j$, $\omega_{ij}$ stands for some average distance between $F_i$ and $F_j$. For the income distribution $F$, we have

$$\begin{aligned} G &= (2\mu)^{-1}\{\int_0^\infty \int_0^\infty |x_1 - x_2| \, dF(x_1) \, dF(x_2)\} \\ &= (2\mu)^{-1}\{\sum_{i=1}^{m} \alpha_i^2 2\nu_i G_i + 2\mu \sum_{1 \leq i < j \leq m} \alpha_i \alpha_j \omega_{ij}\} \\ &= \sum_{i=1}^{m} \alpha_i^2 \nu_i G_i + \sum_{1 \leq i < j \leq m} \alpha_i \alpha_j \omega_{ij} , \end{aligned} \tag{22.27}$$

so that the overall Gini coefficient can be expressed in terms of the individual $\alpha_i$, $\nu_i$, $G_i$ and the $\omega_{ij}$, $i \leq j = 1, \ldots, m$. In particular, if the strata income distributions are nonoverlapping (as in the case of the poor, middle-class, and affluent sectors), then for $F_1 \geq F_2 \geq \ldots \geq F_m$, we have

$$\omega_{ij} = \nu_j - \nu_i \quad \text{for } 1 \leq i < j \leq m , \tag{22.28}$$

so that in this special case (22.27) reduces to

$$G = \sum_{i=1}^{m} \alpha_i^2 \nu_i G_i + \sum_{1 \leq i < j \leq m} \alpha_i \alpha_j (\nu_j - \nu_i) \tag{22.29}$$

and, for the particular case of $m = 3$, (22.29) reduces to (22.21). Thus, for the general case of $m$ ($\geq 2$) and possibly overlapping income distributions, the only extra adjustment needed is to bring in the additional parameters $\omega_{ij}$, $1 \leq i < j \leq m$, which account for the between-sector distances. With this remark, and for the sake of simplicity of presentation, we shall only consider the case of $m = 3$ and nonoverlapping income distributions; a similar picture holds for the general case.

Analysis of income pattern may be done in a parametric setup where $F$ is assumed to be of a specified form (viz., Pareto law) and involves a set $\boldsymbol{\theta} = (\theta_1, \ldots, \theta_r)'$ of unknown parameters, or in a more general nonparametric setup where the functional form of $F$ is not assumed to be given (and it is assumed that $F$ belongs to a general class, say, $\mathcal{F}$). In a parametric model, the $\alpha_i$, $\nu_i$, and $G_i$ may all be expressed in terms of suitable functions of $\boldsymbol{\theta}$, so that the whole analysis may simply be done in terms of suitable parametric constraints on $\boldsymbol{\theta}$. There are, however, some general concerns with such parametric procedures:

(a) In practice, the actual functional form of the income distribution may never be known precisely, and any simple form (such as the Pareto law) may not quite adequately fit the model for the entire range.

(b) As has been mentioned before, some of the socioeconomic factors may affect the different sectors rather differently; as such, in any structural model, with different roles of different input variables, it may be quite counter-intuitive to conceive of a common form of the income distribution for all the strata. That is, merely by varying the associated parameters ($\boldsymbol{\theta}$), it may not be feasible to describe the component income distributions (i.e., $F_P$, $F_M$, and $F_R$) in terms of a single parametric $F$. Inclusion of a large number of parameters may drastically reduce the sensitivity of the parametric models.

(c) With scope for plausible departures from an assumed parametric model (rather differently in different sectors), the issue of robustness is quite an important one. The parametric models thus may not have good robustness properties (against possible departures from the assumed models).

(d) In a parametric framework, to describe a model adequately with a view to studying plausible structural changes, it may be necessary to bring in a large number of parameters ($\boldsymbol{\theta}$). With the increase in the number of such parameters, the usual simplicity of a parametric approach may evaporate fast; moreover, the efficiency of a parametric procedure may drastically go down.

For $F_P$, the income distribution of the poor, the average is $\mu_P = \mu\gamma_P/\alpha_P$, and it lies below the set poverty line $\omega$. Thus, the income gap ratio ($\beta_P$) for the poor people is defined by

$$\beta_P = 1 - \omega^{-1}\mu_P = 1 - \omega^{-1}\mu\,\alpha_P^{-1}\gamma_P\,. \tag{22.30}$$

A crude index of poverty is given by $\pi_P = \alpha_P\beta_P$, while a refined one is formulated in terms of the triplet $(\alpha_P, \beta_P, G_P)$. A popular form [due to A.K. Sen (1976)], based on a set of axioms, is

$$\pi_{PS} = \alpha_P\{\beta_P + (1-\beta_P)G_P\}\,, \tag{22.31}$$

and a more robust version [due to Sen (1986)] is given by

$$\pi_{PS}^* = \alpha_P\{\beta_P^{1-G_P}\}\,. \tag{22.32}$$

An alternative poverty, due to Takayama (1979), is based on the censored (not truncated) income distribution of the poor, and it can be expressed as

$$\pi_{PT} = \alpha_p G_P + (1-\alpha_P\beta_P)^{-1}(1-\alpha_P)(\beta - G_P)\,, \tag{22.33}$$

where $0 \le G_P \le \beta_P$. It is known [viz., Sen (1986)] that

$$\pi_{PT} \le \alpha_P\beta_P = \pi_P \le \pi_{PS}^* \le \pi_{PS} \le \alpha_P\beta_P(2-\beta_P) \le \alpha_P \tag{22.34}$$

for all income distributions. Other forms [viz., Blackorby and Donaldson (1980)] will not be considered here.

For the income distribution $F_R$ (of the affluent people), the average is $\mu_R = \mu\gamma_R/\alpha_R$, and, by definition, $\mu_R \ge \rho$. Thus, a different definition of the income gap ratio is needed here. One way to define $\beta_R$, the income gap ratio of the rich, is to take

$$\beta_R = 1 - \rho/\mu_R = 1 - \rho\alpha_R\mu^{-1}\gamma_R^{-1} = 1 - \rho\alpha_R[\int_\rho^\infty y dF(y)]^{-1}\,. \tag{22.35}$$

In this context, we may recall that wealth (in a form other than income) needs to be transferred into an income form, and there may be real difficulties for accurate assessment of wealth of excessively rich people. From this point of view, $\int_\rho^\infty y dF(y)$ may not be very robust, so that $\beta_R$ in (22.35) may not be robust against such measurement errors as well. The use of a harmonic income gap measure generally leads to a better robustness property [viz., Sen (1988)], and based on this consideration, we may set

$$\beta_R^* = 1 - \frac{\rho}{\alpha_R}\int_\rho^\infty y^{-1}dF(y)\,; \tag{22.36}$$

it is known that $\beta_R^* \le \beta_R$ [Sen (1988)]. For the Gini coefficient $G_R$ (for $F_R$), the same criticism (i.e., lack of robustness against measurement errors) can be labeled to a greater extent, and for this reason a harmonic Gini coefficient ($G_R^*$) has been advocated by Sen (1988). This is defined by

$$G_R^* = E[|Y_1^{-1} - Y_2^{-1}|]/E[Y_1^{-1} + Y_2^{-1}], \tag{22.37}$$

where $Y_1$ and $Y_2$ are independent random variables each having the distribution $F_R$. Then parallel to (22.31) and (22.32), we may consider some indexes of affluence as

$$\pi_{R1} = \alpha_R\{\beta_R + (1 - B_R)G_R\}, \quad \pi_{R2} = \alpha_R\{\beta_R^* + (1 - \beta_R^*)G_R^*\} \tag{22.38}$$

$$\pi_{R3} = \alpha_R\{\beta_R^{1-G_R}\} \text{ and } \pi_{R4} = \alpha_R(\beta_R^*)^{1-G_R^*}. \tag{22.39}$$

From what has been discussed in Sen (1988), we are biased toward the use of $G_R^*$, $\beta_R^*$, and $\pi_{R4}(= \pi_R^*$, say). The situation with the middle-class income distribution is generally more manageable for two basic reasons:

(a) The real income of a person/family in the middle-class group is bounded from below by $\omega$ ($> 0$) and from above by $\rho$ ($< \infty$); within these two bounds, the variation may be more smooth than in the affluent or the poor groups.

(b) There are generally far less difficulties in measuring the real income for the middle-class people than for the rich or the poor ones. Thus, the robustness considerations relevant to the poor or rich sectors may not be that crucial for the middle-class. Note that the average income for the middle-class sector is $\mu_M = \mu_Y/\alpha_M$, and, by definition, $\omega \leq \mu_M \leq \rho$. As such, by analogy with (22.30), we may define $\beta_M$, the income gap ratio for the middle-class people, as

$$\beta_M = (\rho - \mu_M)/(\rho - \omega) = 1 - (\mu_M - \omega)(\rho - \omega)^{-1}. \tag{22.40}$$

With this, an index of middle-classness ($\pi_M$) may be defined as

$$\pi_{M1} = \alpha_M\{\beta_M + (1 - \beta_M)G_M\} \quad \text{or} \quad \pi_{M2} = \alpha_M\{\beta_M^{1-G_M}\}. \tag{22.41}$$

For the general case of $m$ ($\geq 2$) strata with possibly overlapping income distributions, similar indexes can be defined for the individual sectors, and these will usually provide a more comprehensive picture than the alternative ones solely based on $\alpha_1, \ldots, \alpha_m$ or the component Gini coefficients $G_1, \ldots, G_m$. For this reason, we propose to replace the $\alpha$ by the $\pi$, and to employ $(\pi_1, \ldots, \pi_m), (\nu_1, \ldots, \nu_m)$, and $(G_1, \ldots, G_m)$ for a better understanding of plausible structural changes of various socioeconomic factors affecting the original income distribution.

## 22.3 Income Inequality Structures and Response Surface Models

The results in Section 22.2 are incorporated in formulating a stochastic model for which some conventional response surface methodologies can be adopted. For simplicity of presentation, we consider the case of three strata (poor, middle-class, and affluent sectors); the general case can be handled in a similar manner. As has been mentioned before, instead of the overall income inequality measure $G$, we shall deal with the following vector:

$$\boldsymbol{\theta} = (\pi_P, \pi_M, \pi_R; \nu_P, \nu_M, \nu_R; G_P, G_M, G_R), \tag{22.42}$$

where the $\pi$ stand for the individual sector income inequality indexes, the $\nu$ for their relative mean incomes, and the $G$ for the corresponding Gini coefficients. We may also note that the $\nu$ are essentially measures of central tendencies (relative to the overall mean), while the $\pi$ and $G$ are suitable within strata indexes. Although there may be some general concordance between $\pi$ and the $G$, the multi-collinearity problem is not likely to arise here (as the $\pi$ depend on $\alpha$ and $\beta$, in addition to the $G$). This dependent vector $\boldsymbol{\theta}$ is conceivably related to a number of socioeconomic factors (as independent variables or regressors). It is quite clear that a formulation of a suitable response surface methodology may require that all the relevant input variables are capable of being quantified, at least, on an ordinal categorical basis. Actually, for convenience, we assume that these input variables are more or less continuous. Among possible socioeconomic variables, the arbitration of the poverty and affluence lines (i.e., $\omega$ and $\rho$) as well as the usual cost-of-living index (say, $\kappa$) are the most important ones. Again, it may be advantageous to decompose $\kappa$ into suitable components (food and drinks, energy, education, housing, transportation, medical expenses, recreation, etc.) which may have differential effects on the components of $\boldsymbol{\theta}$. In addition, distribution of social welfare or other forms of compensation may constitute relevant factors (especially, if they are likely to vary over time). Other economic factors, such as the prime interest rate, unemployment index, productivity index (for agricultural/industrial and other products), and GNP index may all be quite appropriate for explaining plausible variations in the components of $\boldsymbol{\theta}$ (over time). In due considerations of all such plausible regressors, we conceive of a vector

$$\boldsymbol{\xi} = (\xi_1, \ldots, \xi_q)' \quad \text{(for some } q \geq 1) \tag{22.43}$$

of relevant input (factor) variables, and formulate a regression model:

$$\boldsymbol{\theta} = \boldsymbol{\psi}(\boldsymbol{\xi}, F), \quad F \text{ being the income distribution}, \tag{22.44}$$

where $\boldsymbol{\psi} = (\psi_1, \ldots, \psi_9)'$ and each $\psi_k$ is a functional of the distribution $F$ and a function of $\boldsymbol{\xi}$, $k = 1, \ldots, 9$. The forms of these nine functionals need not be all similar (or linear in nature).

The picture presented above relates to a given time period. In practice, we are interested in the composite picture over a span of successive time periods. For example, these time intervals may be the fiscal years, half-years, or even the quarters, and we may like to study plausible changes over such intervals. In this setup, a change-point model becomes quite relevant. Keeping this in mind, we conceive of an index set

$$T = \{t_1, \ldots, t_n\colon \; t_1 < \ldots < t_n\}, \tag{22.45}$$

where $t_j$ refers to the $j$-th time interval, quantified as its midpoint, $j = 1, \ldots, n$. For the time period indexed by $t_j$, we denote the income distribution by $F_j$ (for the entire population), and let

$$\boldsymbol{\theta}_j = \boldsymbol{\theta}(t_j) \text{ and } \boldsymbol{\xi}_j = \boldsymbol{\xi}(t_j), \; j = 1, \ldots, n. \tag{22.46}$$

We may even allow the form of the functional in (22.44) to be timedependent, and set

$$\boldsymbol{\theta}_j = \boldsymbol{\psi}_j(\boldsymbol{\xi}_j, F_j), \quad j = 1, \ldots, n. \tag{22.47}$$

The functional forms of the components of $\boldsymbol{\psi}_j$ in (22.47) remain stationary (over $j$) or may follow a change-point model. In this case, an abrupt change is likely to occur if the $\boldsymbol{\xi}_j$ or $F_j$ itself undergoes an abrupt change at some point in time—a case that may arise due to some structural changes in the socioeconomic factors affecting the income distributions $F_j$. It is in this general response surface model formulation that we intend to consider a change-point model for plausible structural changes over time. In this context, certain order relations and invariance properties of the income indexes [viz., Sen (1986)] need to be taken into account, and this results in a somewhat different formulation of the problem.

## 22.4 Change-Point Modeling for Income Distributions

Looking at the $\boldsymbol{\theta}_j$ and $\boldsymbol{\xi}_j$, we may gather that the $\boldsymbol{\theta}_j$ are parameters (estimable functionals) of the distribution $F_j$ while the $\boldsymbol{\xi}_j$ are also parametric quantities (depending on some other distributions). The arbitration of the poverty and affluence lines or the adjustment for the cost-of-living index may change the form of the $F_j$ over time and also the $\boldsymbol{\theta}_j$ may vary accordingly. Thus, we may have an implicit functional relationship between the $\boldsymbol{\theta}_j$ and the $\boldsymbol{\xi}_j$. To make this point clear, consider the simple situation where the poverty and affluence lines $\omega_j$ and $\rho_j$ are adjusted by the cost-of-living index, so that denoting by $\kappa_j$ the cost-of-living index at time $t_j$, we have

$$\omega_j = \omega\kappa_j \quad \text{and} \quad \rho_j = \rho\kappa_j \quad \text{for } j = 1, \ldots, n, \tag{22.48}$$

where $\omega$ and $\rho$ are suitable positive numbers. If the income distributions $F_j$ (at time point $t_j$) satisfy the condition that

$$F_j(y) = F(y/c_j), \quad j = 1, \ldots, n, \quad [F \text{ arbitrary}], \tag{22.49}$$

where the $c_j$ are proportional to the cost-of-living indexes $\kappa_j$, then it is easy to show that the income indexes $\pi_P$, $\pi_M$, and $\pi_R$, as well as $\nu_P$, $\nu_M$, and $\nu_R$ and $G_P$, $G_M$, and $G_R$ remain the same over the entire time period. Thus, the parameters $\boldsymbol{\theta}_j$ remain stationary over the span of time. On the other hand, if the income distributions $F_j$ do not satisfy the scale-model in (22.49) and/or the poverty and affluence lines are not adjusted by the proper scale factors, the different components of the $\boldsymbol{\theta}_j$ may be affected rather differently. In this context, it may also be noted that a change in $\alpha_P$ (or $\alpha_M$ or $\alpha_R$) may not necessarily lead to a change in $\nu_P$, $G_P$ (or $\nu_M$, $G_M$ or $\nu_R$, $G_R$) in the same direction, so that $\boldsymbol{\theta}$ may not satisfy a partial ordering with respect to the $\alpha$ or the other parameters. However, the scale-model in (22.49) and the cost-of-living adjustment in (22.48) may work out quite well if the span of the time in $T$ is not too large and there is no abrupt change in some socioeconomic factors affecting the income distributions. As such, we may consider cost-of-living adjusted real income as well as poverty and affluence lines, so that in the ideal situation $\boldsymbol{\theta}$ based on this adjusted distribution would not depend on the cost-of-living index $\kappa$, so that we have

$$\boldsymbol{\theta}_j = \boldsymbol{\theta} \quad \text{for all } j = 1, \ldots, n. \tag{22.50}$$

On the other hand, when a cost-of-living adjustment may fail to induce a scale-equivariance of the income distributions at different time points, the $\pi$, $\nu$, and $G$ may not remain the same; hence, the constancy of the elements of the $\boldsymbol{\theta}_j$ over $j$ may not hold. In this case, the components of $\boldsymbol{\xi}_j$ may provide suitable explanation for the variation in the components of the $\boldsymbol{\theta}_j$. This explains why in (22.47) for the $\psi_j$ being sufficiently smooth (i.e., locally linear), we may actually assume a linear response surface model for the $\boldsymbol{\theta}_j$ in terms of the $\boldsymbol{\xi}_j$ (provided we use the cost-of-living adjusted income distributions). In the sequel, we therefore take the income distributions $F_j$ as the cost-of-living adjusted ones. In passing, we may remark that the elements of $\boldsymbol{\theta}_j$ are regular functions of the distribution $F_j$; hence, when the input variables $\boldsymbol{\xi}_j$ affect the $F_j$ only locally, usual expansion of such functionals ensures the linearity of the elements of $\boldsymbol{\theta}_j$ in terms of the $\boldsymbol{\xi}_j$.

In practice, the distributions $F_j$ as well as the $\boldsymbol{\theta}_j$ and $\boldsymbol{\xi}_j$ are unknown, and we need to estimate them. This does not pose any problem as usually large data sets are available for each of the time periods, and our estimates can be based on them. We denote such estimators by

$$\hat{F}_j, \ \ \hat{\boldsymbol{\theta}}_j = \mathbf{Q}_j = (Q_{j1}, \ldots, Q_{j9})' \quad \text{and} \quad \hat{\boldsymbol{\xi}}_j = \mathbf{U}_j = (U_{j1}, \ldots, U_{jq})' \tag{22.51}$$

for $j = 1, \ldots, n$. It may be quite reasonable to assume that $\hat{F}_j$ is a consistent estimator for $F_j$ (and other asymptotic optimality properties of these sample distributions may also be assumed). We may write

$$\mathbf{Q}_j = \boldsymbol{\theta}_j + \boldsymbol{\epsilon}_j \quad \text{and} \quad \mathbf{U}_j = \boldsymbol{\xi}_j + \boldsymbol{\eta}_j \ , \ \ j = 1, \ldots, n \, , \tag{22.52}$$

where, for each $j$, $(\boldsymbol{\epsilon}_j, \boldsymbol{\eta}_j)$ has a $(9 + q)$-variate joint distribution. As is usually the case, one may have a large sample size pertaining to each time period, so that by an appeal to the classical large sample theory, we may claim that

(i) The stochastic variability of $(\boldsymbol{\epsilon}_j, \boldsymbol{\eta}_j)$ would be small.

(ii) When suitably adjusted by these sample sizes, normalized version of these vectors $(\boldsymbol{\epsilon}_j, \boldsymbol{\eta}_j)$ would have closely some multivariate normal distribution, although this asymptotic distribution may differ from one time period to another.

Based on the above observations, we may therefore assume that for each $j$ ($= 1, \ldots, n$), the conditional distribution of $\mathbf{Q}_j$ given $\mathbf{U}_j$ is closely normal with the mean vector

$$\boldsymbol{\theta}_j + \boldsymbol{\beta}_j(\mathbf{U}_j - \boldsymbol{\xi}_j) \quad \text{[where } \boldsymbol{\beta}_j \text{ is a } 9 \times q \text{ matrix of regression parameters]} \tag{22.53}$$

and a dispersion matrix $\sum_j$. In this setup, borrowing the (approximate) linearity of the $\boldsymbol{\theta}_j$ [as has been discussed after (22.50)], we may regard that the $\boldsymbol{\theta}_j - \boldsymbol{\beta}_j\boldsymbol{\xi}_j$ behaves more or less as stationary, so that the main source of variability comes from the regression $\boldsymbol{\beta}_j\mathbf{U}_j$ (for $j = 1, \ldots, n$). Note that in this setup, both $\boldsymbol{\beta}_j$ and $\sum_j$ are unknown matrices and these can be estimated from the sample data as well. Generally, the matrices $\sum_j$ depend on the sample sizes (say, $N_j$) on which the estimators $\mathbf{Q}_j$ and $\mathbf{U}_j$ are based, and they depend also on some underlying dispersion matrices. Since we assume that these $N_j$ are all large, we may use the classical *jackknifing*, *bootstrapping*, or some other *resampling method* to estimate these $\sum_j$ consistently. On the other hand, the role of the $\boldsymbol{\beta}_j$ is somewhat different, as they enter into the specification of the change-point model (and hence need to be estimated in a more refined manner). As such, for our statistical analysis, we may assume that the $\sum_j$ are all given although they may not be assumed to be homogeneous (i.e., $\sum_j = \sum$, for all $j$). Usually the characteristic roots of the $\sum_j$ are all small (when the $N_j$ are large), but the homogeneity of the $\sum_j$ may demand more restrictive conditions on the underlying dispersion matrices.

From what has been discussed above, we may conclude that in the formulation of our income pattern, structural changes, if any, should be attributed to the regression matrices $\boldsymbol{\beta}_j$, $j = 1, \ldots, n$. In this setup, we therefore confront a *constancy of regression surface model* where the *homoscedasticity* (of the $\sum_j$) is not a part of the model (assumptions). As such, we frame the null hypothesis of no change-point as

$$H_0 : \boldsymbol{\beta}_1 = \ldots = \boldsymbol{\beta}_n = \boldsymbol{\beta} \quad \text{(unknown)} \tag{22.54}$$

and, side by side, we let

$$H_r : \boldsymbol{\beta}_1 = \ldots = \boldsymbol{\beta}_r \neq \boldsymbol{\beta}_{r+1} = \ldots = \boldsymbol{\beta}_n\,, \quad \text{for } r = 1, \ldots, n-1\,. \tag{22.55}$$

Then, the usual change point alternative hypothesis is

$$H^* = H_1 \cup \ldots \cup H_{n-1} = \bigcup_{r=1}^{n-1} H_r\,. \tag{22.56}$$

In testing $H_0$ against $H^*$, we treat the $\sum_j$ as given (but not necessarily homogeneous). For this testing problem, we consider the following:

*I) Pseudo Two-Sample Approaches.* For each $r$ ($1 \leq r \leq n-1$), consider a breakup of the data set into two parts: $\{(\mathbf{Q}_j, \mathbf{U}_j, \sum_j), j \leq r\}$ and $\{(\mathbf{Q}_j, \mathbf{U}_j, \sum_j), j > r\}$. From the first set, by using the classical *weighted least squares method*, we obtain the *weighted least squares estimator* (WLSE) $\hat{\beta}_{(r)}$ of $\beta$. Similarly, let $_{(n-r)}\hat{\beta}$ be the WLSE of $\beta$ from the second set. Let then

$$\mathbf{Z}_r^0 = \hat{\beta}_{(r)} - {}_{(n-r)}\hat{\beta} \quad \text{and} \quad \mathbf{Z}_{(r)} = vec(\mathbf{Z}_r^0)\,, \quad \text{for } r = 1, \ldots, n-1\,. \tag{22.57}$$

Thus, the $\mathbf{Z}_{(r)}$ are $9 \times q$-vectors. Using the normality in (22.53) and the linearity of the WLSE, it is easily seen that under $H_0$ in (22.54), $\mathbf{Z}_{(r)}$ has closely a multi-normal distribution with null mean vector and dispersion matrix (say,) $\boldsymbol{\Gamma}_{(r)}$. The $\boldsymbol{\Gamma}_{(r)}$ can also be consistently estimated from the data set; hence, we assume that these $\boldsymbol{\Gamma}_{(r)}$ are all given. Let then

$$\|\mathbf{Z}_{(r)}\|^2_{\boldsymbol{\Gamma}_{(r)}} = \mathbf{Z}'_{(r)}\boldsymbol{\Gamma}^-_{(r)}\mathbf{Z}_{(r)}\,, \quad \text{for } r = 1, \ldots, n-1 \tag{22.58}$$

and

$$T_{n1} = \max\{\|\mathbf{Z}_{(r)}\|^2_{\boldsymbol{\Gamma}_{(r)}}:\ 1 \leq r \leq n-1\}\,. \tag{22.59}$$

In this context, recall that the $\boldsymbol{\Gamma}^-_{(r)}$ are the generalized inverses of the $\boldsymbol{\Gamma}_{(r)}$ and the characteristic roots of the $\boldsymbol{\Gamma}_{(r)}$ are all small (when the $N_j$ are all large), so that the characteristic roots of $\boldsymbol{\Gamma}^-_{(r)}$ are all large. If the null hypothesis $H_0$ holds, for each $r$ $(= 1, \ldots, n-1)$, (22.58) is bounded in probability, so that $T_{n1}$ is also stochastically finite. On the other hand, if $H_0$ does not hold and $H_r$ holds for some $r$ $(= 1, \ldots, n-1)$, then at least some of these $\mathbf{Z}_{(r)}$ would have a non-null mean vector; hence, $T_{n1}$ will be large. Hence, the null hypothesis $H_0$ is to be rejected in favor of $H^*$ if $T_{n1}$ in (22.59) is significantly large. Thus, our main task is to find out a *critical value* $\tau^{(1)}_{n,\alpha}$, such that

$$P\{T_{n1} \geq \tau^{(1)}_{n,\alpha} \mid H_0\} \leq \alpha\,, \tag{22.60}$$

where $\alpha$ $(0 < \alpha < 1)$ is the desired *level of significance* of the test.

Instead of (22.57), we may also consider

$$\mathbf{Z}_r^{o*} = \hat{\beta}_{(r)} - \hat{\beta}_{(n)} \text{ and } \mathbf{Z}^*_{(r)} = vec(\mathbf{Z}_r^{o*}), \quad \text{for } r = 1, \ldots, n-1 \tag{22.61}$$

denote the dispersion matrix of $\mathbf{Z}^*_{(r)} = \boldsymbol{\Gamma}^*_{(r)}$, and let

$$T_{n2} = \max\{\|\mathbf{Z}^*_{(r)}\|^2_{\boldsymbol{\Gamma}^*_{(r)}} : 1 \leq r \leq n-1\}\,. \tag{22.62}$$

In this setup, parallel to (22.60), we need to find out a critical level $\tau^{(2)}_{n,\alpha}$ for $T_{n2}$, such that $H_0$ would be rejected in favor of $H^*$ whenever $T_{n2} \geq \tau^{(2)}_{n,\alpha}$. We shall consider suitable approximations for these critical levels later on.

*II) Recursive Residual CUSUM Procedures.* With the WLSE $\hat{\beta}_{(r)}$ defined as before in (22.57), let us define the recursive residuals as

$$\mathbf{Y}_r^o = \mathbf{Q}_r - \hat{\beta}_{(r-1)}\mathbf{U}\,, \quad \text{for } r = 2,\dots,n;\ \mathbf{Y}_1^o = \mathbf{0} \tag{22.63}$$

$$\mathbf{S}_r^o = \sum_{k\le r}\mathbf{Y}_k^o \text{ and } \mathbf{S}_{(r)} = vec(\mathbf{S}_r^o)\,, \quad \text{for } r = 2,\dots,n\,. \tag{22.64}$$

Thus the $\mathbf{S}_{(r)}$ are the CUSUM (vectors) of the recursive residuals (using the WLSE), and we denote the dispersion matrix of $\mathbf{S}_{(r)}$ by $\boldsymbol{\Lambda}_r, r = 2,\dots,n$; again these $\boldsymbol{\Lambda}_r$ can be consistently estimated from the data set; hence, we assume that they are given. Let then

$$T_{n3} = \max\{\|\mathbf{S}_{(r)}\|^2_{\boldsymbol{\Lambda}_r} : r = 2,\dots,n\}\,. \tag{22.65}$$

The null hypothesis $H_0$ is to be rejected in favor of $H^*$ whenever $T_{n3}$ exceeds a critical level $\tau_{n,\alpha}^{(3)}$, where $P\{T_{n3} \ge \tau_{n,\alpha}^{(3)} \mid H_0\} \le \alpha$.

These test procedures are formulated by analogy with the usual change-point tests for the classical regression model [viz., Brown *et al.* (1975)]. However, to accommodate possible heteroscedasticity, WLSE have been used instead of the classical least squares estimators. Consistency of these WLSE ensures the consistency of the proposed tests against any fixed alternative covered by $H^*$ in (22.56). The crux of the problem is therefore to find out suitable approximations for the critical levels $\tau_{n,\alpha}^{(j)}$ for $j = 1,2,3$. Given that the sample sizes $N_j$, $j = 1,\dots,n$, are all large (leading to the asymptotic normality in (22.53), for these WLSE, we may as well assume that the asymptotic normality holds, and this is then transmitted onto the $\mathbf{Z}_{(r)}$, $\mathbf{Z}^*_{(r)}$, and $\mathbf{S}_{(r)}$. Thus, we are able to reduce the whole thing to a multi-normal setup. In this context, attention should be paid to two basic points:

(i) The dispersion matrices of these vectors are in general complicated, so that the results for multi-normal distributions with identity dispersion matrix may not always be applicable.

(ii) The number of time points (i.e., $n$) may or may not be small. For large values of $n$, the computational complexities may call for some further asymptotic approximations. In either situation, we shall see that suitable *simulation techniques* work out well. The basis for this simulation study is provided by the weak convergence results for the partial sequence $\mathbf{Z}_{(r)}$, $\mathbf{Z}^*_{(r)}$, or $\mathbf{S}_{(r)}$. In either case, the asymptotic multi-normality results for the finite dimensional distributions follow by an appeal to the classical Cramér-Wold theorem and the central limit theorem for WLSE (which are all linear estimators). In this context, one also needs to establish the *tightness* or *relative compactness* of the stochastic processes constructed from these vectors. When $n$ is small, one does not need to construct such processes, so that the desired simulation results would directly fit with these asymptotic multi-normal laws, and we need to use consistent estimators of the associated dispersion matrices to generate these multi-normal vectors. On the other hand, for large $n$, the process may turn out to be extremely tedious, and suitable Gaussian process approximations (in law) for the associated stochastic processes may provide simpler simulation prospects. In this context, the tightness condition may be verified by standard techniques (applicable for linear processes); hence, we omit the details here. In the context of survival analysis (or in life-testing models), for the weighted Kolmogorov-Smirnov statistics, a similar problem arises, and in Sinha and Sen (1982) some of these simulation studies have been reported in detail. In view of the similarity, we therefore avoid these details. In some specific (simple) cases, one may use some standard results on the so-called *Bessel Processes* [this is particularly the case where the homogeneity of the dispersion matrices (over $j$) can be assumed], and the detailed tables of DeLong (1981) can be used with some advantage. In this context, we may note that there are some simple relations between the critical levels of tied-down Bessel processes over a part of $[0,1]$ and the usual Bessel process on a part of $R^+$ [viz., DeLong (1981) and Sen (1981, Ch. 2)], so that for $T_{n1}$ or $T_{n2}$ such relations can be used to adapt the critical levels from DeLong's tables. If $n$ is large enough then to apply these Bessel process approximations it is not necessary to assume the homogeneity of the dispersion matrices at various time points (all we need that the transformed points of the time arguments are dense on $[a,b]$, for some $0 < a < b < 1$, and this can be justified when these dispersion matrices are not very different from each other).

## 22.5 Forecasting of Economic Structural Changes

The main focus of this study has been on a breakdown of the overall Gini coefficient in terms of a number of component Gini coefficients, relative mean incomes for these sectors, and their relative proportions. It has also been shown in earlier sections that the usual income inequality indexes when suitably modified may provide some further insight in this probe. Thus, given the basic formulation of the (vector-) model in (22.42) and the subsequent analysis in Section 22.4 [viz., (22.48) through (22.56)], it seems quite appropriate to concentrate on a set of pertinent socioeconomic factors [giving rise to the regressors $\boldsymbol{\xi}_j$ (or their estimators $\mathbf{U}_j$)]. The basic issue is therefore to choose these $\boldsymbol{\xi}_j$ in a most judicious manner. This choice is of course dependent on the particular society or community and the major socioeconomic factors affecting the same. Although it may be intuitively appealing to have a large number of components in the $\boldsymbol{\xi}_j$ (to provide more explanation of the impact of various socioeconomic factors), from the statistical analysis point of view, there is a mixed reaction: The larger the dimension of the regressor vectors, the greater should be the sample sizes so that the associated regression matrices can be estimated with comparable precision. The quadratic norms used in (22.59), (22.62) and (22.65) all relate to the dimension of the $\mathbf{Z}_{(r)}$ [or $\mathbf{Z}^*_{(r)}$] and $\mathbf{S}_{(r)}$. The larger the value of $q$, the greater will be the (stochastic) variability of these norms, so that the critical levels $[\tau^{(j)}_{n,\alpha}]$ will be larger. This automatically points out that unless all the explanatory regressors are informative, the power or sensitivity of the tests in Section 22.4 may not increase with the increase in $q$. The situation is quite comparable with the usual chi-squared (goodness of fit) tests where an increase in the degrees of freedom may not necessarily lead to an increase in the power (unless the noncentrality increases at a commensurate rate). Thus, in choosing $q$ and the subset of the regressors, sufficient care must be taken to ensure that unnecessary loss of efficiency does not arise owing to redundant regressors or because of omission of some relevant ones.

For forecasting purposes, the above findings are quite important. First of all, if $q$ and $\xi_j$ are chosen properly then a change in the structural form may be studied in terms of these explanatory regressors, and the model in Section 22.4 can be used with advantage. On the other hand, it may be such that some factor (or regressor) may be quite insignificant up to a certain time point and, then due to some extraneous factor, it may become quite an important one at some stage. Thus, if no attention is paid to this factor (based on its past history), a forecasting formula may not work out that well. Thus, in a forecasting situation, one needs to pay attention to a possible change point model with respect to the regressors $\xi_j$; a progressive examination of these factors with a view to eliminating the redundant ones and accommodating new important ones should lead to a better forecasting.

Second, in Section 22.4, to eliminate the redundant variation due to plausible scalar adjustments, the income distributions were taken as cost-of-living adjusted ones. It provided us with a reasonable way of achieving a linear model (c.f. Section 22.4) when the regressors are not too different from each other (i.e., there is no abrupt change in their realizations over the time period considered). In the context of forecasting, this cost-of-living adjustment for the $F_j$ should also be examined carefully. If a cost-of-living adjustment fails to bring two distributions (say, $F_j$ and $F_{j'}$) close to each other, the approximate linear expansion in Section 22.4 may not be that appropriate; hence, a forecasting based on this model may not work out that well. This may be particularly important when an existing important regressor phases out of the system and a new one enters into the scheme and the cost-of-living adjustment as adapted might not have paid due importance to this new factor. Nonlinear regression models may be necessary for the forecasting problem if no such cost-of-living adjustment is made to the income distributions and the regressors may vary considerably over time. Finally, in forecasting for a very near future time period, the methodology in Section 22.4 can be used with greater confidence. However, as the time gap between the time domain under study and the projected time increases, the effectiveness of forecasting models may decrease drastically, especially, in the context of structural changes. Nevertheless, the decomposition considered in Sections 22.2, 22.3, and 22.4 provide us with pertinent insight into the structure of such income patterns (and inequalities), and a forecasting based on such a decomposition should be much more effective than the usual one based on the overall Gini coefficients or some other conventional measure of income inequality.

This section concludes with a remark on the WLSE used in Section 22.4 (and in this one too). The justification for the WLSE is based on the asymptotic multi-normality result in (22.53). This is generally quite appropriate. If, however, (22.53) is not that appealing (but still the linearity of regression may be tenable), one may use some robust estimators instead of the WLSE. For the constancy of regression relationships over time, suitable $M$-tests based on recursive $M$-estimators have been considered by Sen (1984), Hušková (1988), and Hušková and Sen (1989), among others, and these may be considered here too. However, in all the other works, homoscedasticity has been a vital part of the basic model; while in the current setup, there may be good evidence that such an homoscedasticity assumption may not hold. Thus, the effect of heteroscedasticity of the original model on the performance characteristics of such recursive $M$-tests for change-points or $M$-estimators in forecasting remains to be studied. The basic reason for using the WLSE instead of the ordinary LSE is to take into account the possible heteroscedasticity in the model, and this approach is likely to generate sufficient interest in other areas too.

## References

Blackorby, C. and Donaldson, D. (1980), Ethical indices for the measurement of poverty. *Econometrica*, **48**, 1053-1060.

Brown, R.L., Durbin, J., and Evans, J.M. (1975), Techniques for testing the constancy of regression relationships over time (with discussion). *Journal of the Royal Statistical Society*, **B-37**, 149-192.

DeLong, D. (1981), Crossing probabilities for a square root boundary by a Bessel process. *Communications in Statistics: Theory and Methods*, **A-10**, 2197-2213.

Hušková, M. (1988), Recursive $M$-test for detection of change. *Sequential Analysis*, **7**, 75-90.

Hušková, M., and Sen, P.K. (1989), Nonparametric tests for shift and change in regression at an unknown time point, pp. 71-85 in P. Hackl (ed.), *Statistical Analysis and Forecasting of Economic Structural Change*. Berlin: Springer-Verlag.

Sen, A.K. (1976), The measurement of poverty: An axiomatic approach. *Econometrica*, **44**, 219-232.

Sen, P.K. (1981), *Sequential Nonparametrics: Invariance Principles and Statistical Inference*. New York: John Wiley.

Sen, P.K. (1984), Recursive $M$-tests for the constancy of multivariate regression relationships over time. *Sequential Analysis*, **3**, 191-211.

Sen, P.K. (1986), The Gini coefficient and poverty indexes: Some reconciliations. *Journal of the American Statistical Association*, **81**, 1050-1057.

Sen, P.K. (1988), The harmonic Gini coefficient and affluence indexes. *Mathematical Social Sciences*, **16**, 65-76.

Sen, P.K. (1989a), Affluence and poverty indexes, pp. 1-5 in S. Kotz and N. L. Johnson (eds.), *Encyclopedia of Statistical Sciences, Supplement*. New York: John Wiley.

Sen, P.K. (1989b), Change-point problem relating to the poverty structure, pp. 329-344 in P. Hackl (ed.), *Statistical Analysis and Forecasting of Economic Structural Change*. Berlin: Springer-Verlag.

Sinha, A.N. and Sen, P.K. (1982), Tests based on empirical processes for progressive censoring schemes with staggering entry and random withdrawal. *Sankhya*, **B-44**, 1-18.

Takayama, N. (1979), Poverty, income inequality and their measures: Professor Sen's axiomatic approach reconsidered. *Econometrica*, **47**, 747-760.

# Author Index

# Subject Index

# List of Authors

Yngve Abrahamsen, University of St. Gall, St. Gall, Switzerland

Masanao Aoki, School of Engineering and Applied Science, University of California, Los Angeles, USA

Lyle D. Broemeling, University of Texas Medical Branch, Galveston, USA

Juanita H. Chin Choy, Northrup King Inc., Minnesota, USA

Lennart Claesson, Department of Economic Statistics, Stockholm School of Economics, Stockholm, Sweden

Peyton J. Cook, University of Tulsa, Oklahoma, USA

Francis X. Diebold, University of Pennsylvania, Philadelphia, USA

Jean-Marie Dufour, Department of Science and Economics, Montreal, Canada

Clive W.J. Granger, Department of Economics, University of California, La Jolla, USA

Peter Hackl, Institut für Statistik, University of Economics, Vienna, Austria

Andrew C. Harvey, Department of Statistical and Mathematical Sciences, London School of Economics, London, UK

David F. Hendry, Nuffield College, Oxford, UK

Marie Hušková, Department of Statistics and Probability, Charles University, Prague, ČSFR

Walter Krämer, Institut für Statistik, University of Dortmund, FRG

Wilhelm Krelle, Institut für Gesellschafts- und Wirtschaftswissenschaften, University of Bonn, FRG

Maozu Lu, Department of Economics, University of Southampton, UK

G. S. Maddala, Department of Economics, University of Florida, Gainesville, USA

Spyros Makridakis, European Institute of Business Administration, Fontainebleau, France

Grayham E. Mizon, Economics Department, Southampton University, Southampton, UK

Werner Ploberger, Institut für Ökonometrie, Technical University, Vienna, Austria

Peter M. Robinson, Department of Economics, London School of Economics, London, UK

Bernd Schips, University of St. Gall, St. Gall, Switzerland

Irene Schlüter, Institut für Statistik, University of Dortmund, FRG

Wolfgang Schneider, Christian Albrechts University, Kiel, FRG

Pranab K. Sen, Department of Biostatistics, University of North Carolina, Chapel Hill, USA

Mariane Streibel, Institut of Mathematics and Statistics, University of São Paolo, Brazil

Timo Teräsvirta, Department of Economics, University of California, La Jolla, USA

Hiroki Tsurumi, Department of Economics, State University of New Jersey, Rutgers, New Brunswick, USA

Anders H. Westlund, Department of Economic Statistics, Stockholm School of Economics, Stockholm, Sweden

## THE INTERNATIONAL INSTITUTE FOR APPLIED SYSTEMS ANALYSIS

is a nongovernmental research institution, bringing together scientists from around the world to work on problems of common concern. Situated in Laxenburg, Austria, IIASA was founded in October 1972 by the academies of science and equivalent organizations of twelve countries. Its founders gave IIASA a unique position outside national, disciplinary, and institutional boundaries so that it might take the broadest possible view in pursuing its objectives:

*To promote international cooperation* in solving problems arising from social, economic, technological, and environmental change

*To create a network of institutions* in the national member organization countries and elsewhere for joint scientific research

*To develop and formalize systems analysis* and the sciences contributing to it, and promote the use of analytical techniques needed to evaluate and address complex problems

*To inform policy advisors and decision makers* about the potential application of the Institute's work to such problems

The Institute now has national member organizations in the following countries:

**Austria**
The Austrian Academy of Sciences

**Bulgaria**
The National Committee for Applied Systems Analysis and Management

**Canada**
The Canadian Committee for IIASA

**Czechoslovakia**
The Committee for IIASA of the Czechoslovak Federal Republic

**Finland**
The Finnish Committee for IIASA

**France**
The French Association for the Development of Systems Analysis

**Federal Republic of Germany**
Association for the Advancement of IIASA

**Hungary**
The Hungarian Committee for Applied Systems Analysis

**Italy**
The National Research Council

**Japan**
The Japan Committee for IIASA

**Netherlands**
The Netherlands Organization for Scientific Research (NWO)

**Poland**
The Polish Academy of Sciences

**Sweden**
The Swedish Council for Planning and Coordination of Research (FRN)

**Union of Soviet Socialist Republics**
The Academy of Sciences of the Union of Soviet Socialist Republics

**United States of America**
The American Academy of Arts and Sciences

MIX
Papier aus verantwortungsvollen Quellen
Paper from responsible sources
FSC® C105338

If you have any concerns about our products,
you can contact us on
**ProductSafety@springernature.com**

In case Publisher is established outside the EU,
the EU authorized representative is:
**Springer Nature Customer Service Center GmbH**
**Europaplatz 3, 69115 Heidelberg, Germany**

Printed by Libri Plureos GmbH
in Hamburg, Germany